Designing Intelligent Machines

Volume 2

Concepts in Artificial Intelligence

The two volumes of this book were produced as the major components of the third-level undergraduate course *Mechatronics: Designing Intelligent Machines*, written by a Course Team at The Open University, UK. They are:

Volume 1: ***Perception, Cognition and Execution***

Edited by George Rzevski

Volume 2: ***Concepts in Artificial Intelligence***

By Jeffrey Johnson and Philip Picton

Designing Intelligent Machines

Volume 2

Concepts in Artificial Intelligence

By Jeffrey Johnson and Philip Picton

Butterworth-Heinemann in association with The Open University

OXFORD LONDON BOSTON
MUNICH NEW DELHI SINGAPORE SYDNEY
TOKYO TORONTO WELLINGTON

MILTON KEYNES

BUTTERWORTH-HEINEMANN LTD, Linacre House, Jordan Hill, Oxford OX2 8DP, England, UK

OXFORD LONDON BOSTON
MUNICH NEW DELHI SINGAPORE SYDNEY
TOKYO TORONTO WELLINGTON

in association with

THE OPEN UNIVERSITY, Walton Hall, Milton Keynes MK7 6AA, England, UK

First published in the United Kingdom by the Open University in serial form for Open University students and staff 1994.

This edition first published in the United Kingdom 1995, and reprinted with corrections 1999.

Edited, designed and typeset by The Open University.

Printed and bound in the United Kingdom by the Alden Press, Oxford, United Kingdom.

This text forms part of an Open University course. If you would like to know more about Open University courses, please write to the Course Reservations Centre, PO Box 724, The Open University, Walton Hall, Milton Keynes, MK7 6ZS, United Kingdom.

British Library Cataloguing in Publication Data

A record is available from the British Library

ISBN 0-7506-2403-5

Cover: Computer art created by Dr Paul Margerison using a Silicon Graphics IRIS workstation as part of his PhD research in the Design Discipline at the Open University. The images were created by random sampling of previously drawn images and subsequent interpolation.

PREFACE

George Rzevski

This textbook is aimed at undergraduate and postgraduate students and those working in industry who wish to learn the fundamentals of a branch of engineering called ***mechatronics***.

The name was coined in the 1970s to acknowledge an urgent need to integrate two engineering disciplines – ***mecha***nics and elec***tronics*** – with a view to developing and manufacturing mechanical machines controlled by means of electronic circuits. Since then the control and communication technologies have advanced beyond recognition and are now dominated by the software and hardware of digital computers and by embedded artificial intelligence. The name, therefore, may now be considered to be somewhat restrictive. It is, nevertheless, widely used.

The second part of the title – ***designing intelligent machines*** – emphasizes that this book covers new aspects of mechatronics, that is, how to specify and design machines capable of smart sensing, planning, pattern recognition, navigation, learning and reasoning.

The book consists of two independent volumes. Volume 1 covers the fundamentals of mechatronics and discusses the design of machine perception, cognition and execution. Volume 2 is concerned with the concepts of artificial intelligence needed for the design of machines with advanced intelligent behaviour.

Each volume has an 'Overview' which provides the reader with the orientation needed when approaching the study of an unfamiliar and multidisciplinary subject, and provides the rationale for the inclusion and ordering of the topics.

These two volumes were written as the major components of a package of distance learning material for the Open University undergraduate course *Mechatronics: Designing Intelligent Machines*. The contributors to these two volumes were part of an interdisciplinary Course Team, brought together to integrate the disciplines and techniques underlying mechatronics. This Course Team has also generated complementary components of the course, which include video tapes, software, a home experiment kit, study guides and course assessments. More detailed information on the course is given overleaf.

The Open University 'Mechatronics' Course

The two volumes of this book were produced as the major components of the undergraduate third-level course *Mechatronics: Designing Intelligent Machines* by a Course Team at the UK Open University.

Complementary components of the undergraduate course include video tapes, a home experiment kit, software, study guides and course assessments. Video tapes provide students with an opportunity to watch state-of-the-art mechatronic systems in action, to listen to interviews with leading designers of intelligent machines, and to use visual aids to clarify more advanced concepts. The home experiment kit is used to build a scanner and a small vehicle connected by an infra-red link to the student's own personal computer. A variety of computer programs and programming environments enable students not only to simulate and experiment but also to design new vehicle behaviours, including autonomous navigation.

If you would like a copy of *Studying with the Open University*, please write to the Central Enquiry Service, P.O. Box 200, The Open University, Walton Hall, Milton Keynes, MK7 6YZ, United Kingdom. Enquiries regarding the availability of supporting material for this and other courses should be addressed to: Open University Educational Enterprises Ltd, 12 Cofferidge Close, Stony Stratford, Milton Keynes, MK11 1BY, UK.

Course Team Chair and Academic Editor

Professor George Rzevski, The Open University, UK

Authors

Chris Bissell
Chris Earl
Jeffrey Johnson
George Kiss
Anthony Lucas-Smith
Phil Picton
Joe Rooney
George Rzevski
Alfred Vella
Paul Wiese

Supporting staff

Geoff Austin (Academic Computing Service)
George Bellis (Project Officer)
Pam Berry (Text Processing)
Phillippa Broadbent (Print Production)
Jennifer Conlon (Secretary)
Roger Dobson (Course Manager)
Ian Every (Academic Computing Service)
Ruth Hall (Graphic Designer)
Garry Hammond (Editor)
John Newbury (Staff Tutor)
Christopher Pym (Course Manager)
Janice Robertson (Editor)
John Stratford (BBC Producer)
John Taylor (Graphic Artist)
Helen Thompson (Academic Computing Service)
David Wilson (Project Control)
Bill Young (BBC Producer)

External assessor

Professor Duc-Truong Pham, of University of Wales, College of Cardiff.

Acknowledgements

The Course Team wishes to acknowledge the contributions made in the development of the course by: Mike Booth of Booth Associates; Professor John Meleka; and Dr Memis Acar, Professor J. R. Hewit, Paul King and Dr. K. Bouazza Marouf of Loughborough University of Technology. The Course Team is also indebted to Stuart Burge, Douglas Leith, Don Miles and Peter Steiner, and the many students who participated in the piloting exercises for the course material.

Contents of Volume 2

CHAPTER 5 Scheduling

CHAPTER 6 Reasoning

CHAPTER 9 Intelligent control

CHAPTER 10 Computer vision

CHAPTER 11 Integration

OVERVIEW OF VOLUME 2

Jeffrey Johnson and Philip Picton

Volume 2 provides the theoretical background for the implementation of concepts of artificial intelligence (AI) in engineering design and mechatronics. Our goal has been to explain the ideas to those who have no previous knowledge of the subject, but at the same time to give sufficient technical information to be useful at the operational level. But AI is a huge subject, and it is impossible to cover all the details in a single introductory book. This means that some subjects are discussed in depth, while others are raised to set the wider context and to guide readers to the more specialist literature.

This book was written in the context of the development of the SmartLab software and home experiment kit which the Open University provides to all students on its Mechatronics course. Many of the examples given here come from our work on SmartLab, and hopefully this has kept us sufficiently close to our goal of presenting theoretical concepts in an applications-oriented way. We are both practising engineers, and we appreciate that the ultimate value of theoretical ideas is how well they can be applied to solving practical problems. In this we encourage our students to adopt the European connotation of 'engineer' of an ingenious and creative problem solver as implied by the Latin root of the word. It is our belief that engineering requires the highest levels of intellect, exploiting the new problem-solving paradigms enabled by new technologies of information processing.

We do not adopt a purist approach to 'Artificial Intelligence' and we are not here involved in the fierce debates about the possibilities of creating artificial brains or human-like robots. Although we enjoy speculating on this as much as anyone else, the approach adopted in this book will follow that attributed to Edsgar Dijkstra: 'the question of whether a computer can think is no more interesting than the question of whether a submarine can swim'. Although we have concentrated on the main ideas of AI in the context of solving practical engineering problems, AI raises some important philosophical, social and scientific issues. We hope that after reading this book you will be in a better position to make up your own mind on this.

The book starts off in Chapter 1 with a general introduction to artificial intelligence. This is followed in Chapter 2 by quite a detailed discussion of *pattern recognition* within an engineering context. Chapter 3 introduces one of the most important topics in AI, namely *search*. Search is important because many problems can be construed as searching the universe of all possibilities for something that can be considered to be a solution to a problem. Many interesting problems do not have solutions in the sense that a mathematical equation has a solution, and deciding if a particular answer is acceptably good involves many theoretical subtleties. Artificial intelligence has given new insights into the nature of search, and produced new approaches such as simulated annealing and genetic algorithms. The theme of search permeates the whole book.

In Chapter 4 we explain the new computational paradigm of *neural networks*, which learn from data rather than being programmed. In Chapter 5 we consider the problems of *scheduling* which intelligent machines must solve in order to decide where they should be and when, and what they should be doing when they get there. Chapter 6 introduces machine *reasoning*, including traditional logical and non-deterministic approaches such as probability and fuzzy logic. Chapter 7 shows how logical reasoning can be implemented in computers using *rule-based systems*.

Chapter 8 introduces the idea of machine *learning* which is widely felt to lie at the heart of machine intelligence. Alan Turing, one of the founding fathers of AI, knew that if machines are to achieve the levels of intelligence to which we aspire, then they must be capable of learning for themselves.

The remaining chapters are concerned with implementation. Chapter 9 considers the subject of *intelligent control*, in which the ideas developed in earlier chapters are applied to the benchmark problem of balancing an upside-down broom on a wheeled trolley. The ideas of the earlier chapters can also be applied to less well-defined problems with more uncertainty such as a robot negotiating a path through an unknown landscape. Chapter 10 discusses the problem of *computer vision*, which involves machines abstracting useful information from images. Chapter 11 shows how, of many possibilities, all the techniques discussed in the book can be implemented and integrated in a simple architecture called the *blackboard system*.

After reading the book, readers should feel that they know the main issues in artificial intelligence as far as they apply to practical engineering design, and that they are sufficiently familiar with the basic techniques to use them in practice. However, there is no substitute for hands-on experience, and we hope that readers will try out some of these ideas for themselves, as our students do with SmartLab.

This book is the outcome of many meetings and discussions with our colleagues on the Mechatronics course team. Their input has been invaluable. The book would not exist without the academic editor and course team chair, George Rzevski, who had the original idea. Also it would not exist without the efforts of Roger Dobson who, like all Open University course managers, had the impossible task of interfacing academics to reality. We would especially like to thank the course editor, Garry Hammond, who often transformed our impenetrable English and mathematical formulae into the current much improved form.

We would be pleased to receive constructive criticism from readers, but we hope that most will find the book a useful introduction to some profound and important ideas that will characterize the engineering of the future.

CHAPTER 1 INTRODUCTION

1.1 Artificial intelligence in engineering

This book, Volume 2 in the series, sets out to explain how the fruits of fifty years' research into artificial intelligence (AI) can be applied to make intelligent and better machines. We have two objectives: the first is to explain the theory of the mainstream ideas in AI, and the second is to show how these ideas can be applied in practical engineering situations.

Artificial intelligence is a young discipline which has had some spectacular successes, and some equally spectacular failures. The failures have mostly been due to underestimating the complexity of apparently simple problems combined with a belief that brute computer power *ought* to be able to solve any problem. The successes have been due to human ingenuity, scientific analysis, good engineering practice, and sometimes good luck. Out of all this experience, engineering principles are emerging which can be used to guide engineers who have to tackle problems of ever-increasing complexity in an increasingly competitive world.

After reading this chapter you should:

- be aware of the distinction between *strong* and *weak AI*;
- be aware of the term *cognitive science*;
- be aware of the concepts of *human–computer systems* and *human–computer interaction*;
- know the benefits of building more intelligence into machines;
- realize that there are currently limits to how much intelligence can be built into machines;
- know the five critical features of AI and the ten enduring characteristics suggested by Schank;
- be aware of how AI can be applied in the study of perception, cognition and execution.

As in Volume 1, key terms are picked out in bold type when they are first introduced.

1.2 Strong AI, weak AI and cognitive science

AI has two main camps. The proponents of ***strong AI*** believe that it will be possible to build machines with human-like intelligence. By contrast, the proponents of ***weak AI*** believe that machines can exhibit what might be called intelligent behaviour, but that there are limits which mean that machines will always be intellectually inferior to humans. Within the weak AI camp there is an active research area called ***cognitive science*** which uses computers to model human behaviour with the intention of learning more about human beings. A very important aspect of this is the relationship between humans and computers. Whether or not it is possible to build human-like machines, we certainly build ***human–computer systems*** which involve both machines and people. For such systems to function properly it is just as important to engineer the human part of the system as it is to engineer the physical computing part of the system. It is also essential to understand how the human and physical subsystems interface to each other, and the last twenty years have seen a huge increase in research into ***human–computer interfaces*** (HCIs).

In this book we will be mainly concerned with explaining the concepts of AI as they relate to physical systems. We are interested in engineering problems which either have no conventional solution, or which can be solved better by the application of information processing and computation. We will not attempt to give an absolute definition of 'intelligence'. Rather, we will explain some recent ideas and computational paradigms coming under the umbrella of AI. Readers can decide for themselves if machines which embody these are intelligent.

1.3 Why build intelligence into machines?

The main reasons for building more machine intelligence into machines are that they may be cheaper to build, cheaper to maintain, more reliable, or able to overcome problems which other engineering approaches cannot. This question is discussed in some detail in Volume 1. We will summarize the conclusions as:

- machine intelligence offers new possibilities;
- machine intelligence can give better solutions to problems;
- software is relatively inexpensive to mass produce;
- software can often be changed more easily than hardware.

1.4 How much intelligence can be built into machines?

The simple answer to this question is that no one knows. Engineering is the adaptation of general scientific knowledge to particular situations and problems. The engineer often has to fill in the details which the scientist has not provided, and in some cases the engineer has to become a scientist in order to fill in essential pieces of theoretical knowledge. Applied AI is like this. Often the engineer will have to do research to find how a general technique can be applied to the particular problem. Although research is by its nature unpredictable, there are some guidelines.

The first guideline is that *no machine works without some level of human supervision*. In other words, *all machines are part of a human–computer system.* Suppose the intelligent part played by machines in any particular system can be quantified between 0% (no machine intelligence) and 100% (no human intelligence, which we claim is impossible). Then the cost of increasing the machine component of the system's intelligence can be sketched as shown in Figure 1.1.

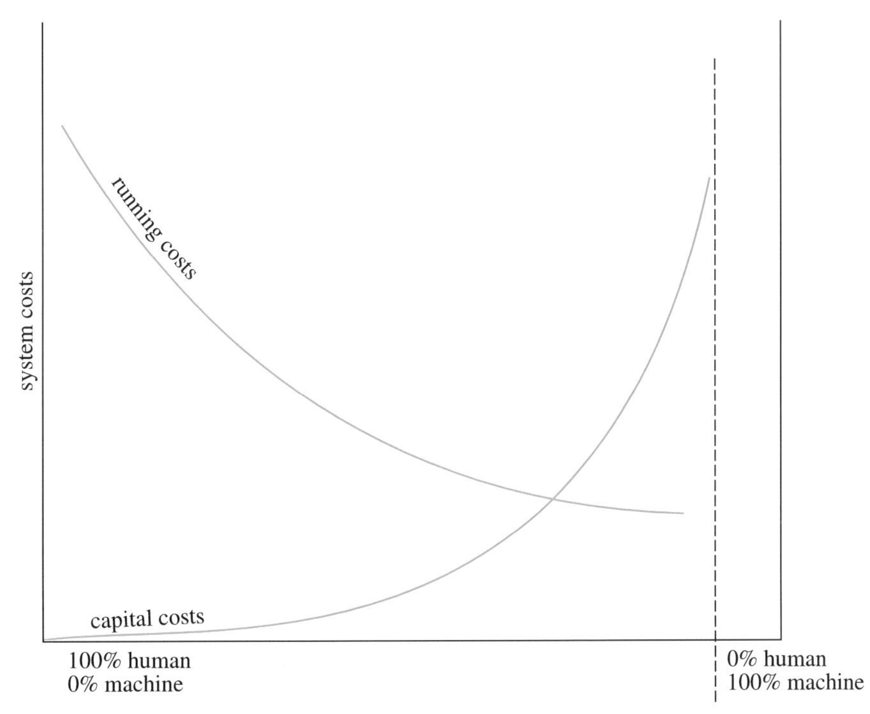

Figure 1.1
The running costs of systems combining human and machine intelligence decrease with the introduction of greater machine intelligence, but the capital costs increase.

The first principle that the engineer can infer from this is that it gets progressively more expensive to replace human intelligence with machine intelligence. In general, putting greater intelligence into machines gets increasingly expensive. The design engineer should therefore be aware that there will always be a cut-off where the investment in going just a bit further cannot be justified.

The information technology revolution of the past thirty years has seen many mundane human tasks being taken over by computers. It has been cost effective to begin the process of introducing machine intelligence into systems which require data processing and the handling of large amounts of information. Yet although the check-out counters at supermarkets are now intelligent enough to recognize the groceries passed and do all the 'special offer' calculations on our bill, we still have human operators at the till at the time of writing. The human tasks require only a relatively low educational attainment, and yet the operators possess a level of intelligence which cannot be built into the check-out machines. This is the subtle human intelligence which, among other things, smiles engagingly at the irate customer, has a kind word for the lonely pensioner, and alerts security to deal with the shifty potential shoplifter.

1.5 What is artificial intelligence?

In giving an answer to the question 'What is AI, anyway?', Roger Schank writes:

> Artificial Intelligence is a subject that, due to the massive, often quite unintelligible, publicity it gets, is nearly completely misunderstood by people outside the field. Even AI's practitioners are somewhat confused with respect to what AI is really about.

Schank suggests five critical features of AI:

- communication
- internal knowledge
- world knowledge
- goals and plans
- creativity.

He also suggests ten 'enduring characteristics' of AI (we have added the comments in parentheses):

1 representation
2 decoding (real systems encoded into and from machine representations)
3 inference

4 control of combinatorial explosion
5 indexing (for recalling knowledge)
6 prediction and recovery
7 dynamic modification (including learning)
8 generalization
9 curiosity
10 creativity.

Most of these are of immediate importance in designing intelligent machines. A major theme throughout this book will be the problem of *representing* the machine and its environment. If we want to *reason* and make *inferences* we will have to *encode* these data in a *symbolic language*. The resultant *knowledge base* will have to be *indexed* in a way which allows information to be extracted quickly and efficiently. Alan Turing (1912–54), one of the founding fathers of machine intelligence, realized right from the outset that for a machine to become 'intelligent' it would have to *learn*, and to *generalize* from specific information in order to acquire new knowledge. *Prediction* is an important requirement for intelligent machines since they must know what is possible and choose between the options. Similarly, it is highly desirable that a machine that has made a mistake should be able to detect that mistake and *recover* from it.

The characteristics of *curiosity* and *creativity* are rare in contemporary machines. In one sense we expect intelligent machines to be 'curious' through the actions of their sensors, which are constantly seeking to know 'what's out there'. Higher levels of curiosity may appear in the next few generations of machines. A robot might muse along the lines of 'although I have been doing this job for years with my right arm, I wonder if I could do it better with my left leg', and might thus spontaneously improve its performance. Whether we want our machines to be creative, and possibly unpredictable, remains an open question.

Although there is a substantial body of applicable knowledge arising from AI, the subject remains young with many more questions than answers. One of the major failures of AI to date is the inability of machines to understand natural languages such as English, French, Hebrew, Arabic or German. Certainly we can feed the electrical signals from microphones into computers, but we cannot make machines which abstract the kind of information that humans do so effortlessly with our ears and brains. Vision is another area in which progress in AI has been slow despite many hundreds of man-years of research effort. So we can conclude that computers are rather poor at the cognitive functions which lie at the heart of much of our human intelligence. However, the story is not entirely negative, and in this book we will show how AI and information engineering can be applied in the practical design of intelligent machines.

1.6 How is AI applied to engineering in practice?

AI can make major contributions to designing intelligent machines in the areas of perception and cognition, and also to actuators such as 'intelligent' grippers on robot arms.

1.6.1 AI in perception

As seen in Volume 1, effective perception is essential in the design of intelligent machines. In particular, these machines need to have *sensors* which provide information on internal aspects and on the environment. Many sensors deliver information which is not useful in itself. Sometimes sensors produce 'noisy' data with uncertain interpretation. Sometimes information from many sensors must be combined in order to give useful information. And sometimes the cognition subsystem requires information in symbolic form. AI has developed many principles and techniques for processing noisy sensor information and synthesizing it into useful symbolic forms of known reliability. Techniques which are particularly useful in perception include:

- pattern classification
- neural networks
- image interpretation: computer vision, sonar, radar
- data fusion
- learning.

1.6.2 AI in cognition

Once a machine has reasonably reliable information about itself and its environment it must constantly be making decisions as to what to do in the long and short terms. Thus a machine must be able to *model* itself within its environment and predict the possible states of both itself and the environment. Techniques which are particularly useful in cognition include:

- reasoning:
    - representation
    - logical reasoning
    - knowledge-based systems
    - fuzzy logic
- scheduling and planning:
    - representation
    - activity planning

critical-path analysis
path planning
emergency planning

- problem solving:

  heuristics

- learning.

1.6.3 AI in execution

Assuming that a machine has established its goals and has a plan to enable it to achieve them, it must *execute* that plan. Execution usually involves the machine *moving*, either the whole machine as in the case of an autonomous vehicle, or parts of the machine as in the case of a robot gripper.

Some intelligent machines use human beings in the execution stage. For example, machines to detect drugs and explosives use human customs inspectors to take the appropriate action once a contraband substance has been detected. The link between the cognition subsystem and the human being is often implemented through graphic user interfaces (GUIs). The sensing–cognition–execution loop is then closed using devices such as the keyboard, the mouse, and so on. The area of human–computer interfaces is becoming increasingly important as machines gain more intelligence and greater functionality.

Sometimes intelligence is distributed throughout machines, with parts such as grippers having their own processing ability. Distributed systems may have a central controller, or the overall behaviour may be allowed to emerge from the interacting subsystems. Often, distributed intelligence is limited and dedicated to specific tasks. This may allow subsystems to be controlled by relatively high-level commands such as 'pick up the block' or 'spray the panel'. The intelligent execution subsystem must interpret such commands and 'unpack' them in the context of the knowledge or model it has of its tasks or function. The final result of this will, in general, be to activate switches which power motors and other actuators.

In this volume AI in execution appears mainly under the heading of 'intelligent control'. This involves decision-making using various kinds of reasoning, including rule-based systems, fuzzy logic, and even neural networks.

From the viewpoint of this book, reasoning in execution differs from reasoning in cognition by its motivation. In cognition the goal is to find appropriate information. In execution the goal is to control actuators to make things happen. Thus the techniques used in cognition and execution may be similar to those listed in Section 1.6.2.

1.7 The principles behind the applications

Apart from applying AI-derived solutions to engineering, AI has also given us methodological knowledge which can be applied in engineering design.

One of the main lessons we learn from AI research is that it can be hard to find an appropriate way of *representing* things inside a computer. Often this means having to analyse things very carefully and abstract their parts and relationships. Also, we sometimes have choices in methods of representation, and we can look to AI for principles which help us choose between the possibilities.

Another important lesson learnt from AI is that 'brute force' computer power cannot solve a large class of important problems. Our computers can store and process huge amounts of information, but for many problems the *search* for particular outcomes or solutions must be guided by ***heuristics***, i.e. rules of thumb based on trial and error, which usually work, but don't always; i.e. they give reasonably good results most of the time.

Machine reasoning is a major area of research in AI and has resulted in successes such as *knowledge-based systems, fuzzy logic* and *non-monotonic reasoning*. To build intelligent machines we must have a good understanding of what it means for a machine to reason.

Learning is another major area of AI which we can exploit in designing intelligent machines. Building everything into machines once and for all makes them expensive and inflexible. Machines that can learn tend to be cheaper, more adaptable, and in principle able to improve their performance.

The building of intelligent machines, especially *robotics*, is itself a mainstream research area in AI. From this the subject of *intelligent control* has evolved, which deals with situations in which conventional control theory is not applicable. These include open-loop control, where the machine has inadequate feedback information or lacks the reference information required to use conventional control techniques. These new approaches to control are often based on logic rather than more traditional continuous mathematical techniques such as calculus.

In the rest of this book we will examine these topics in greater detail. Our goal is for you to understand the principles of AI sufficiently well to be able to apply them to real engineering problems.

Reference

Schank, R. C. (1990) 'What is AI, anyway', in Partridge, D. and Wilks, Y. (eds) *The Foundations of Artificial Intelligence*, Cambridge University Press.

CHAPTER 2
PATTERN RECOGNITION

2.1 Introduction

Pattern recognition is fundamental to perception and cognition in intelligent systems. A machine's sensors can generate a huge number of combinations of inputs over short periods of time. To be useful it is necessary to transform these data into one of a set of known classes by recognizing patterns in the data. In perception, for example, a microphone delivers a waveform corresponding to sound. Until parts of the wave which correspond to words such as 'yes' and 'no' are recognized, these data may have little or no value. Pattern recognition occurs also during cognition, for example, when a machine has to decide what to do next. Some patterns of data from the sensors combined with data in memory will require one action, other patterns will require other actions.

Human beings are astoundingly good at pattern recognition; so good that the pioneers of machine intelligence severely underestimated how difficult it would be to represent patterns of any complexity. For example, can you see a pattern in the following configuration of dots?

● ●● ●●● ●●●● ●●●●● ●●●●●● ●●●●●●●

Figure 2.1
Is there a pattern in the dots?

You probably see a pattern of one dot, two dots, three dots, etc., and you probably see that the dots form a straight line.

Human pattern recognition is used extensively in large and small systems. For example, security systems include human guards whose job includes looking at television monitors and recognizing unusual patterns (perception), deciding if this pattern is an emergency (cognition), and raising the alarm if it is (execution). In industrial systems human quality controllers have to look at assemblies and recognize unusual configurations (perception), decide if the configuration is outside the specification (cognition), and physically move rejected parts (execution). In many similar cases the execution could be performed by machines if only one could automate the pattern recognition. In some systems this has already happened; for example, most large aircraft are controlled much of the time by an autopilot.

Although human pattern recognition can be very good, it is unreliable. People get tired, lose concentration, miss things, and make mistakes. Also it is sometimes not possible to use human pattern recognition, especially in environments in which the fragile human body cannot survive. For these reasons automating pattern recognition has been an active research area for many years.

Figure 2.2 shows that human beings have the remarkable ability to recognize patterns which are not explicit in the sensory data. The 'sun illusion' shows how a circle can emerge from a set of straight lines. There is nothing in any individual line to suggest a circle, but together they produce what is called a ***subjective contour***.

Figure 2.2
A subjective contour in the sun illusion.

Where does the circle come from? Something in our brains seems to allow it to recognize a circle in the lines, although this remains something of a mystery. As Figure 2.3 shows, the explanation that subjective contours are due to edge detectors in the brain is inadequate (edge detectors are defined in Chapter 4 of Volume 1). There the subjective contours are orthogonal to the edge segment data and in most places there is no explicit edge data for an edge detector to respond to.

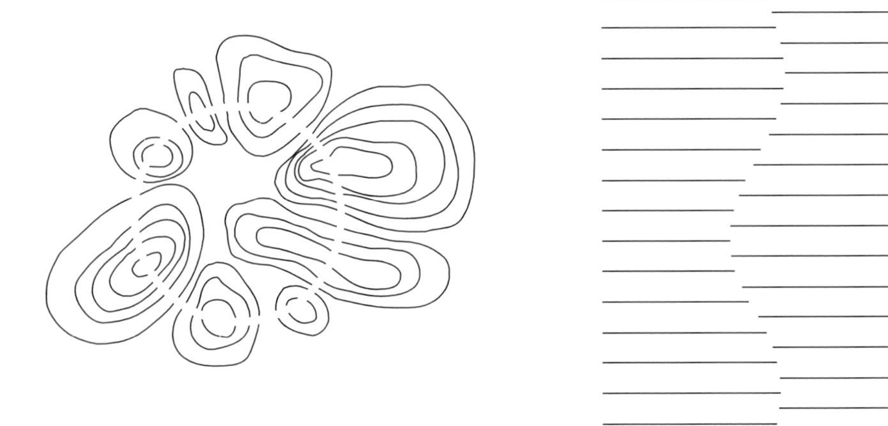

Figure 2.3
Subjective contours cannot be explained by simple edge detectors: here the edges are orthogonal to the lines and there is no explicit edge data in these directions.

Figure 2.4 shows that just a few points can give the strong illusion of lines and geometric shapes, even when there are no explicit edge data at all. These phenomena are not well understood and they suggest that automating pattern recognition may be very hard: how can we get a machine to see patterns which are not explicit in the data?

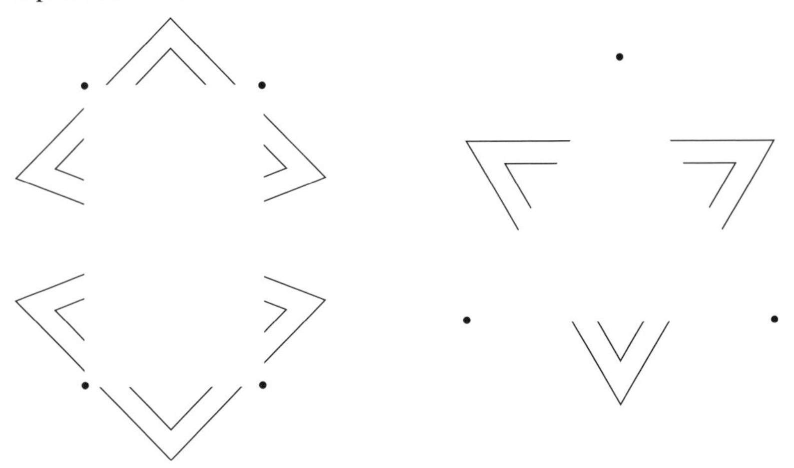

Figure 2.4
A few dots are sufficient for us to perceive lines and geometric shapes in images.

Figure 2.5 shows shapes that can be perceived in patterns of dots. The first of these can be recognized as the letter **A**. When you see this shape it ***symbolizes*** the first letter of the Roman alphabet. You have a tremendous amount of implicit information about the symbol. You can equate it with the symbol ***A*** for some purposes, and sometimes you equate it with ***a***. You know it can be a word by itself, and you can even make a sound which corresponds to the symbol: 'a' to rhyme with 'hay'.

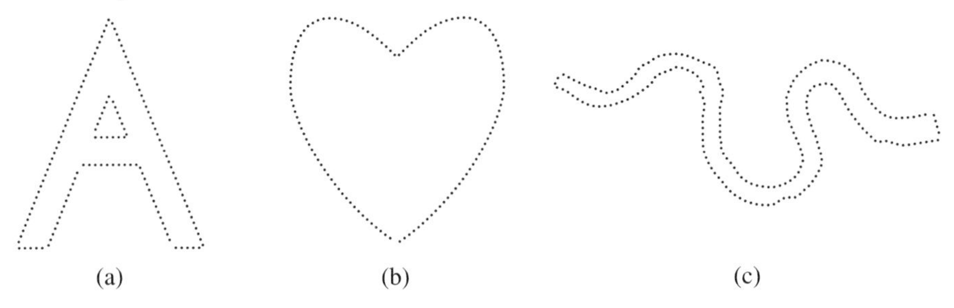

Figure 2.5
Patterns of dots as shapes and symbols.

The second shape can be perceived to be a heart shape. You have a tremendous amount of implicit information associated with this symbol. It symbolizes romantic love and 'affairs of the heart', and you have probably seen bumper stickers giving the message **I ❤ N Y** which can easily be read as 'I love New York'. The third shape is not a symbol that will be recognized by everyone. In fact it represents the River Thames between Tower Bridge in central London and the Thames Barrier in the east. This shape will easily be recognized by many Londoners as symbolic of the capital.

Symbols form the link between explicitly sensed information, and *a priori* information stored in a different form within a machine.

What information is there in the following patterns of letters?

M C A R N C   N F D I B U S P O J D T

E H T O I S   _ _ _ _ _ _ _ _ _ _ _ _

Both of these patterns use symbols to represent information, but recognizing the symbols does not complete the information abstraction process.

In the first pattern a word can be abstracted by applying the rule 'read the symbols from left to right as top-bottom pairs'. Thus the first two letters are M and E, the second two letters are C and H, and so on. In this case the pattern has ***atomic parts*** (letters) with a spatial ***relation*** on them.

The second pattern is also read from left to right, but this time we have used the schoolboy code which replaces a letter by the letter that follows it in the alphabet: N replaces M, F replaces E, D replaces C, and so on. Here we have used the order relation on the letters of the alphabet to abstract the pattern, and this involved pre-existing knowledge.

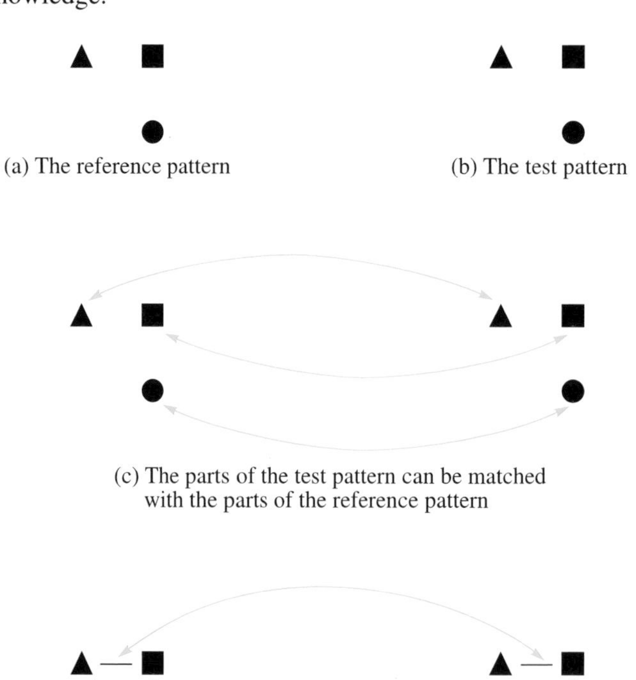

(a) The reference pattern

(b) The test pattern

(c) The parts of the test pattern can be matched with the parts of the reference pattern

(d) The relationships between the parts of the pattern can be matched

Figure 2.6 Matching a test pattern against a reference pattern.

Consider the patterns shown in Figure 2.6 (a) and (b). The first is a *reference pattern*, a standard against which others will be compared. The second is a test pattern. Is the test pattern the same as the reference pattern? Using our human pattern recognition system most of us would instantly say that the two patterns are the same. However, how might this be implemented on a computer? As shown in Figure 2.6, this pattern recognition problem can be split into two. First, we test to see if the test pattern has the same number of parts as the reference pattern, and try to *match* them. Here we can match the triangles, squares and circles, so the patterns are the same as far as the set of their parts is concerned, Figure 2.6(c). Second, we ask if those parts are *assembled* in the same way. In other words, do the patterns have the same relationships between their parts? In this case they do, as shown in Figure 2.6(d): the triangle is to the left of the square in both (shown by the solid line —), the square is above the circle in both (shown by the dotted line ······), and the triangle is above and to the left of the circle in both (shown by the dashed line - - - -). Since the parts match and the relationships between the parts match, we could define the patterns to be the same.

Other considerations have to be taken into account when matching patterns. For example, which of the patterns in Figure 2.7 can be matched with the reference pattern of Figure 2.6(a)?

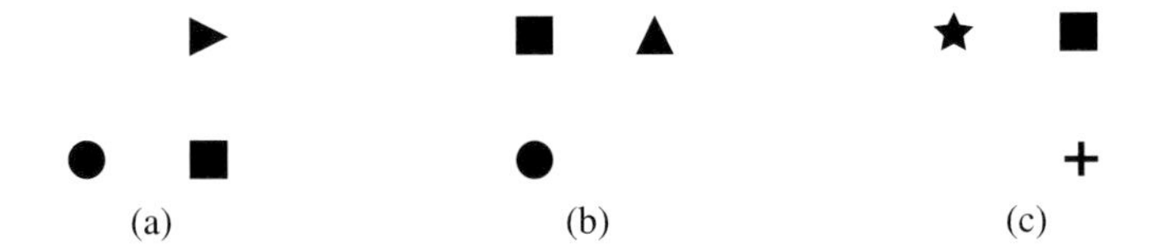

Figure 2.7
Can these patterns be matched with the reference pattern, Figure 2.6(a)?

The first of these, Figure 2.7(a), is the reference pattern rotated 90° clockwise. The triangular, square and circular parts can be matched as before. If the reference pattern is rotated, the relationships between the parts can be matched. Whether or not rotating the reference pattern makes any difference depends on the application. Similarly, the second pattern (Figure 2.7b) is the same pattern as the reference pattern, but this time it is flipped about a vertical axis. In the third case the patterns are similar by having three parts, but they are different because the parts are different. In an application in which the precise nature of the parts matters, these patterns are different. On the other hand, if the precise nature of the parts is irrelevant, or if it is known that stars are equivalent to triangles, and crosses are equivalent to circles, then the patterns are the same. In Section 2.3 we will show how graph theory can make these concepts more precise, as a step towards implementing them on machines.

Apart from matching patterns of objects in space, we often need to match patterns of numbers. For example, consider the following observation of sensor inputs corresponding to a machine's pressure and temperature taken under normal and abnormal running conditions:

Normal	Abnormal
(0.812, 0.423)	(0.714, 0.518)
(0.823, 0.433)	(0.622, 0.444)
(0.720, 0.302)	(0.719, 0.483)

Given these data, could one conclude from a subsequent observation (0.721, 0.310) whether the machine was operating normally or not? And what of the observation (0.705, 0.530)?

Intuitively, (0.721, 0.310) looks more similar to the last 'normal' value and so is likely to indicate normal running. On the other hand, (0.705, 0.530) looks more similar to the 'abnormal' values and so indicates abnormal running. Section 2.5 will show how these data can be systematically analysed as the basis of automated pattern recognition.

Machine pattern recognition attempts to automate the process of finding patterns in both relational and numerical data. Invariably it does this by having a set of patterns of known type stored in its memory, and new patterns are compared with these. If the new pattern *matches* one of the known patterns it is classified as being of that type (or class). The known patterns may come from a variety of sources, but usually they come from ***models***, or a process of ***training***.

In order to train a pattern recognition system it is necessary to start with a set of ***training data***, which consists of input–output pairs. The input data in some way characterize an object type or class. The output is usually the code corresponding to the object type or class.

Once trained, the pattern recognition system will be expected to recognize objects from input data it has not encountered before. For example, it may have been trained on a number of handwritten characters: the input data consist of the pattern of black/white dots from an optical scanner and the output is a symbol such as the letter **L** (Figure 2.9 in the next section). The objective of such a system is to recognize other handwritten characters which are sufficiently similar to the trained **L**. We speak of the system ***generalizing*** from its training data. This means that, given particular examples during training, the system is expected to generalize to other examples which it has not seen before.

This chapter will develop the theoretical foundations of pattern recognition and illustrate these with some particular techniques. It will also consider in some depth how it is possible to train a pattern recognition system and, most importantly, how one conducts rigorous tests to determine success rates.

2.2 Theoretical foundations

2.2.1 What is a pattern?

A ***simple pattern*** is defined to be a set of atomic parts assembled by a relation. A ***pattern*** is a set of parts (possibly simple patterns and parts of patterns) assembled by a relation. A part of a pattern is an *atom* or *atomic feature* if it can be perceived independently of the rest of the pattern.

In the case of the 'sun illusion' (Figure 2.2) there are 16 straight lines. Each can be sensed independently of the rest of the pattern, and so they are atomic features. These atomic features are assembled by a spatial configuration relation which arranges them as the spokes of a wheel. This relation can be expressed precisely by the angles between the lines and their positions relative to their neighbours.

The circle in the sun illusion is an ***emergent feature*** which human perception abstracts from the pattern. It can be extremely difficult to program machines to recognize such subtle illusions. However, there are many emergent features which can be detected quite easily. For example the sides of a square do not possess the 'squareness' property, but it is easy to test if four lines satisfy the relational requirements for them to form a square.

The circle illusion is an example of what psychologists call a *Gestalt*, which comes from the German word meaning 'form'. There are some patterns which can only be sensed as a whole and whose existence breaks down when one of the parts is removed. For example, the property of being the letter L is not possessed by either its vertical or horizontal strokes, I and _ respectively. ***Gestalt patterns*** are those in which the whole is more than the sum of its parts, i.e. their emergent features cannot be recognized in the absence of any of the supporting parts.

2.2.2 Patterns and operators

In his book on patterns and operators, Jean-Claude Simon (see References), a pioneer of pattern recognition, suggests that the subject is best understood in terms of:

computational complexity

the properties of representation spaces

the properties of interpretation spaces.

Computational complexity is discussed in some depth in Chapter 3 on search. Simon writes that 'pattern recognition is first and foremost a battle against complexity'. This means that the obvious approach to pattern recognition where the input pattern is compared with all possible matches is usually not feasible. In general, the computational demands are too great and they always will be.

Therefore we have to try to find methods which are computationally feasible. In general this means devising heuristics (procedures which usually work but are not guaranteed to do so).

Representation is one of the fundamental problems in designing intelligent machines: how is the information and knowledge that underlies intelligence to be stored in the machine? In this chapter two major classes of representation will be considered. The first represents patterns as objects and their relationships. For example, a square is a set of four lines subject to a set of spatial relationships. The second representation uses numbers in multidimensional spaces to represent patterns. For example, coins can be classified by a two-dimensional space of their diameters and their weights as discussed in Chapter 4 of Volume 1.

In general, it is more difficult to handle relational patterns and usually we try to find a representation in terms of numbers. This can mean finding transformations from one representation to another.

As far as intelligent machines are concerned, Simon presents pattern recognition in terms of spaces and operators (Figure 2.8).

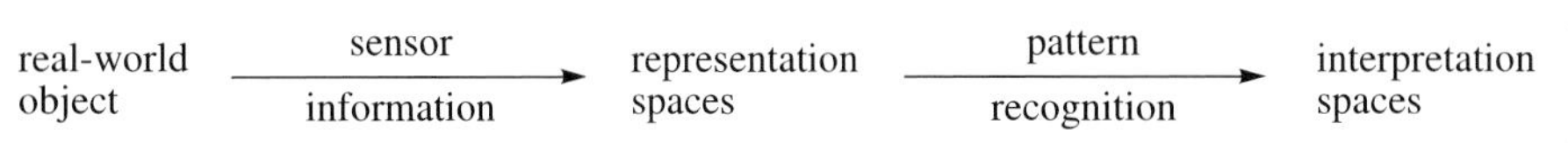

Figure 2.8
The fundamental diagram of pattern recognition, after Simon (1986).

For example, in Figure 2.9 the handwritten pattern which is to be interpreted as the character **L** could have a physical representation as dots in a rectangular array after being digitized by a document scanner. These data in the representation space form the input to the pattern recognition system. This is shown as an operator which takes the bit-mapped representation to a symbolic interpretation as the letter **L**. In fact, in the machine it is more likely to be represented by the ASCII code for the letter **L**, i.e. the binary number 01001100.

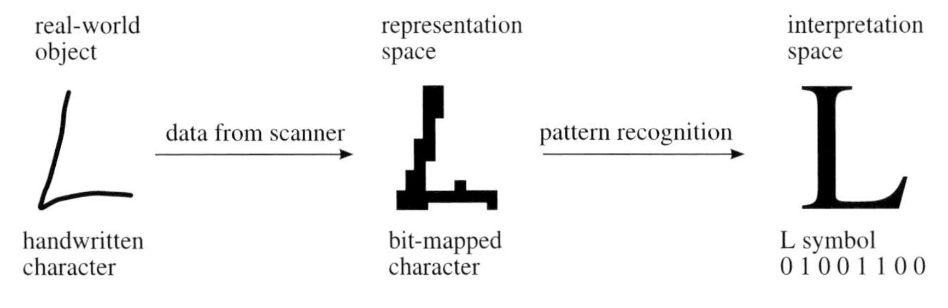

Figure 2.9
The representation and interpretation of a handwritten character.

A ***representation*** of an item observed by a machine's sensors is a string of elements in a finite alphabet representing (or coding) it. At the machine level the representation will usually be a string of bits written in the machine's memory; for example, the 0/1 bits representing black/white picture elements in the bit-mapped image of the handwritten character **L**.

The possibility of multiple interpretations is one reason for the difficulty of automating pattern recognition. For example, in Figure 2.10 the interpretation of the central shape can be either H as in THE, or A as in CAT. Here the interpretation depends on the context.

Figure 2.10
The representation /-\ can be interpreted as H and A.

A single representation can have several different ***interpretations***; these can be made either by a human being or by a machine, and it is important that the two make the same interpretation; e.g. the symbol L or its ASCII code 01001100.

Identification is defined as the action of giving a particular interpretation to a representation. This is the objective of pattern recognition, e.g. interpreting the bit-mapped image of the handwritten character as the symbol L.

A ***feature*** is defined to be the result of a partial identification, e.g. one stroke of a character, one phoneme in a spoken word, an edge or texture in a visual image. The term ***initial*** or ***primitive feature*** is sometimes used in connection with the initial description of a representation. This underlines the fact that there is always a lowest level of data in pattern recognition. In general, primitive features are identifications of the information represented by the sensor inputs.

In many pattern recognition systems the first step is to identify primitive features which together form part of a higher level representation in a hierarchy of identification. For example, a system might first recognize strokes as image primitives, and then recognize configurations of strokes as characters.

It is most important in pattern recognition that the primitives are robust and easy to recognize without confusion. Although subsequent reasoning can correct errors in pattern recognition, it is very hard to do so when the initial data are ambiguous.

2.2.3 Invariance

In pattern recognition we want the identification to be invariant to some things but not to others. Figure 2.7 showed three patterns and asked if they were the same as the reference pattern in Figure 2.6(a). In the case of Figure 2.7(a), if the pattern recognition allows the reference to be rotated the answer is yes. But if it is not *invariant* to rotation (i.e. rotation matters), then the patterns are not the same. Similarly, if the reference pattern can be flipped over then it can be matched with Figure 2.7(b). If the pattern recognition need not be invariant to the precise nature of the parts, then Figure 2.7(c) can be matched with the reference pattern of Figure 2.7(a).

Consider an optical character reading (OCR) machine. We would want it to work whatever the angle of the text, i.e. it should be ***invariant*** to rotation. However, complete rotational invariance would mean that, for example, the symbol + might get confused with the symbol ×. Here it is required that the overall pattern recognition should have rotational invariance but that the system can decide the orientation of the text and treat each character in a rotation-dependent way.

There are many invariances that are of interest in pattern recognition. They include:

invariance to sensor errors and noise

invariance to sensor position and orientation

invariance to signal strengths received by sensors

size invariance

colour invariance

speed invariance

distance invariance.

When specifying a pattern recognition system it is important to specify the required invariances.

2.3 Relational patterns and graph matching

A ***pattern*** is defined to be a set of parts assembled by a relation. The relation can be very complicated, and it is often made up of many subrelations.

An ***n-ary relation***, R, on the set of elements $\{a_1, a_2, \dots a_n\}$, is defined by a proposition concerning these elements which can be judged true or false, and an operational procedure for making that judgement. The related set is written $<a_1, a_2, \dots, a_n; \mathrm{R}>$.

For example, in Figure 2.11 the set {block 1, block 2, block 3} is assembled by the ‘arch’ relation defined as follows. Here $n = 3$ since there are three components being related. The proposition defining the ‘arch’ relation can be stated as a set of relationships such as:

(a) Block x and block y have equal length.

This means there are two blocks which have the same length. For the arch in Figure 2.11 we can choose $x = 1$ and $y = 2$. Substituting these values, relationship (a) reads as: 'block 1 and block 2 have equal length', which can be operationally tested by measuring.

(b) Block *x* must stand vertically and block *y* must stand vertically.

This requires that the blocks designated x and y must stand vertically. This can be tested by measuring their height against their base. Then we find that 'block 1 stands vertically' is true, and that 'block 2 stands vertically' is true.

(c) Block *x* stands to the left of block *y* with a gap between them not exceeding the length of block *z*.

To test this we look to the left of block 2 to see if block 1 can be found there. It can. Then we measure the distance between blocks 1 and 2. In order to test the rules we try block 3 as block z. By measuring block 3 it can be found that the distance between block 1 and block 2 is less than the length of block 3.

(d) Block *z* stands above block *x* and above block *y*, touching both.

This means that block 3 must touch both block 1 and block 2. It does. It also means it must be on top of block 1 and block 2, which it is.

In this way it has been shown that the pattern in Figure 2.11 satisfies the conditions of the 3-ary relation which defines the arch relation. So, by definition, the blocks form an arch. Any set which satisfies all the conditions (a) to (d) is defined to be of the 'arch' pattern.

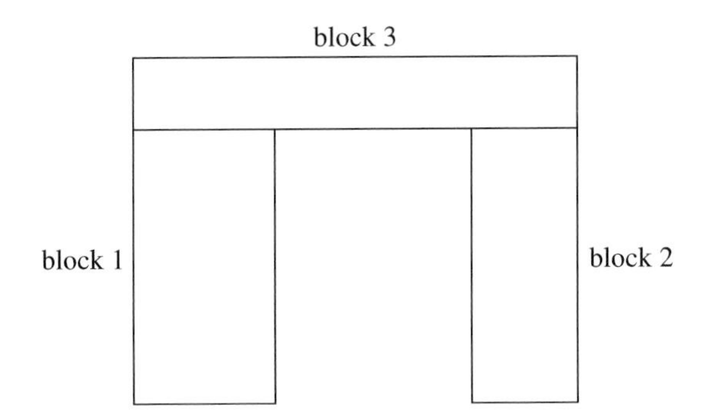

Figure 2.11
An arch pattern <block 1, block 2, block 3; arch relation> assembled from three blocks.

The compound proposition which defines a relation can be hard to understand when it is written in words, and it is common for it to be represented by a ***graph*** such as Figure 2.12, if this is possible. In a graph the ***vertices*** represent objects and the ***links*** represent relationships between pairs of objects.

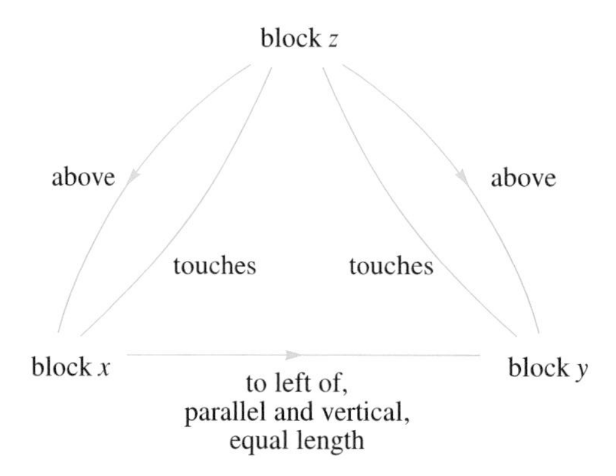

Figure 2.12
A relational pattern.

Given a set of learnt patterns, pattern recognition can be considered to be the process of comparing new patterns with those in memory. If a pattern is identified with one in memory it is identified as being of that class.

One method of establishing whether a test pattern is the same as a known pattern is to compare their graphs: if the graphs are the same then the patterns are the same.

Consider the configuration in Figure 2.13. Should it be recognized as an arch? Let block x = block 4, block y = block 5, and block z = block 6. Then it has the same graph as that in Figure 2.11, and by the definition given it is an arch. But it is clearly not the *same* arch as that in Figure 2.11, even though it satisfies the definition.

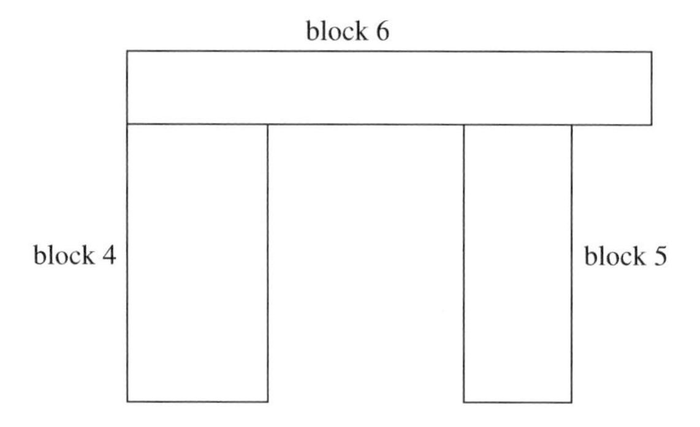

Figure 2.13
An arch with an overhang.

The difference between the two arches is that the second has an overhang on the right. Is this what was required? The answer to this question depends on the purpose of the pattern recognition. If this *is* what is required, then the pattern recognition has worked. If it *is not* what was required, then the definitions of the relations are not precise enough. They might be made more precise by adding the conditions:

(e) All the blocks x, y and z must have proportions between 3:1 and 10:1.

(f) The ends of the top block, z, should be in line with the outsides of the side blocks x and y.

The second arch then fails to be an arch of the pattern of the first. This suggests the following definition. Two patterns are defined to be ***identified*** if:

1 the elements of the first are identified with the elements of the second in a specified way, and
2 the relations of the first are identified with the relations of the second in a specified way.

In practice this means that the engineer has to give rules for matching the vertices in the graphs and rules for matching the links in the graphs.

A ***graph isomorphism*** is defined by two one-to-one mappings. The first maps each vertex of one graph to a unique vertex of the other. The second maps every link in the first graph to a unique link in the second. The mappings have to satisfy the requirement that if link L_i is mapped to link L'_i, then the ends of link L_i are mapped to the ends of link L'_i. Thus if the vertices of L_i are a and b, and the vertices of L'_i are a′ and b′, then a is mapped to a′, and b is mapped to b′.

Figure 2.6 illustrated a graph isomorphism. The mapping between the vertices was given in 2.6(c). The mapping between the links was given in Figure 2.6(d).

One of the major ways of establishing pattern matches is to require that the graphs of the patterns are isomorphic.

In principle the letters p, q, d and b could be considered to be isomorphic, as could the numbers 6 and 9. Indeed the symbols are sometimes written $\underline{6}$ and $\underline{9}$ to make sure they are not identified incorrectly. The underlining is a convention which establishes a relationship between the whole pattern and an ***external reference*** (in this case the horizon). This suggests that the definition for two patterns to be identified requires a third condition:

3 Any relations of the first pattern with external reference objects must be identified with the relations of the second pattern with the external reference objects in a specified way.

The vertices of the configuration in Figure 2.14 can be put into one-to-one correspondence with those of Figure 2.11. However, the blocks do not satisfy our rule (d); block 3 is *under* the other two so there are no links to express block 3 being above block 1 and above block 2. No isomorphism can be established between the graphs, and this shape fails to be recognized as an arch. (Strictly speaking the relations ‘above’ and ‘below’ require an external reference, but this is not developed here for reasons of simplicity.)

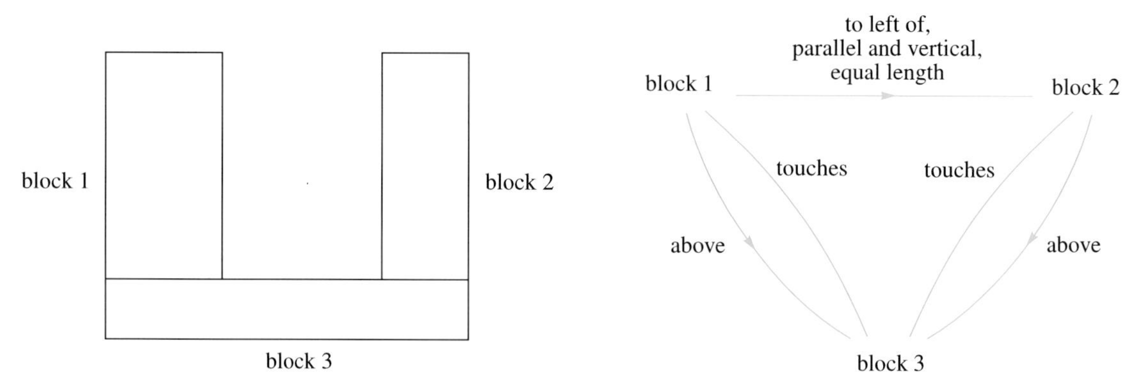

Figure 2.14 An inverted arch.

In pattern recognition we have to allow for the possibility that parts of the pattern are missing, or not quite right. For example, the configuration in Figure 2.15 is almost the same as that in Figure 2.11, but it just fails to meet the requirement that block 3 touches block 2. In this respect it is a ***near miss.***

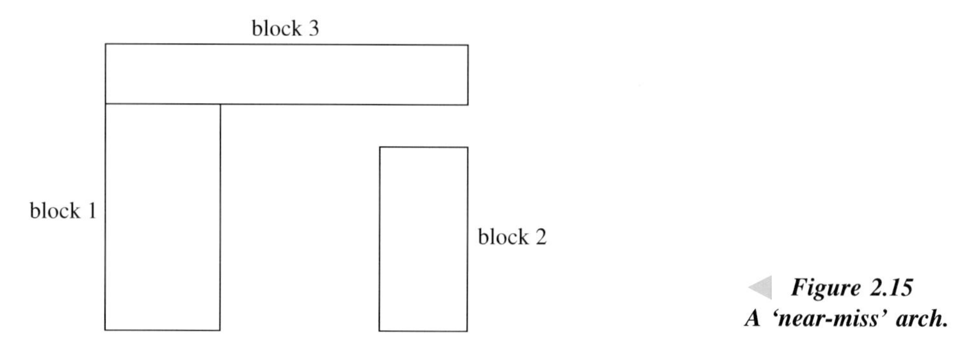

Figure 2.15
A 'near-miss' arch.

The graphs for the arch in Figure 2.11 and that in Figure 2.15 are given in Figure 2.16. The 'near miss' is expressed by the absence of a link in the second graph.

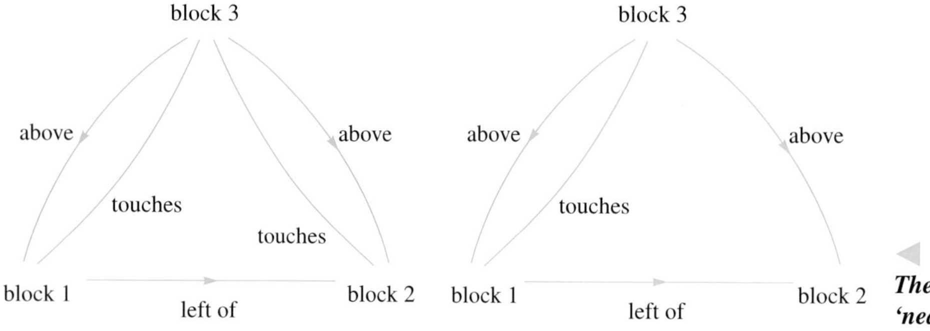

Figure 2.16
The absence of a link in a 'near-miss' pattern.

The graphs give a homogeneous way of representing the multitude of relations between elements of patterns. They allow heuristics to be expressed such as identifying patterns which are near misses by just one link. However, care is needed in this because some links are essential to the integrity of a pattern while others may be more expendable.

When patterns get very complicated their graphs can have many vertices and links. Comparing graphs can be very expensive in terms of computation due to problems of complexity. (Using the notation developed in the next chapter, the complexity is usually of order $O(n^2)$.)

Relational pattern representations might be considered to be *models* of the things they represent. In principle, one could build a pattern recognition based on general propositions. This kind of ***model-based*** pattern recognition is different from the training-based pattern recognition discussed in the next section.

2.4 Hierarchical structure in pattern recognition

Pattern recognition is in general too complex to be performed in a single operation. Usually it begins with *low-level* pattern recognition of the primitive features. These primitives are selected by the criteria:

> they must be robust: easy and reliable to detect
> they should appear in a large class of patterns
> there should not be too many primitives.

In the last two respects, we seek primitives which have properties similar to the alphabet: every English word can be made up from the alphabet, but there are only 26 primitives.

Once primitive features are detected, another level of pattern recognition assembles them to form *higher-level* structures. For example, in an OCR system the primitives might be the strokes of the letters. Intermediate structures might be assemblies such as the configuration L which is found in the letters B, D, E and L.

At the next level of assembly the system might recognize a letter such as E as the assembly of configurations, as in Figure 2.17.

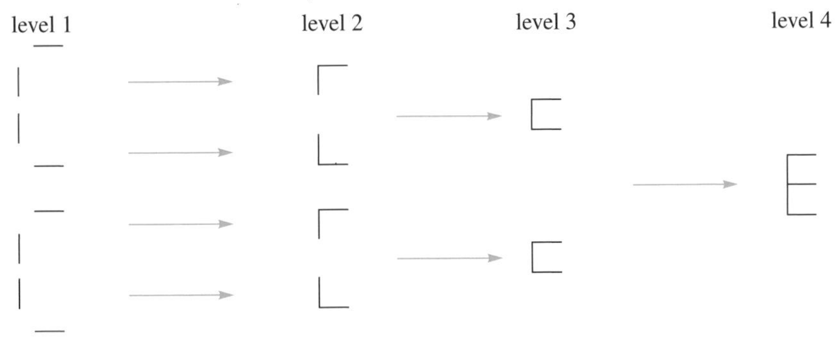

level 1 level 2 level 3 level 4

Figure 2.17
A pattern recognition hierarchy.

The great advantages of using hierarchies in classifications are:

breaking down the process reduces the computational complexity;

this approach can handle missing data.

This last requirement is very important. In many pattern recognition applications one does not have all the data. For example, in computer vision pieces of one object may be occluded by parts of another. Within hierarchical classification we can accept that a configuration such as

|_
|_

might be an E which has lost its top stroke.

Hierarchical pattern recognition is illustrated by computer vision in Chapter 10.

2.5 Data transformation in pattern recognition

Sometimes the initial representation of a pattern does not lend itself to simple methods of automatic pattern recognition. For example, in Chapter 10 there is a discussion of the problem of recognizing insects in digital images.

A digital colour image is effectively three arrays of numbers, one array representing the intensity of red in the image, one array representing the intensity of green in the image, and one array representing the intensity of blue. In other words, each picture element (***pixel***) in a horizontal and vertical grid has three numbers assigned to it, one each for its red, green and blue intensities.

These data can be used to put coloured dots on a computer screen, and our eyes and brains can abstract structure from these mosaics to recognize shapes and configurations. It is difficult to program a computer to achieve the same degree of subtle pattern recognition which our biological vision system performs so effortlessly.

Simply matching the pixels in an image of a reference insect with a test image would be hopeless because the test insect may be in a different position with a different orientation in a different pose. Abstracting the shape of the insects is a quite complex operation which gives poor results in such images.

In Chapter 10 we suggest a very simple solution to the problem of recognizing insects in the images. First, the background is made a single blue colour which is easy to detect. Then six colours are defined* which can be determined from the

* The details do not matter here. In fact the six colours are obtained by partitioning the two-dimensional red–green sub-space of three-dimensional RGB space using a rectangular box classification similar to that described later, in Section 2.6.4.

red–green–blue values of each pixel. In this way most of the pixels can be classified as being blue, red, green, yellow, ochre (a browny orange), black, and white. Blue pixels are background pixels and are ignored. The others are insect pixels and are counted to give a list of six numbers for each insect. For example, in Table 2.1 the ladybird has the most red pixels, the wasp has the most yellow pixels, the greenbottle has the most green pixels, and the blowfly has the most black pixels.

This is an example of a ***data-to-data transformation*** in which the original domain (digitized video image data, 768×576 pixels $\times 2^{24}$ possible colours) has been transformed to another domain (lists of six numbers). For the four insects shown the pattern recognition then becomes very easy, based on the predominant pixel colour.

TABLE 2.1 THE RESULT OF TRANSFORMING DATA FROM THE VIDEO IMAGE DOMAIN TO A NUMERICAL LIST DOMAIN

	Red	Green	Yellow	Ochre	Black	White
Ladybird	12872	554	423	291	9107	223
Wasp	2314	3590	12753	1686	8386	2846
Greenbottle	842	7931	3192	63	14925	3287
Blowfly	1801	4033	432	231	24722	24

The previous example illustrates a very important technique in pattern recognition: if pattern recognition is difficult in one representation domain, then seek a transformation into another representation domain in which the desired pattern recognition may be easier.

Finding appropriate domains and transformations is a creative activity, and is one point at which pattern recognition becomes as much an art as a science. However it is an art that requires the engineer to understand the science of data representation, and the many possible ways of representing things and transforming them within machines.

An important class of pattern recognition problems concerns the recognition of waveforms, since this is the way that many sensors deliver their data. Although these usually have characteristic forms, there may be considerable variation between them. This is illustrated in Figure 2.18 (over page) in which there are three examples of a waveform with a large peak at A, followed by a lesser peak at B, followed by a double peak at C and D, followed by a lesser peak at E.

Although humans can see the similarities in this kind of waveform, it is very difficult to program computers to make the match. For example, although the concept of 'peak' seems fairly clear, the A-peak in (b) is actually a double peak, something that happens quite often in sensor output waveforms. A similar observation can be made for the B-peak in (c). In fact, recognizing peaks is itself a major pattern recognition problem.

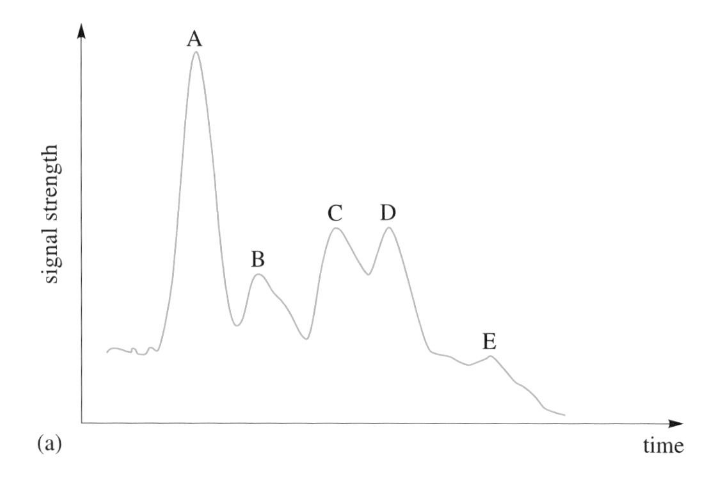

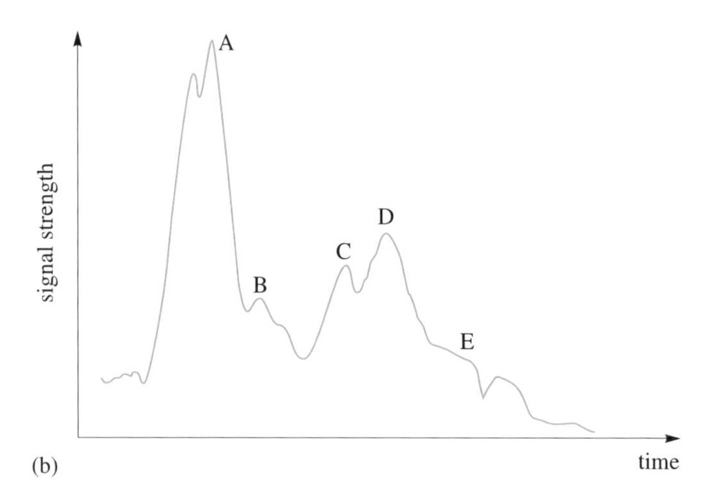

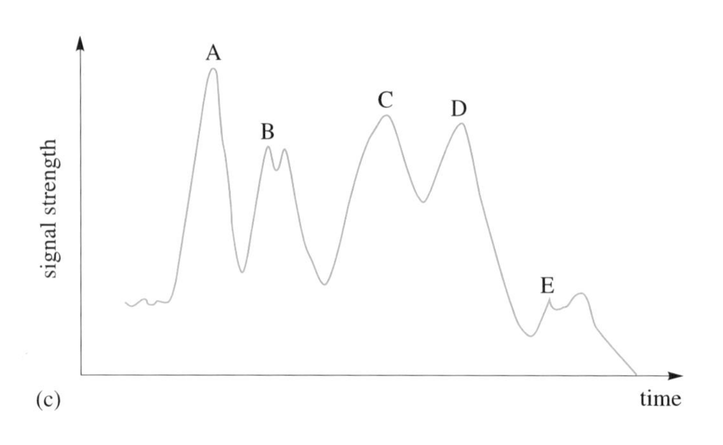

Figure 2.18
Three similar waveforms from a sensor.

Superimposing one waveform on another will not be very informative because the peaks are out of phase, e.g. the C-peak of (c) corresponds more or less to the D-peaks of (a) and (b).

This waveform representation does not lend itself to methods of matching the waves which can be programmed easily on a computer. In such circumstances engineers attempt to ***transform*** the representation from one domain to another in which it is easier to detect the pattern of interest.

An example of such a transformation, described in Volume 1, is the ***Fourier transform***. The Fourier transform converts the data from the time domain to the frequency domain. Recall that periodic signals can be represented by coefficients in the series

$$f(t) = a_0 + a_1 \sin(2\pi f_1 + \phi_1) + a_2 \sin(2\pi f_2 + \phi_2) + a_3 \sin(2\pi f_3 + \phi_3) + \dots$$

from which we can draw a ***magnitude spectrum*** and a ***phase spectrum***, with frequency as the horizontal axis, and the values of a_i and ϕ_i as the vertical axes, respectively.

To illustrate this, consider the sampled waveforms of Figure 2.19 which show voltage varying through time. These waves are said to be in the ***time domain***, and are typical of the data produced by many sensors.

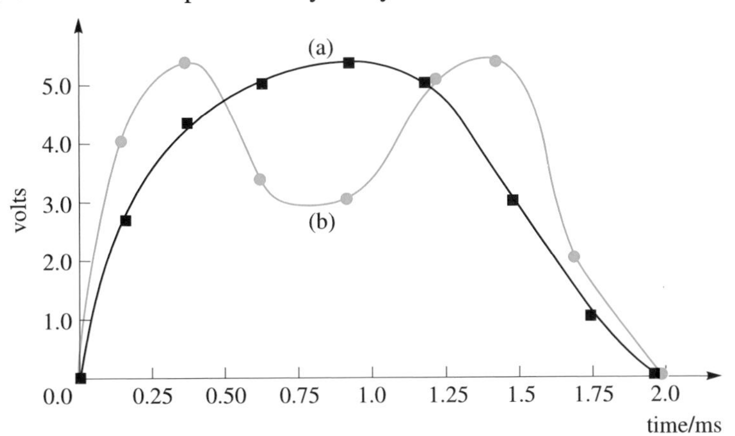

Figure 2.19
Waveforms such as those produced by sensors in the time domain.

Although it is sometimes possible to analyse waveforms in the time domain, it can be difficult when the data are 'noisy', with erroneous spikes and other distortions.

Using the Fourier transform, we can obtain a representation in the ***frequency domain*** as shown in Figure 2.20 (over page). In this case the comparison of the waves can be performed by comparing the five pairs of numbers (vertical bars) representing the magnitudes. In general, similar waves will have similar spectra, or parts of their spectra will be similar.

Transforming the data from the original domain to the spectral coefficients of the Fourier domain can make it much easier to classify the waveforms. Because they offer a more tractable representation than the waveforms from which they are derived, Fourier transforms hold out the possibility of easier pattern recognition operators.

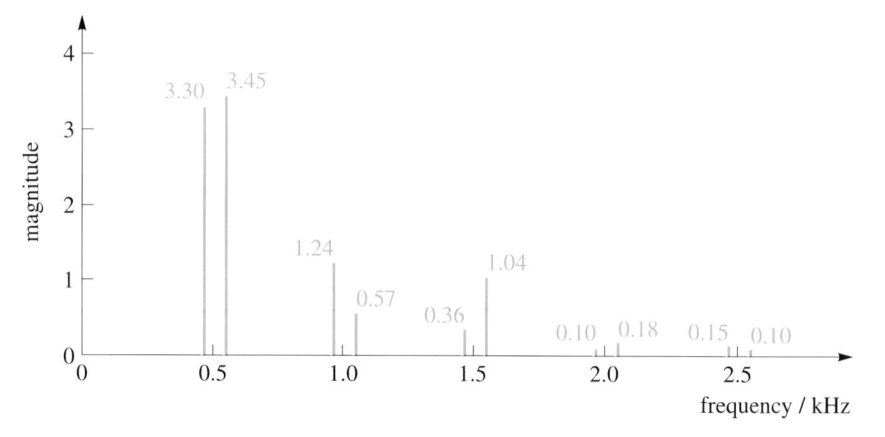

Figure 2.20
The waveforms of Figure 2.19 transformed from the time domain to the frequency domain. The left lines of each pair are the spectral components of (a), the right lines are the spectral components of (b) in Figure 2.19.

Data transformations in pattern recognition frequently reduce the problem to that of comparing sequences of numbers. For example, the insect recognition problem becomes that of comparing the counts of pixels with specified colours. Often the sequences of numbers are considered to be the coordinates of points in multidimensional space, and the pattern recognition problem is transformed into the problem of classifying points in multidimensional space, which we consider next.

2.6 Pattern recognition using multidimensional data

2.6.1 Representing items in multidimensional data spaces

It is very common for the representation of an item to be a set of numbers in multidimensional space. Sometimes this is a consequence of data coming as numbers from n sensors in the form $(x_1, x_2, \ldots, x_n)$, and sometimes data are transformed into this form in order to exploit the many classification techniques for such data.

For example, as discussed in the previous section, computer images of insects can be represented by the numbers of pixels of various colours in the image. This results in each image of an insect being represented by six numbers, $(n_{\text{red}}, n_{\text{green}}, n_{\text{yellow}}, n_{\text{ochre}}, n_{\text{black}}, n_{\text{white}})$. Pattern recognition then occurs in the six-dimensional space of numbers of coloured pixels.

To illustrate pattern recognition based on multidimensional data, consider a hypothetical machine to be used by customs officials to detect explosives of a certain kind. In general, such a machine would have many numerical inputs, but here it will be supposed that there are two, x_1 and x_2. Suppose that in ten trials it is found that suitcases containing explosives give the set of responses shown in Table 2.2.

TABLE 2.2 DATA FROM SUITCASES THAT CONTAIN EXPLOSIVES

Suitcase number	Measurement of x_1	Measurement of x_2
1	25	83
2	29	94
3	10	50
4	15	75
5	25	79
6	23	85
7	19	90
8	27	56
9	26	65
10	17	77

These data can be plotted as a two-dimensional scatter diagram as shown in Figure 2.21, in which the points cluster together in the top-left corner. Suppose a new measurement is taken with $x_1 = 45$ and $x_2 = 80$; does this suitcase contain explosives or not? At first sight one might think that this point in the representation space is close enough to the others to indicate the presence of explosives. However, more information is needed.

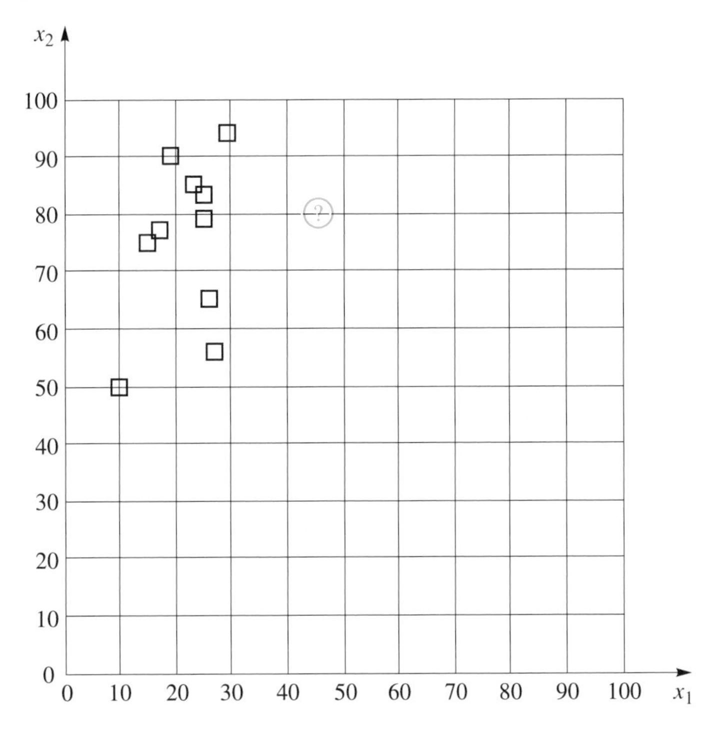

Figure 2.21
The representation points for suitcases containing explosives, plotted as a two-dimensional scatter diagram.

The pattern recognition problem here involves discriminating between two sets: suitcases which contain explosives and suitcases which do not contain explosives. These are the real-world objects the system is trying to recognize. All suitcases presented to the system are represented by a pair of numbers, (x_1, x_2). The objective of the system is to give suitcases the identification 'explosives present' or 'explosives not present'. Let us suppose that the machine encodes these interpretations on a piece of wire attached to an alarm bell with 9 V (bell rings) meaning explosives are present and 0 V (bell silent) meaning explosives are not present. Should the sensor data (45, 80) make the bell ring?

More data are required in order to complete the design of this machine, namely pairs of numbers which are typical of the class of suitcases which do not contain explosives. Suppose a series of trials gives the data in Table 2.3.

TABLE 2.3 DATA FROM SUITCASES *NOT* CONTAINING EXPLOSIVES

Trial number	Measurement of x_1	Measurement of x_2
1	55	83
2	29	30
3	40	50
4	55	75
5	82	49
6	23	45
7	49	90
8	87	56
9	56	25
10	57	63

When these points are plotted in Figure 2.22, it can be seen that the point (45, 80) is actually closer to the samples which do not contain explosives. This suggests that the suitcase with values (45, 80) should be classified as not containing explosives.

The main idea behind this approach to pattern classification is that a point in the representation space will be assigned to the class to whose samples it is 'closest'. This is equivalent to the assumption that the representation space can be ***classified*** so that every point in the representation space is associated with one of the classes of interpretation. This is illustrated in Figure 2.23, in which the partition of the representation space is based on a rather simple procedure which examines the closest pairs of samples between the classes.

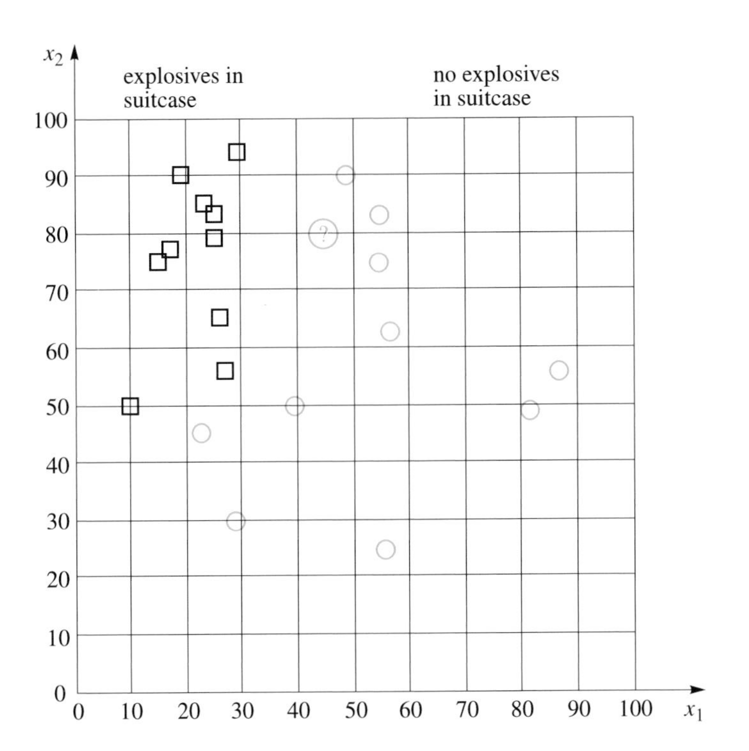

Figure 2.22
The representation points for suitcases containing and not containing explosives, plotted as a two-dimensional scatter diagram (squares correspond to suitcases containing explosives, circles correspond to suitcases which do not contain explosives).

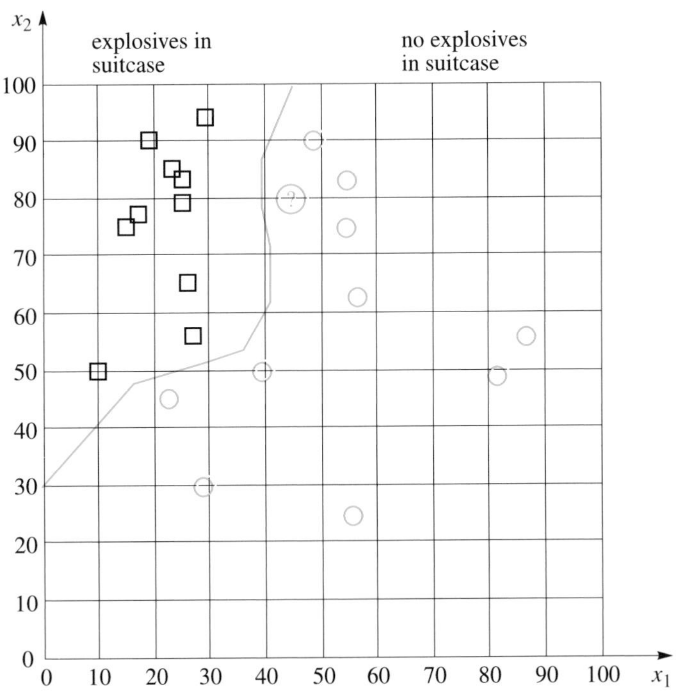

Figure 2.23
Partitioning the representation space to facilitate classifying data points and identification.

This discussion illustrates a general approach to pattern recognition, which involves making the representation space well defined, and finding some criteria that partition it in a way which is consistent with the recognition classes. Then the representation of any new pattern can be located as a point in the partitioned representation space and its class determined.

2.6.2 Multidimensional pattern classification

The general idea behind classification using multidimensional spaces is as follows:

(A) Partition the representation space into classes of points, with each class associated with one of the identifications.

(B) This *a priori* classification allows any new data point to be mapped into the representation space and associated with an identification.

(C) This establishes a pattern recognition.

Although it is very useful to see pattern recognition in this way, it highlights two of the main problems of pattern recognition using multidimensional data:

Problem 1: For some data a simple partition of the representation space may not exist.

Problem 2: When a partition of the representation space does exist, in general there are many ways to partition the sample data in a representation space.

The first of these is often overlooked by those using pattern recognition, but it is most important to test this fundamental property. In the extreme, the data collected may be irrelevant to the classes. For example, suppose the variable x_1 had been the cost of the suitcase and the variable x_2 had been its size in cubic metres. Let data be collected such as that in Table 2.4.

TABLE 2.4

Suitcase number	Cost	Volume / m^3	Explosives found?
1	£37	0.36	no
2	£28	0.39	yes
3	£30	0.36	no
4	£33	0.40	no
5	£27	0.36	no
6	£40	0.39	yes
7	£45	0.38	yes
8	£37	0.36	yes
9	£48	0.42	no
10	£40	0.39	no
11	£40	0.34	no

TABLE 2.4 - Continued

Suitcase number	Cost	Volume / m^3	Explosives found?
12	£45	0.33	yes
13	£29	0.39	no
14	£45	0.40	no
15	£45	0.36	no
16	£30	0.34	yes
17	£39	0.37	no
18	£35	0.33	no
19	£47	0.32	no
20	£45	0.43	yes

When these data are plotted as a graph (Figure 2.24), the representation space cannot naturally be partitioned into classes of suitcases which do contain explosives versus those which do not. The two classes are all mixed together for these variables. In fact things are worse still: there are two points in the representation space which correspond to both explosives found and explosives not found (labelled 'multiple-class points'). This is a profound weakness in these data, and the engineer is well advised to seek indicators which better discriminate the data.

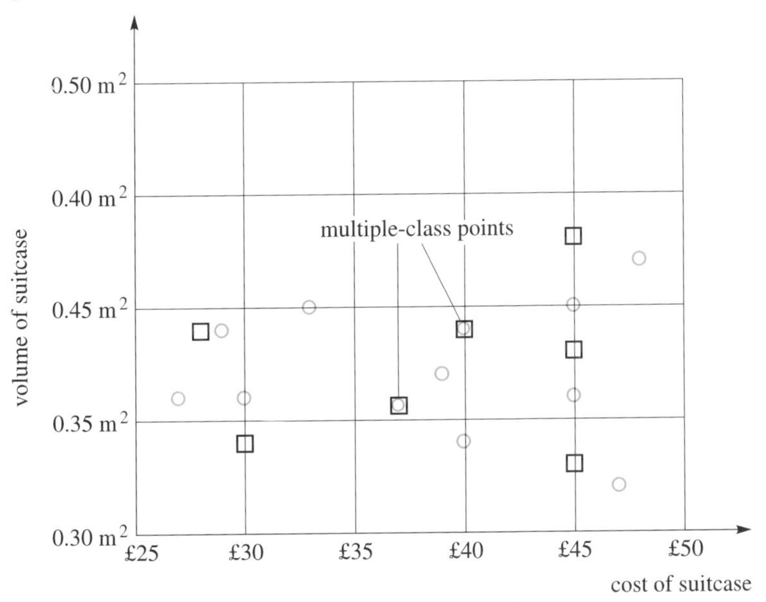

Figure 2.24
The scatter diagram for cost and volume of suitcases shows that these variables give poor discrimination between suitcases which contain explosives and those which do not.

Problem 2 means that selecting one method of partitioning the representation space will give one set of results, while a different method may give different results. Since it is assumed that an item belongs to a single class, irrespective of the method used, the case of contradictory data cannot be resolved by a 'better' classifier; ideally, the source of contradiction should be found.

Engineers sometimes overlook the necessity of testing to ensure that the assumptions underlying a particular classifier are satisfied by the data, and sometimes they use inappropriate classifiers which give poor results for their system in its environment. Such problems could be detected by rigorous testing, as discussed in Section 2.9.

2.6.3 Classifying multidimensional spaces by statistical methods

Figure 2.25 shows a two-dimensional scatter of data points obtained from a remotely-sensed satellite image. The image is a grid of pixels. The dimensions are two spectral bands: x_1 = Band 3 (red, very strong vegetation absorbency) and x_2 = Band 4 (near infra-red, high land–water contrasts, very strong vegetation reflectance). The numbers on these scales are ***pixel greyscales***, and they vary between 0 (no light) and 255 (measurement instrument saturated).

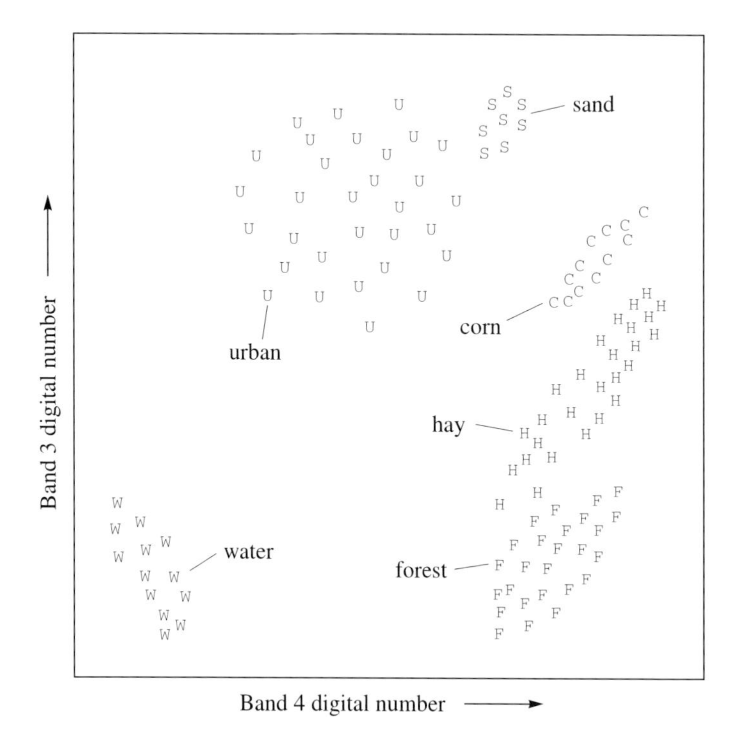

Figure 2.25 Pixel observations on a scatter diagram.

The classes are

S	Sand	F	Forest
C	Corn	W	Water
H	Hay	U	Urban

To obtain this scatter diagram a number of pixels from each of the classes is sampled to give ***training points*** (x_1, x_2) from which the system will 'learn' and generalize.

As can be seen, a class such as 'Water' is quite distinct from the others. However some classes such as 'Hay' and 'Forest' are not so clearly separated.

The objective in this application is to recognize the correct class for every pixel in the image (about half a million) on the basis of a few hundred samples taken on the ground (often called ***ground truth***). Since ground truth samples are time-consuming and expensive to collect, this approach attempts to optimize the information they contain. This example and the subsequent discussion of various ways of partitioning the representation space are taken from Lillesand and Kiefer's standard textbook, *Remote Sensing and Image Interpretation* (see References list).

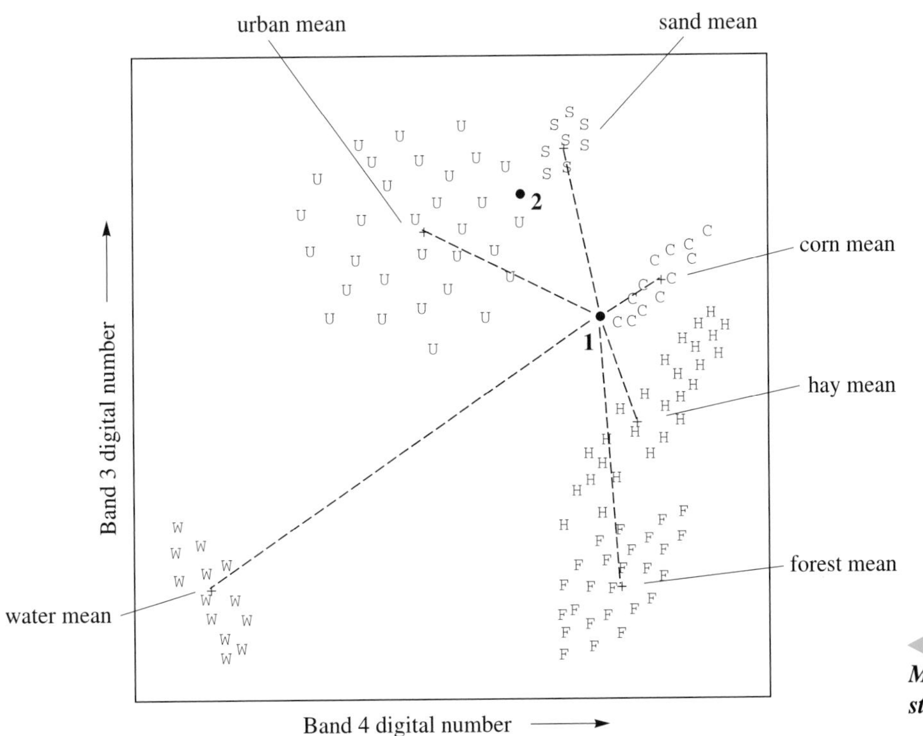

Figure 2.26
Minimum distance to means strategy.

The ***minimum distance to means method*** of partitioning the data space is illustrated in Figure 2.26. The idea here, and in many other approaches, is to abstract statistical measures from all the pixels in a given class, and use these for classification. One of the simplest measures is the ***mean*** greyscale value in each of Band 3 and Band 4. The means of each class are shown by crosses in Figure 2.26.

A test pixel, such as that labelled **1** in Figure 2.26, can then be compared with these means. In this case it is closest to the Corn mean, and so pixel **1** is identified as belonging to the Corn class. Similarly, pixel **2** is closest to the Sand mean, and is identified as a Sand pixel.

This approach is simple and computationally undemanding. However it does not take into account the statistical properties of the distributions in each of the classes. In particular, some classes are much 'tighter' than others. For example, the class of Sand pixels is grouped much closer together than the class of Urban pixels. A second look at Figure 2.26 suggests that pixel **2** has been misclassified. Although it is further from the mean of the Urban pixels than it is from the mean of Sand pixels, the former spread much more than the latter and pixel **2** is actually closer to the sampled Urban pixels than it is to any particular Sand pixel.

Other classification techniques attempt to overcome this by considering the *variance* of the distribution, which measures the spread of the distribution. For example, the ***maximum likelihood*** method of classification assumes that the data points are sampled from a *normal distribution*. One of the main features of a normal distribution is that it is symmetric about its mean. Another important feature is that a normal distribution can be modelled by two statistics, the mean and the variance. The details of this are beyond this text, but these assumptions mean that the ideal distributions of the classes appear as shown in Figure 2.27(a). These have equiprobability contours as shown in Figure 2.27(b).

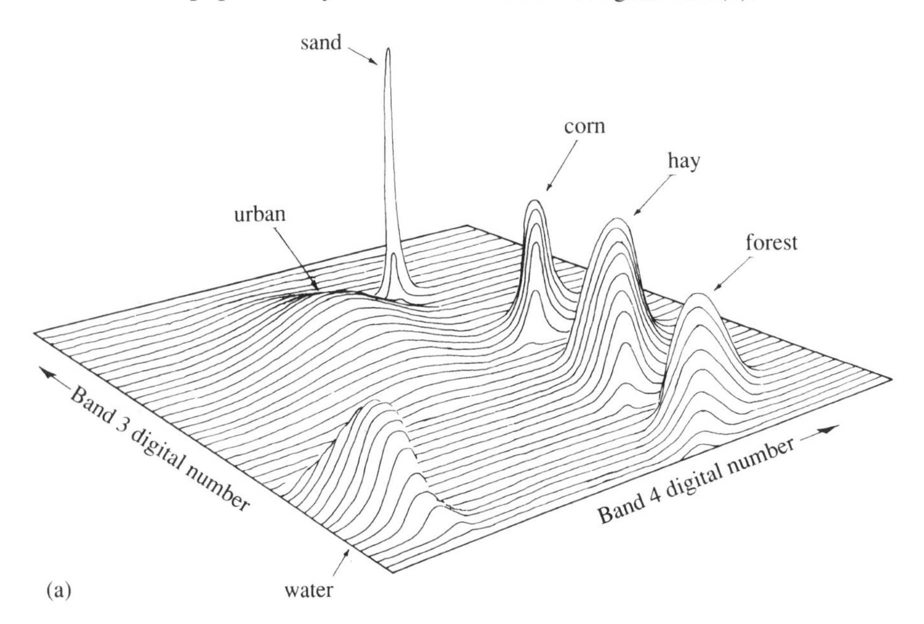

Figure 2.27(a)

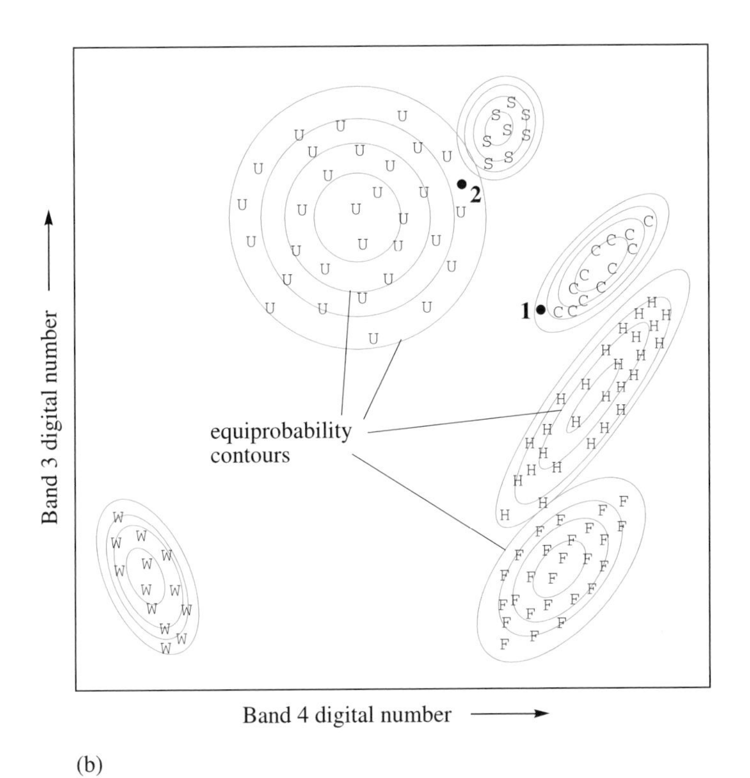

Figure 2.27(b) Probability density functions (a), and equiprobability contours (b), defined by a maximum likelihood classifier.

The maximum likelihood method uses the training data to calculate the necessary statistics and 'calibrate' the probability model based on normal distributions. Once these statistics have been calculated, the method classifies a point (x_1,x_2) according to the greatest probability for each of the classes at this point. Thus in Figure 2.27(b), point **1** has the greatest probability of being a Corn pixel. This method assigns pixel **2** to the class Urban, which accords much better with intuition.

The maximum likelihood method has the disadvantage that it is computationally expensive and too slow for all but the simplest real-time applications. It also has the disadvantage that it systematically misclassifies points and introduces error.

There are other statistical techniques, such as Principal Component Analysis and Factor Analysis, which attempt to give summary information in the multidimensional data by projecting them onto axes in a way which accounts for the maximum variance. In general, they have demanding data requirements. They are also expensive and slow, and may be unsuitable for real-time applications.

2.6.4 Rectangular box classification

A very simple approach to classifying multidimensional spaces involves defining intervals for each dimension. In the case of two dimensions these intervals define a rectangular box, as shown in Figure 2.28(a).

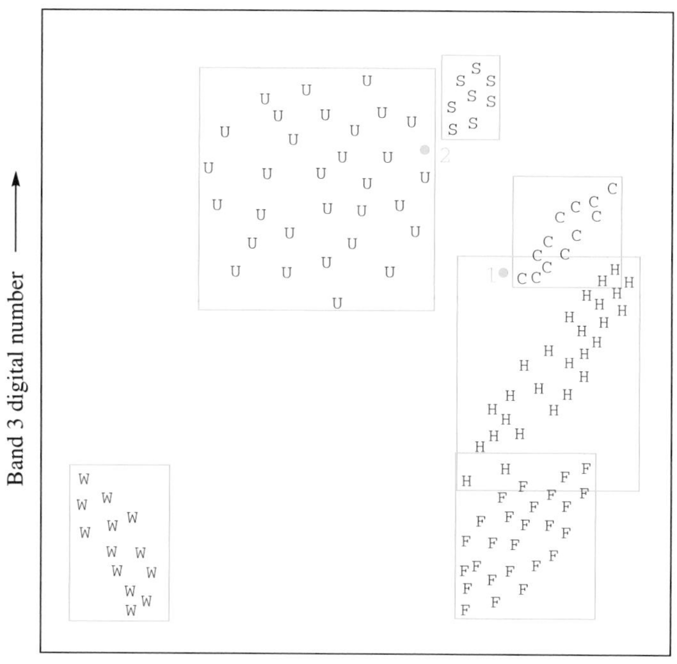

(a)

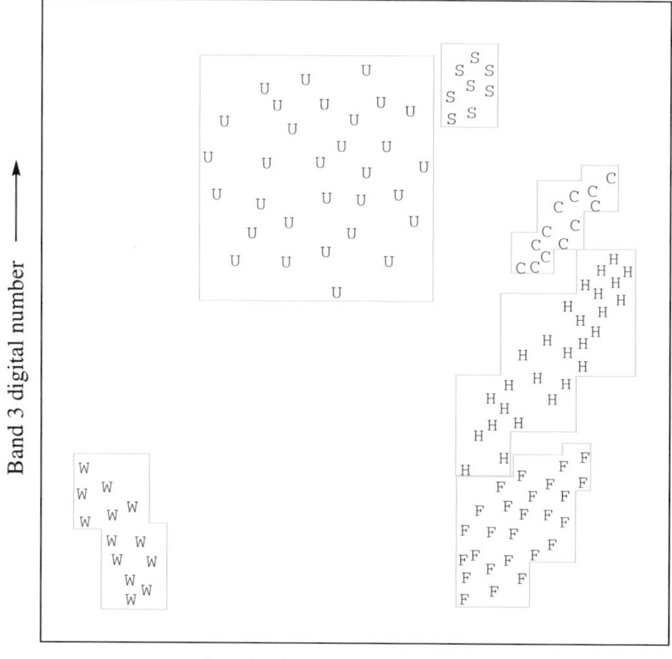

(b)

Figure 2.28
(a) Rectangular box (parallelepiped) classification strategy; (b) strategy employing stepped decision boundaries.

The technique is very inexpensive computationally and suitable for pattern recognition in which the classes are widely separated, as in the case of Water, Urban, Sand, Corn and Forest. This technique becomes problematic when the classes intersect, as in the case of Hay and Corn, and Hay and Forest. Sometimes this can be overcome by a refinement which uses stepped decision boundaries, as in Figure 2.28(b).

Lillesand and Kiefer call this technique ***parallelepiped classification***, but we prefer the term ***rectangular box***.

2.6.5 Non-metric classification for chalk–cheese systems

One of the great dangers of representing pattern data by a sequence of numbers such as $(x_1, x_2 \ldots, x_n)$ is that it is almost irresistible to assume that this is a point in a metric space, i.e. a space in which a meaningful *distance* can be attributed between every pair of points.

Sometimes things are referred to as being 'as different as chalk and cheese'. For illustration, suppose one dimension of a representation is *chalk* and the other dimension is *cheese*, and that for the variables concerned there is no natural equality between the two. For example, density might be measured along the cheese axis and purity might be measured along the chalk axis.

A 'distance' can be calculated between the points $(1, 2)$ and $(4, 6)$ as

$$\sqrt{(1-4)^2 + (2-6)^2} = \sqrt{9 + 16} = 5$$

But this is exactly the same as the distance between the points $(1, 2)$ and $(6, 2)$. So, as far as this measure of distance is concerned, a difference of 5 along the chalk axis can be 'traded' against a difference of 3 along the chalk axis and a difference of 4 along the cheese axis. In fact the assumption that the distance metric exists is equivalent to assuming that one unit of cheese/density equates to one unit of chalk/purity. This could lead to some very odd results!

When the dimensions of a multidimensional representation space have no natural trade-offs, it will be defined to be ***non-metric***. How then can the points in non-metric space be classified? The answer to this question lies in understanding that the concept of distance is related to that of ***closeness***, and that the required classification depends upon ***relative closeness***.

In the following, the symbol $|a|$ means the *absolute value* of a. It has the magnitude of a irrespective of whether the sign of a is positive or negative, so that $|a|$ is always positive. For example, $|-7| = |7| = 7$.

In a non-metric representation, let a point $\mathbf{p} = (x, y, z, \ldots)$ be defined to be ***closer*** to the point $\mathbf{p}_c = (x_c, y_c, z_c, \ldots)$ than to the point $\mathbf{p}_d = (x_d, y_d, z_d, \ldots)$ when

$$|x_c - x| \leqslant |x_d - x|, \quad |y_c - y| \leqslant |y_d - y|, \quad |z_c - z| \leqslant |z_d - z|, \quad \text{and so on.}$$

In other words, point $\mathbf{p}$ is closer to $\mathbf{p}_c$ than it is to $\mathbf{p}_d$ if it is closer on *every* dimension $x, y, z, \ldots$ (Figure 2.29).

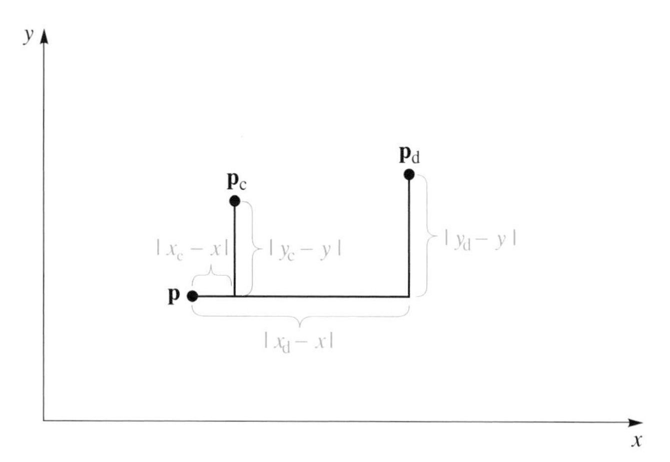

(a) Point $\mathbf{p}$ is closer to $\mathbf{p}_c$ than $\mathbf{p}_d$ because $|x_c - x| < |x_d - x|$ and $|y_c - y| < |y_d - y|$.

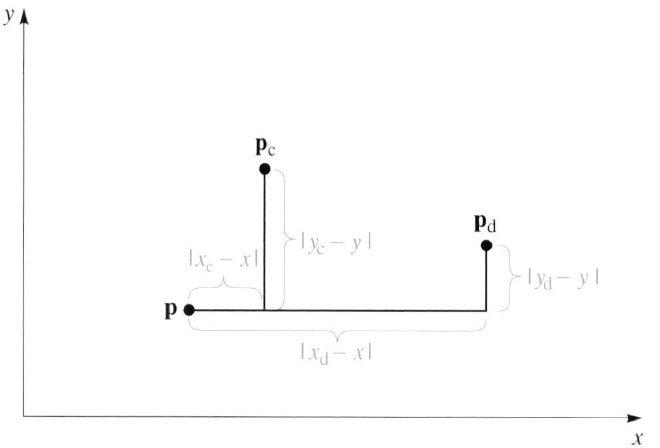

(b) The 'distances' between are non-comparable because $|x_c - x| < |x_d - x|$ but $|y_c - y| > |y_d - y|$

Figure 2.29 Relative closeness and non-comparability in a non-metric classification space.

This definition of closeness means that it is not always possible to say that a given point is closer to one of two others. Such pairs of points are said to be ***non-comparable***.

Let an ***identification point*** $\mathbf{p}_c$ be a point which is associated with identifications for class C. Typically, identification points come from *training data*, i.e. known examples of the patterns which are used to 'train' the system.

Then every point $\mathbf{p} = (x, y, z, \ldots)$ which is 'closer' to the identification point $\mathbf{p}_c = (x_c, y_c, z_c, \ldots)$ than any other identification point will be identified with class C.

This definition means that for any identification points $\mathbf{p}_c$ and $\mathbf{p}_d$ the representation space will be partitioned into three parts: those that are closest to $\mathbf{p}_c$, those that are closest to $\mathbf{p}_d$, and those that are not closest to either, as shown in Figure 2.30.

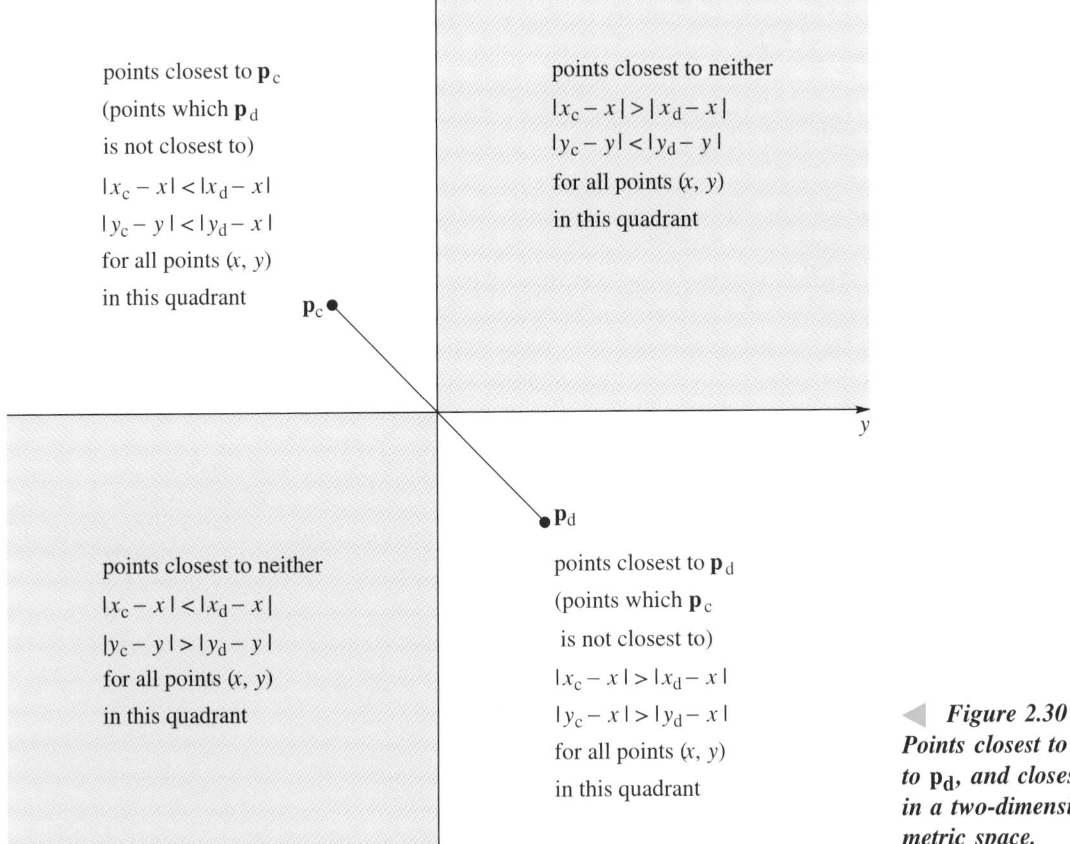

Figure 2.30
Points closest to $\mathbf{p}_c$, closest to $\mathbf{p}_d$, and closest to neither in a two-dimensional non-metric space.

For a set of identification points, pairwise comparison gives the points which are *not* closest to a given identification point, **p**. This is illustrated in Figure 2.31, in which the identification points around **p** establish which points are not closest to **p** and which points cannot be classified by examination of **p**. The remaining points are closest to **p**.

This method establishes a partial classification of the multidimensional space, i.e. some points can be classified as being closest to one of the identification points, but some cannot. This classification does not require a distance function which trades off values on one dimension against another, and so it is appropriate for chalk-and-cheese representation spaces. Non-classification occurs when a test point is closest to an identification point on one dimension but closer to another identification point on another dimension. The more general problem of what to do with the areas of non-classification is discussed next.

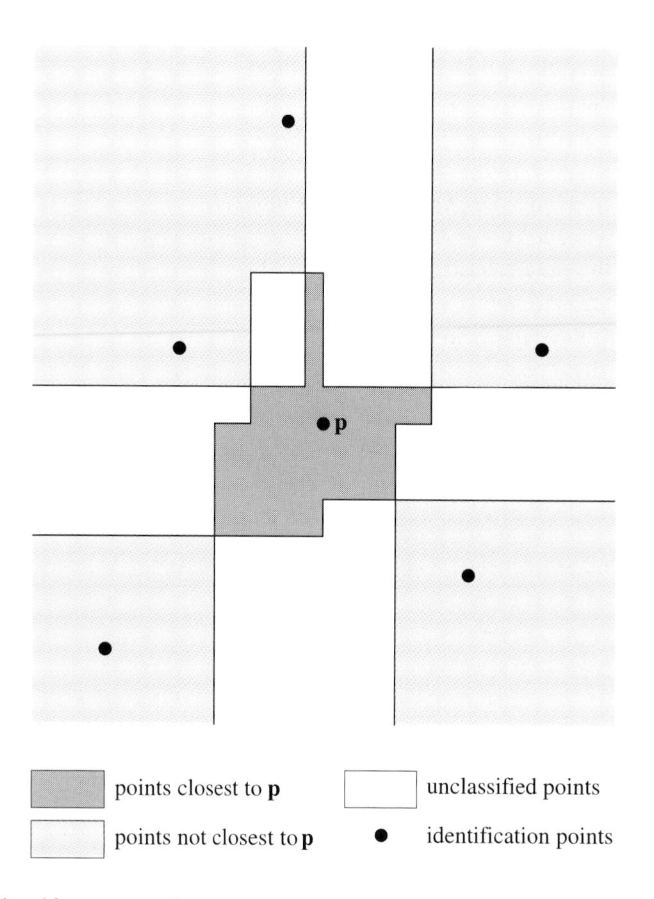

Figure 2.31
The points closest to p in a 2-D non-metric classification space. Unshaded areas are unclassified points.

To classify a test point we can proceed as follows. Given a set of identification points $\mathbf{p}_a$, $\mathbf{p}_b$, $\mathbf{p}_c$, $\mathbf{p}_d$, ..., it is required to know which, if any, is closest to the test point **p**.

For each dimension we calculate the minimum distance between **p** and all of the identification points. We can then test each identification against these minimum distances. If a particular identification point is closest on all dimensions, it is the closest to **p**. If none of the identification points is closest on all dimensions, then **p** cannot be classified by them.

The main problem with the non-metric approach is that it may systematically misclassify if the training data contain errors. It is therefore only suitable for applications in which there is a low cost of misclassification, or for which we can be certain that the training data are correct. However, the systematic nature of errors resulting from faulty training data is an advantage because their consistency makes them relatively easy to detect and remove.

In practice, non-metric classification makes the analyst address some difficult questions. On what basis can the data points be assumed to be *separable*? Intuitively, sets of data points such as those of Figure 2.25 are considered to be separable if a line can be drawn between the set of their points. If the separating line is straight, then the data points are said to be *linearly separable*, as discussed in more detail in Chapter 4 on Neural networks.

The problem of separating a data set into two classes is illustrated by the pathological example shown in Figure 2.32(a). In this case the two classes spiral round each other in a way that makes standard classification techniques inappropriate. This space can be partially classified by the non-metric classification discussed in this section. It can also be classified by the process of *dilation*. Classification by dilation involves, for example, defining a square of sides of one unit around each data point. Assuming none of the squares intersect, all the points in a square are assigned to the data class of the dilated point. The process is then repeated by defining a square of two units around each data point. Again all the points within these squares are assigned to the class of the 'dilated point'. Eventually the dilated squares from different classes meet and the dilation has to cease on that dimension.

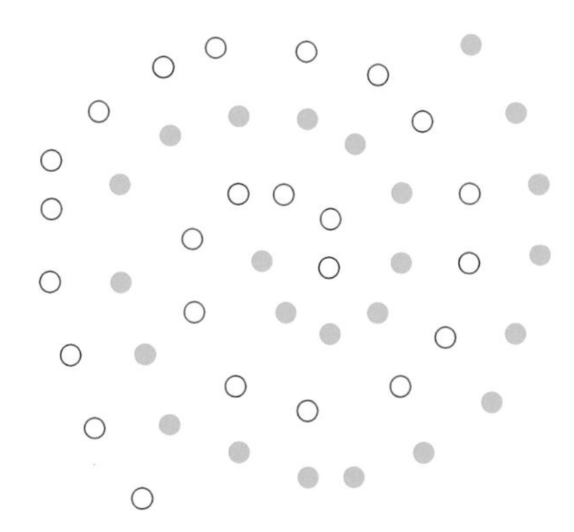

(a) The double spiral problem

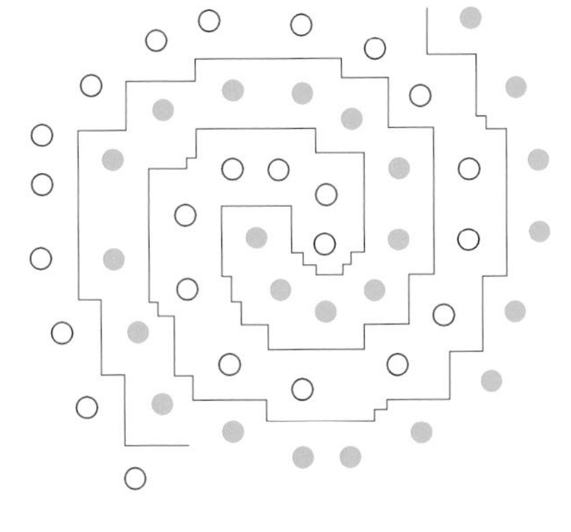

(b) The non-Euclidian dilation solution

Figure 2.32
(a) The double spiral problem. (b) Johnson's non-Euclidean dilation solution.

Classification by dilation gives good results for the spiral data, as shown in Figure 2.32(b). It is a variant of the rectangular box classification discussed in Section 2.6.4. It has the disadvantage that dilating by the same amount on different dimensions requires justification for chalk-and-cheese systems. It is easier to take into account the need for dilations due to different distributions along the dimensions, but the problem of trading chalk and cheese still holds.

Figure 2.32 highlights an interesting property of the separating data dimensions, namely that their importance is *local*. The separation of the points at the bottom and top of the spirals depends only on the vertical data dimension, while the points at the sides depend only on the horizontal dimension. Thus the data points are behaving differently, and treating them as a homogeneous population without taking into account the different roles played could lead to misleading results. For example, it may be possible to construct populations of black and white spiralled data points which are normally distributed and have the same means and variances, even though they are different classes!

In this chapter we cannot resolve all the problems that have been raised, but they serve to show that classification can be very subjective and can sometimes depend on the method chosen and the nature of the data.

2.6.6 Neural networks as pattern classifiers

All the methods of pattern recognition discussed so far have drawbacks of one kind or another. A particular problem is that they do not filter out irrelevant data.

In the example of suitcases which might contain explosives, variables such as the cost and size of the suitcase were not useful for the classification. In that example these were chosen to make that point, but in general we do not know how much information there is in the data for any given dimension.

The emerging technology of neural networks, discussed in detail in Chapter 4, automatically filters out irrelevant data by altering weights on connections which involve those data. For this and other reasons, neural networks are sometimes proving to be very powerful classifiers for multidimensional data.

2.7 Multiple classifications and fuzzy sets

In all classifications there is the problem of those elements 'at the edges'. This is an artificial problem caused by insisting that the observed world can be conveniently partitioned into non-intersecting classes. Traditional set theory requires that an element either belongs to a set or that it does not. Similarly, Boolean logic requires that a proposition is either true or false.

But, as illustrated in the Escher engraving reproduced in Figure 2.33, there can be a gradation of set membership from 'very strong membership' to 'very weak membership'. For example, the first four rows of birds at the top of the picture are

Figure 2.33
M. C. Escher's 'Sky and Water I' (1938).

clearly bird shaped. The next row is less clearly bird shaped. The black shape at the centre of the picture could be said to be bird shaped, but then again it could be said that it is not bird shaped. How can one decide? Fuzzy set theory and fuzzy logic do not force this kind of decision. These are discussed in detail in Chapter 6 on reasoning.

Arguably, classifying things into mutually exclusive sets is a non-problem which has its roots in a scientific tradition which has obtained success from dividing things up into mutually exclusive parts. Indeed, methods based on partitions, such as the maximum likelihood, result in systematic misclassification, as shown in Figure 2.34.

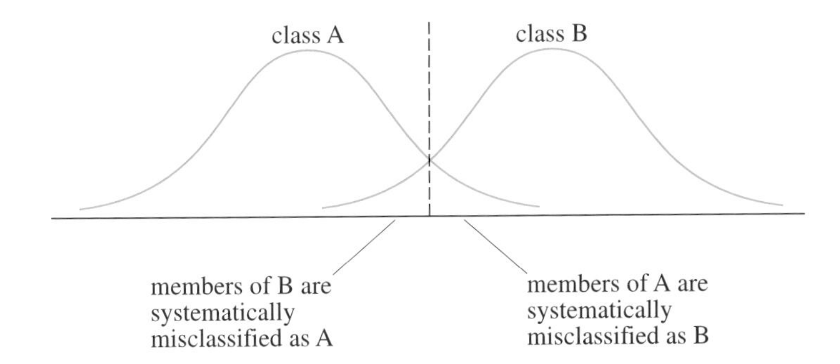

Figure 2.34
Forced classification may lead to systematic error.

Usually, when classifying a representation space some of the classes are disjoint, while a few classes intersect. Sometimes it is better to record that an item belongs to class A *or* class B without attempting to decide between them.

Pattern recognition occurs in perception when a machine has to transform raw data from its sensors into a form of information that it can operate on. In a simple machine the result of the classification is fed directly to an actuator. For example: inputs belonging to class A result in a control sequence 1; inputs belonging to class B result in a control sequence 2; inputs which belong to class C result in control sequence 3; and so on. In such a case it is necessary for the machine to decide which of the available control sequences to invoke. A decision must be made, even if it turns out that the classification was incorrect.

In more sophisticated machines the classification information may be used during reasoning. For example: *if* the inputs currently belong to class A and the machine is in state B and the environment is in state C, *then* invoke control sequence 2.

When the machine is using the classification information for reasoning it may not be necessary to force a classification decision. This is especially the case when the machine is using *fuzzy reasoning* as explained in Chapter 6. A ***fuzzy classification*** is one which assigns a value between zero and one to the set membership. Choosing one of the fuzzy options to make a particular action is sometimes called *defuzzification*, as explained in Section 6.3. As noted earlier, it may be better to allow that an outcome is A or B but *not* C or D. This kind of partial defuzzification may give better information when the data are fed into a knowledge-based system.

As an example, in the case of the 'explosives in the suitcase' example developed at the beginning of this chapter, the machine may decide that a given suitcase belongs to the set of 'suitcases with explosives' with a weighting of 0.3 and that it belongs to the set of 'suitcases without explosives' with a weighting of 0.7. The *decision* as to what to do about this will depend on many things. For example, it may be that very few suitcases reach a value as high as even 0.1 for containing explosives. In such a case it may be decided to adopt procedures which require that a suitcase with a weighting of 0.1 or more should be opened, even though the expectation of finding explosives in these suitcases may be one in a million or less.

This discussion brings us to the important question of the costs of failure and success in pattern recognition.

2.8 Errors: non-recognition versus misclassification

In pattern recognition there are two related measures of failure. The first is the ***proportion of misclassification***, i.e. the number of times the system assigns an item to the wrong class. The second is the proportion of non-classifications in which the system cannot assign an item to a class.

Misclassification is usually a more serious error than non-classification since the former comes with no indication that something is wrong.

In pattern recognition systems which require the match to exceed a threshold there is a trade-off between non-recognition and misclassification. In general, increasing the match threshold makes the recognition criterion more severe, and so reduces the number of misclassifications. On the other hand, it is likely to increase the number of non-classifications due to more borderline cases failing to meet the more rigorous criterion.

In any pattern recognition application the engineer should take into account the cost of misclassification, and should design into the system procedures for handling misclassification or their consequences, and for handling non-classifications.

2.9 Rigorous procedures for training pattern recognizers

Pattern recognizers which are trained from data present the engineer with the problem of deciding which of the available data to use for training, and which to use to test the trained system.

Training data consist of ***input–output pairs***. Usually these are obtained by experiment and it is assumed that the given output is the correct pattern class for the inputs. When one is training it is very tempting to hold back for testing those pairs which seem to give the best results. To yield to such temptation is a grave error which invariably leads to poor system behaviour in the future.

Suppose then that n training pairs are available to train a pattern recognition system. The engineer must find reasoned answers to the following questions:

> What is the smallest value of n which will give a reliable interpretation of the test results?
>
> How many of the n pairs should be used to train the system?
>
> How many of the n pairs should be held back to test the trained system?
>
> How should the test pairs be selected?
>
> How can we interpret the results of testing the system?

Suppose that all the n training pairs were used to train a system. Suppose also that these pairs form an ***unbiased sample*** of all possible representation–identification pairs. Usually this means that they are selected at random.

After training, if m of these were correctly classified, then the proportion of failures is $(n-m)/n$. This is likely to be the best performance this system will achieve, and it may perform much worse on unseen data. As discussed in the previous section, the engineer should know the cost of misclassifications and whether this upper limit on the recognition rate of m/n is sufficiently high for each identification class.

In general, one would not want to use methods which did not have a ratio of m/n very close to unity, i.e. one would have to justify the use of pattern recognition techniques which significantly misclassified their training data.

Assuming that a pattern recognition technique can train sufficiently well on its training set, how well does it ***generalize*** to other data? Usually this is an empirical question. The method to test this is to divide the training data into two sets: one to use for training and one to use for testing. In general, the more data used to train the system the better it will perform. This suggests that one wants to hold back as few input–output pairs as possible for testing.

Statistical sampling theory tries to answer the question of generalization. Suppose you asked someone if they like champagne and they answered 'no'. Since 100% of your sample answered 'no', could you deduce that 100% of the population do not like champagne? Obviously not. Suppose you asked a second person and they answered 'yes'. Could you deduce that 50% of the population like champagne and that 50% do not? What if you asked ten people and five said 'yes' and five said 'no'? Intuitively, the more people you ask the more confidence you have that the proportion of their answers represents the proportions of the whole population. This kind of reasoning lies at the heart of the problem of how many training pairs should be held back for testing.

Statistical sampling theory is beyond the scope of this book. It is based on the idea that the result of any particular experiment is a sample from the whole population of all experimental outcomes. This theory has been used to prepare Table 2.5.

The column headed 'Experimental results' shows the number of failures that occurred in a series of trials. (For example, we might be testing the classification of a neural network and observe that it misclassifies one time in ten.) The second column headed 'Estimated failure rate' is simply the observed proportion of failures. However, it is unlikely that this reflects quite precisely the true underlying failure rate, which might be somewhat lower than that observed or, more critically, *rather higher.* For instance, for an observed failure rate of once in ten trials, the actual underlying failure rate might be a little lower than 0.1, or possibly higher than that: 0.2 is certainly plausible, 0.4 too, ... 0.6? ... but 0.7 or higher seems scarcely credible. The next four columns quantify these notions: they give 90%, 95%, 99% and 99.9% *upper confidence limits* for the underlying failure rate, based on the observed experimental results.

TABLE 2.5 UPPER CONFIDENCE LIMITS FOR THE UNDERLYING FAILURE RATE

Experimental results	Estimated failure rate	Upper confidence limits for underlying failure rate			
		90%	95%	99%	99.9%
1 in 10	0.1	0.34	0.39	0.50	0.62
10 in 100	0.1	0.15	0.15	0.19	0.22
100 in 1000	0.1	0.11	0.12	0.12	0.13
1 in 100	0.01	0.038	0.047	0.065	0.089
10 in 1000	0.01	0.015	0.017	0.020	0.024
1 in 1000	0.001	0.0039	0.0047	0.0066	0.0092
2 in 1000	0.002	0.0053	0.0063	0.0084	0.0112
5 in 1000	0.005	0.0093	0.0105	0.0131	0.0164

After one failure in 10 trials, while it might be tempting to interpret this as '10% failure', Table 2.5 shows how dangerously conservative such an extrapolation might be – one can 'only' be 99% confident that the true failure rate is less than one-half. In some contexts, even this level of confidence is unsatisfactorily low.

If only ten failures are observed in 100 trials, the estimated underlying failure rate would be the same at 10%, and because the conclusions are based on a much more extended experiment, much more confidence can be attached to a low underlying rate; but even then you can see that an underlying failure rate more than double this (22%) is still just credible. If only 100 failures are observed in 1000 trials (again, an estimated 10%), the 99.9% upper confidence limit (which by any standards must be deemed quite high!) for the underlying failure rate is just 13%.

Other rows in the table tell a similar story. But notice that when the estimated failure rate is very low (say, 0.001 after observing 1 failure in 1000 trials) then, despite the size of the experiment, 'reasonable' upper confidence limits for the underlying failure rate might still be as high as five or seven times the estimate.

To illustrate the interpretation of Table 2.5, consider a machine which tests for explosives. Suppose that in ten trials of suitcases containing explosives the system failed once, giving an estimated failure rate of 0.1. The underlying failure rate is not as low as this, but these data can be interpreted as meaning that we can have 99% confidence that the underlying failure rate is less than 0.5. Put another way, a person contemplating trying to smuggle explosives could be 99% confident that they would get caught at least half of the time, and this would probably be sufficient deterrence, making the machine viable from a detection viewpoint. However, this is only part of the story. After the machine has identified a suitcase as containing explosives, it can be assumed that a customs officer will open the case and conduct a more detailed search. This costs time and money. It is therefore important to know how many false alarms the system generates. Suppose 1000 suitcases not containing explosives are tested, and there are five failures. This means that we can be 95% confident that the rate of false alarms is 0.0105 or less. At this level of confidence, opening about 1% of suitcases as a result of false alarms might be considered acceptable.

These figures should be studied and understood by those who intend to build pattern recognition systems. It is perhaps surprising that just one or two observed errors can give rise to such large statistical ranges.

For some pattern recognition purposes misclassifications can be very expensive, and the feasibility of the system depends on the proportion of misclassification being very low.

In recent years there has been a great interest in classifiers such as the neural networks described in Chapter 4. Unfortunately, there has been a tendency to overlook rigorous statistical methods and to quote 'success' rates which are nonsense. For example, some engineers quote 99% success to mean that their system has correctly classified 99% of its training data. In fact this means that their system has misclassified 1% of the training data which suggests that it will fare worse on unseen data. How much worse? We cannot say unless the pattern recognition system is tested rigorously.

Having to generate so much data for training and testing pattern recognition systems can be a daunting task. Some people optimize the use they get from the available data by adding 'noise'. This means that they add small random values to the inputs to obtain data points in the representation space which are close to the given sample, and which it is assumed will correctly have the same identification.

To test the ***generalization*** of pattern recognition from its training data, it is necessary that the training data and the test data be kept separate. As discussed in this section, some several hundred data points may be required to test a pattern recognition system. In an ideal world we would have a similar number of data points for each identification class. Smaller numbers of data points can of course be used and they may be found to give good generalization on testing.

The designation of an input–output pair as training or test data must of course be done at random. In general, one would make a random selection for each identification class. Randomness is essential, otherwise it is very easy to cheat 'just a little' in order to get good laboratory results – and pay the price of machines which do not function well in the field.

Engineers could be responsible for some very expensive mistakes if exaggerated performance figures were misguidedly quoted. However, if you are not familiar with statistical methods, you may find that all these figures can be very confusing. If so you should at least learn this: ***when it is important that a machine achieves a given rate of performance you must ensure that the tests are properly designed from a statistical viewpoint. If you cannot do this yourself, you should consult a qualified statistician.****

2.10 Conclusion

This chapter has given an overview of pattern recognition as it relates to the design of intelligent machines. As a result you should understand:

- that human abilities in perception may make pattern recognition appear more easy than it is;
- that pattern recognition is fundamental in perception;
- that pattern recognition is important in cognition;
- that pattern recognition usually requires input–output pairs of training data;
- that a pattern is a structured set of objects;
- what a representation space is;
- what an interpretation space is;
- that an identification is the action of giving an interpretation to a representation;
- what features and primitive features are;

* Readers may be interested to learn that the authors took their own advice. We had this section checked by Dr Trevor Lambert and Dr Fergus Daly of the Open University's Statistical Advisory Service who supplied Table 2.5 and its commentary. We are very grateful for their advice and help.

- how relational patterns can be represented by graphs;
- how relational patterns can be recognized using graph matching;
- how graph matching can be modified for 'near misses';
- that transforming the representation can facilitate recognition;
- that the Fourier transform is useful in pattern recognition of waveform data;
- how patterns can be represented as points in multidimensional spaces;
- how multidimensional spaces can be classified for pattern recognition;
- that chalk-and-cheese spaces have no metric;
- that spaces can be classified by order relations;
- that there may be problems when classifying multidimensional spaces;
- multiple classification with uncertainty is better than incorrect classification;
- multiple classifications can be represented using fuzzy sets;
- the important distinction between rejection (no classification) and error (using classification);
- the importance of rigorous statistical sampling methods to find error rates for a classifier;
- how complex pattern recognition is achieved by hierarchical pattern recognition.

Pattern recognition is a recurrent theme throughout this volume, and in later chapters we will build on the theory and methods developed in this chapter.

References and further reading

Hopgood, A.A. (1993) *Knowledge-Based Systems for Engineers and Scientists*, CRC Press, London.

Lillesand, T.M. and Kiefer, R.W. (1979) *Remote Sensing and Image Interpretation*, John Wiley & Sons, New York.

Simon, J.C. (1986) *Patterns and Operators: The foundations of data representation*, North Oxford Academic, Kogan Page, London.

CHAPTER 3
SEARCH

3.1 Introduction

The meaning of the word ***search*** in an everyday context is well known – it means to look for something. If you had to search for a key in your house, for example, you would decide (possibly subconsciously) on a strategy to adopt. You might decide to look everywhere in a systematic way, starting in one room and looking on the floor, in all the cupboards and under the furniture, then moving on to the next room and repeating. Alternatively, you could just look in selected places where you think there is a strong likelihood of finding a key – in pockets, drawers, etc.

The two strategies just described are ways of searching a set, or 'space', of possibilities. They can be classified as:

exhaustive search: where potentially the whole space is examined, and

heuristic search: where some 'heuristics' or knowledge acquired through experience is used to restrict the search to a smaller space.

Intelligent machines must constantly monitor their status in terms of existing goals and plans in the context of new information provided by their sensors. At every moment they must review what they are doing, and when unexpected events make the current course of action inappropriate, they must find another. To do this they must determine the space of all possible actions, and *search* it for the most appropriate action.

In the early days of AI it was thought that many problems would be solved by the ability of computers to examine many alternatives very quickly. For example, consider all the possible sequences of moves in the game of chess. In principle the problem of winning at chess could be considered to be that of searching the set, or ***search space,*** of all alternative moves to find one that does not lose. Methods which examine every alternative in the search space until it is exhausted are often said to work by ***brute force***.

It was soon found that exhaustive brute force methods are impractical for many problems: existing computers are just not powerful enough to search the space of all possibilities in a reasonable time. For example, a robot controller which took ten hours to predict an imminent collision would not be practical.

Throughout the relatively short history of electronic computers, there has been an amazing increase in the power of machines over time. This has meant that some problems can be solved by brute force on the new machines, and people sometimes think that *every* problem will be solvable by brute force when the right generation of computers comes along. This is a profound error.

Consider the problem of finding your key. Suppose the search was not restricted to your house, and your key could be in any house in the country. Suppose you could search ten houses per day. Then to search all the houses in Britain would take you over a thousand years. You could ask a friend to help, and so increase your search ability. But what if your key might be anywhere, in any country of the world? You could recruit more friends. But then, what if your key could be anywhere in the universe? No matter how you increased your army of key searchers, you would never be able to search the space of all possible places for your key. An infinity of possible hiding places cannot be exhaustively searched by a finite number of people. And an infinite search space cannot be exhaustively searched by a finite machine (which all computers are).

This is the nub of the search problem in AI. For many interesting and important problems, exhaustive search is not an option and *it never will be*. To understand why this is so, it is necessary to consider the issue of ***computational complexity***.

Some computer procedures are inherently more demanding than others. For example, a program that has to decide 'which of a set of characters is a vowel' will work in time proportional to the number of characters to be classified. On the other hand a program which has to calculate the distances between a set of n cities has to perform $n \times (n-1)$ computations, i.e. $n^2 - n$. When n is large, say 100, the $n = 100$ term becomes insignificant compared with the $n^2 = 100 \times 100 = 10000$ of the squared term. This leads to a rough and ready measure of computational complexity, called the ***'Big-O' notation***.

Suppose the time a computation takes on a given machine is related to the size of the data set it acts on. Let n be the number of data items to be processed. If the time taken to process these data can be expressed as a polynomial such as $an^4 + bn^3 + cn^2 + dn + e$ then the algorithm is said to have ***polynomial complexity***. In general, the highest term is far more significant than the lower terms, and these are ignored, to give a complexity of an^4. The constant a reflects the power of the machine: for a machine with half the power the constant would be $2a$. Since we are interested in the complexity of the algorithms and not the particular machines they are run on, these constants are ignored to make the measure of complexity machine-independent. So for this algorithm the run time is of the 'order' of n^4, which is written $O(n^4)$ in the Big-O notation. There are other measures of computational complexity, but this is one of the most widely used.

Consider a machine that is planning its movements ahead in time. Suppose for any given state and time it can examine ten subsequent states for the next time. Then suppose that for each of these it can examine ten more. Then to plan ahead to time $t = 1$ takes 10 computations, to plan ahead for time $t = 2$ takes 10×10

computations, to plan ahead for time $t = 3$ takes $10 \times 10 \times 10$ computations, and to plan ahead for $t = n$ takes 10^n computations. Thus the order of complexity for this planning program is $O(10^n)$, which is ***exponential***.

Since the 1950s, computers have become, very roughly, ten times more powerful (or faster) every five years (Figure 3.1). This spectacular increase in power over the last forty years has enabled many new applications of computing in many fields. It is tempting to be euphoric and to suppose that this can go on forever. It cannot, because there are physical limits to computation determined by physical constants such as the speed of light. But, *even if it could go on forever*, there are problems whose inherent complexity makes them impossible to solve in any practical time scale.

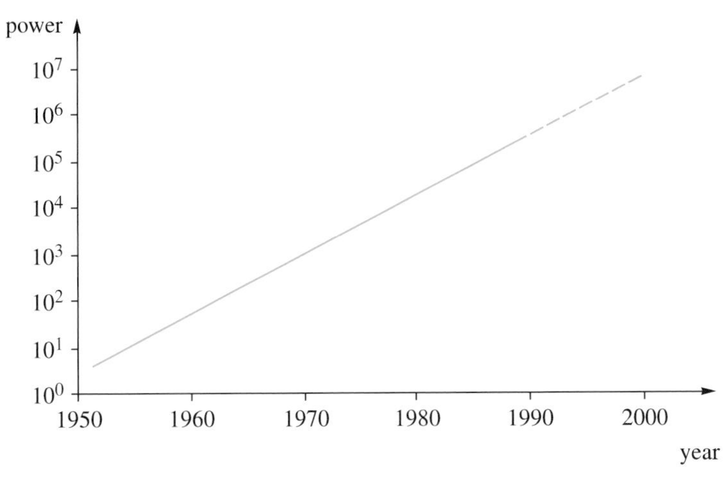

Figure 3.1
The relative increase in computer capabilities since 1950 (schematic).

Consider the machine above with exponential complexity. Suppose today that it can look n moves ahead. A machine with 10 times the power will allow us to look $n + 1$ moves ahead in the same time. With good fortune this extra computer power will take about five years to develop. What about $n + 2$ moves ahead? By the same argument it will take ten years for the necessary computer power to evolve. To look ten moves ahead will require waiting for fifty years, even assuming the spectacular rate of increase in computer power that has been seen in the last forty years. This is outside the time scales of most engineering projects.

An exponential algorithm such as this is *non-polynomial*, and its complexity is of a different order to polynomial algorithms. There is a large class of problems which are not known to have algorithms of polynomial complexity to solve them. One of the best known of these is the *travelling salesman problem* which appears in Chapter 5 of this volume on scheduling. These problems are said to be *non-polynomial indeterminate,* and the algorithms in this class are often called ***NP-algorithms***.

The improvements in computer power seen to date make little impact on these NP-algorithms: the limitations on computer power relative to algorithms are absolute. Like the speed of light, they are a fact of life which will not change.

The triumph of artificial intelligence is to have developed methods for obtaining practical solutions to problems which do not yield to brute force. They do this by considering the nature of the spaces being searched and developing strategies which may not give the best result every time, but give acceptable sub-optimal results most of the time. This approach is said to be ***heuristic***.

For any search problem, the ***search space*** is the set of possible solutions. The subset of the search space which contains actual solutions is called the ***solution space***.

For example, the search space for the problem of finding two dominoes whose spots add up to twenty is all the possible pairs of dominoes. The solution space is the set of pairs:

(4/4, 6/6), (4/5, 5/6), (4/6, 5/5).

Note that here there is more than one solution, and they are all equally acceptable.

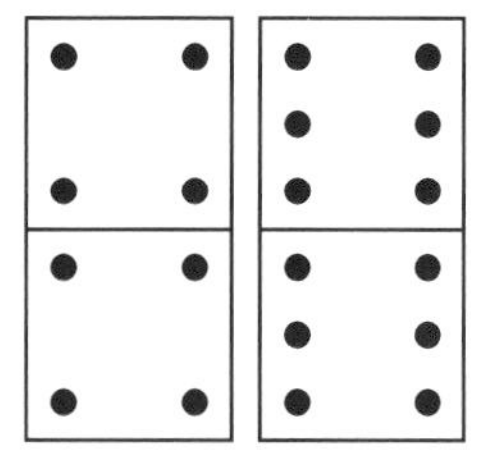

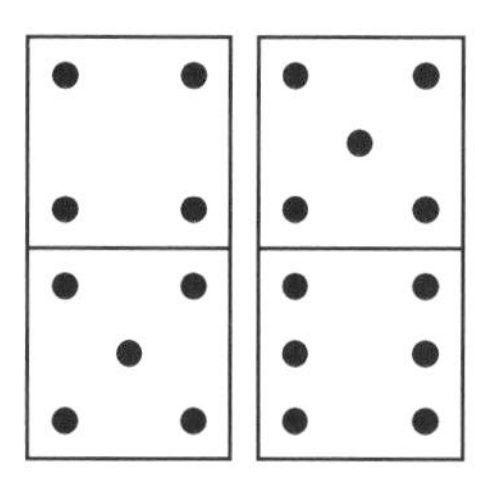

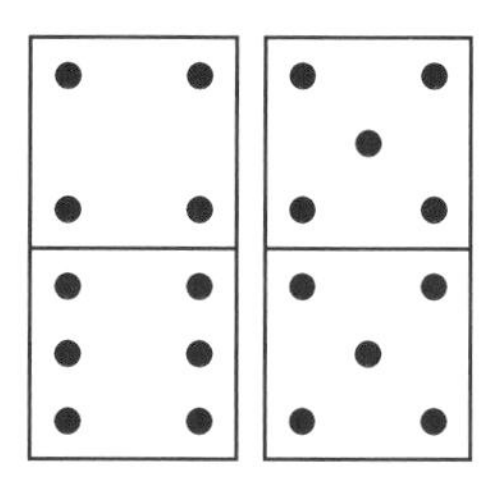

Figure 3.2
A solution space – pairs of dominoes with spots adding up to 20.

For another example, consider the problem of finding a path for a vehicle if we know that its power consumption depends on the load it carries and the landscape it encounters. In general, the path is non-linear with some discontinuities, as might happen, for example, when it leaves the road to take a short cut over rough ground, or when it has to cross a river. The *optimum* solution to this problem, it will be supposed, is that which uses the least fuel, subjects the vehicle to the least mechanical stress, and carries the greatest load in the least possible time. In general, with problems like this one does not know in advance whether such an optimal solution exists. Indeed we may not know whether or not a solution exists at all, i.e. it may be that it is impossible for the vehicle to find a path between its origin and its destination which can be traversed using the fuel that is available. The machine may run out of fuel, fail to find a solution, and be stranded.

The difference between these two examples is that the first involves searching a finite set of combinations, and any particular combination is either a solution or not a solution. Sometimes *domain knowledge* can guide such a search. For example, we can reason that to make twenty spots on a pair of dominoes, the smallest number of spots on any one of a successful pair must be eight, since the largest number of spots available is twelve on the double six. Later in this chapter we will see how such heuristics can reduce the number of combinations which have to be examined, and so make the search space smaller or easier to search.

In the second case the problem is well defined, but we may have sparse or imprecise knowledge about the domain. In such a case one rarely aspires to finding the 'best' solution, and is satisfied with sub-optimal solutions which have acceptable statistical properties in the long run. In general, we try to reduce the risk of failing on any given trial, accepting the consequence of smaller rewards and losses. This is a better strategy for both physical and financial survival. In this second case it may be impossible to find the optimal solution, even when it exists. So the strategy of taking the best that you can find given the time or resource available usually ensures the best long-term outcome.

A major area of search addresses the question of how an *infinite* search space can best be searched by a *finite* machine in a *finite* time. To understand the difficulty of this, consider a spacecraft which is prospecting for minerals on an unexplored planet. Each point on the planet has a financial value associated with it according to the minerals there. For simplicity, consider a search restricted to just one dimension. Let us suppose that the machine can fly long distances, but cannot do ground surveys and analysis when flying. To do this it must land, collect samples, and analyse them. What is the best way for the machine to search this environment in order to optimize the value of its findings?

In order to do any surveying the spacecraft must land at least once. Since it knows nothing about the planet it may as well land anywhere. After taking some samples it will have an idea if this is a promising place to stay. Even if the results here are very good, they may be better elsewhere. This suggests that the spacecraft should try elsewhere, as illustrated in Figure 3.3. In the absence of prior knowledge, anywhere is as good as anywhere else. So the spaceship could take off and land somewhere else. A great many search techniques address the question of where that somewhere else should be. The underlying problem is that, by hypothesis, it is impossible to sample every place. This leads to the problem of selecting places to search which give a 'good enough' spread over the search space, and so improve the expectation of overall return in the long term.

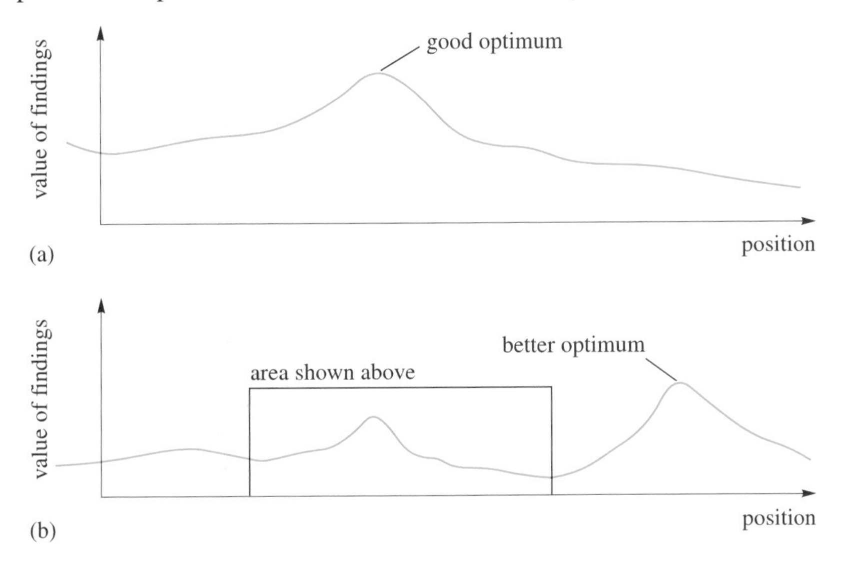

Figure 3.3
(a) The search surface for the surveyor spacecraft (in one dimension); (b) there is always more of the space which has not been searched, and may have a better optimum.

In this chapter we will elaborate on the technical ideas underlying search in the context of finding solutions to problems and optimization. In general, optimization involves mathematical functions, and these lead to *calculus-based search*, the concept of *hill climbing*, and special forms of hill climbing such as *gradient descent*. Other methods that are discussed relate to methods for 'getting around' the search space to sample it adequately, such as *simulated annealing* and *genetic algorithms*.

The combinatorial nature of search is expressed through *search trees*, and the ideas of *breadth-first search*, *depth-first search*, and *best-first search* are discussed, all of which are attempts to speed up the search.

3.2 Tree search

For a long time research in artificial intelligence has included work on game playing, particularly of chess. At each turn a player has to select the next move based on what possibilities lie ahead. The player is therefore searching through the space of all possible next moves looking for a favourable outcome. Part of the skill of the player lies in being able to look several moves ahead. Computers are quite good at this, and can evaluate moves in terms of the possible outcome several moves later.

The method used by a chess-playing computer is called a ***tree search***. The tree is constructed from a ***root node*** which represents the current state or position. A number of ***branches*** spring from the root node, themselves ending at nodes. Each branch represents a possible decision. The branches terminate when there are no further decisions to be made, either because a dead end has been reached or a solution has been found.

To illustrate this, let's take an example of an autonomous vehicle (AV) again, which has the possibility of moving a fixed distance forwards, backwards, right or left. These moves will be denoted as directions N, S, W or E respectively. Now imagine that the AV is in an environment, as shown in Figure 3.4. This environment consists of objects (the squares), with paths in between the objects. The task set for the AV is to get out through the exit. Assume that the AV has some form of internal representation or map of its environment.

0	0	[1]	0		0 = empty space
[1]	**0**	[1]	0	← exit	**0** = current position
0	0	0	0		[1] = object
0	[1]	[1]	[1]		

Figure 3.4 Environment of the AV.

Wherever it starts, the AV has to make a decision about which way to travel. This can be represented by a tree structure with a root node and four branches, corresponding to N, E, S and W, as shown in Figure 3.5.

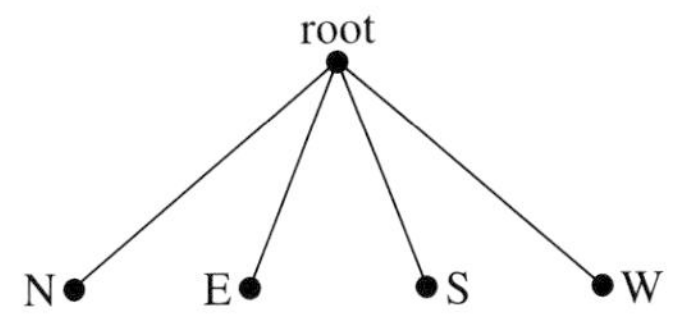

Figure 3.5
Initial search tree for the AV.

In this example, the paths are so limited that the AV can only ever choose to go either of two ways. In addition, one of those options is usually to go back in the opposite direction to the way that it has just travelled, which means that it could oscillate between two squares by executing the movements N, S, N, S etc. for ever. A simplified tree can therefore be drawn with branches showing only the moves that take the AV to a *new* position. This is shown in Figure 3.6 for this particular example.

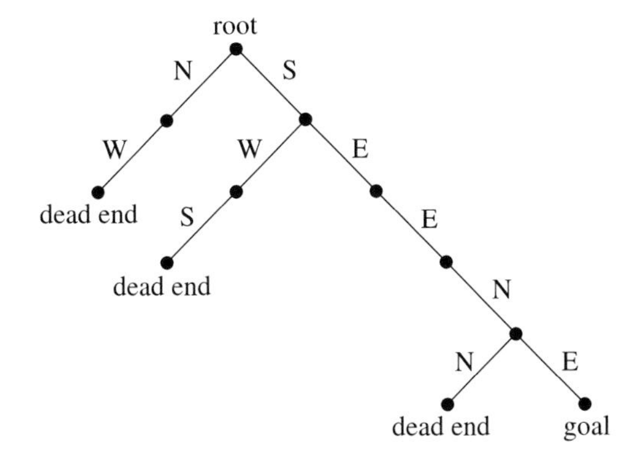

Figure 3.6

The start is indicated at the top of the tree, and the goal is at the bottom. The figure shows the complete tree, with all of the moves that are possible from this one particular starting position. From this you should be able to see that this tree search is an exhaustive search.

Now this is a relatively simple tree. It's not hard to appreciate that sometimes these trees are enormous, so methods have been developed to avoid searching the whole tree, and better still to avoid having to construct the whole tree. These include ***depth-first***, ***breadth-first*** and ***best-first searches*** that will be described in the following sections. However, it still may be the case that the solution lies at the very tip of the very last branch that is searched, so that the search can still be exhaustive. These methods speed up the search by trying to find the solution before the whole tree is searched, and require some heuristic knowledge about the tree.

3.2.1 Depth-first search

Consider the tree shown in Figure 3.7, where the nodes are labelled A to G. The search starts at node A and finishes when node G, the goal, is found.

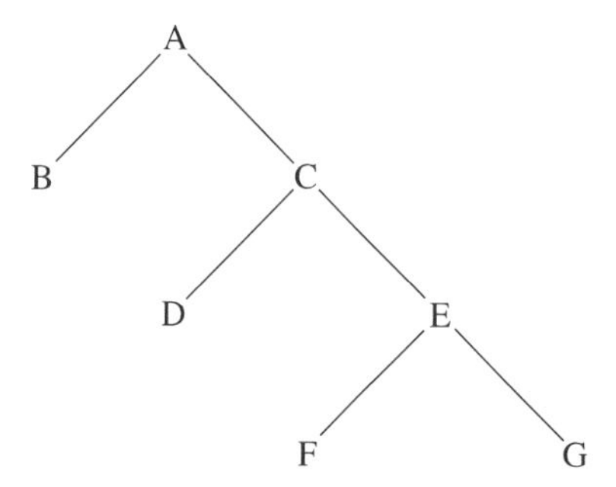

Figure 3.7
Example of a search tree.

A depth-first search follows the rules:

(a) If there is a branch, take the left one first. (This could equally well be the right one. One must decide which and then stick to it.)
(b) If a left branch turns out to be a terminal node, go back to its parent node and take the right branch.

Now let's use these rules. A depth-first search would proceed as follows:

1. Starting from A, examine the left branch (or child) from the parent node. This leads to B.
2. If B is the goal, stop. It's not.
3. As the branch terminates at B, go back to its parent node, A, and take the right branch which leads to C.
4. If C is the goal, stop. It's not.
5. As C has branches, examine the left branch which leads to D.
6. If D is the goal, stop. It's not.
7. As the branch terminates at D, go back to its parent node, C, and examine the right branch which leads to E.
8. If E is the goal, stop. It's not.
9. As E has branches, examine the left branch which leads to F.
10. If F is the goal, stop. It's not.
11. As the branch terminates at F, go back to its parent node, E, and examine the right branch which leads to G.
12. If G is the goal, stop. It is.

In this example, a depth-first search has not managed to speed up the search at all – it is still exhaustive. However, a simple re-ordering could make all the difference, as shown in Figure 3.8.

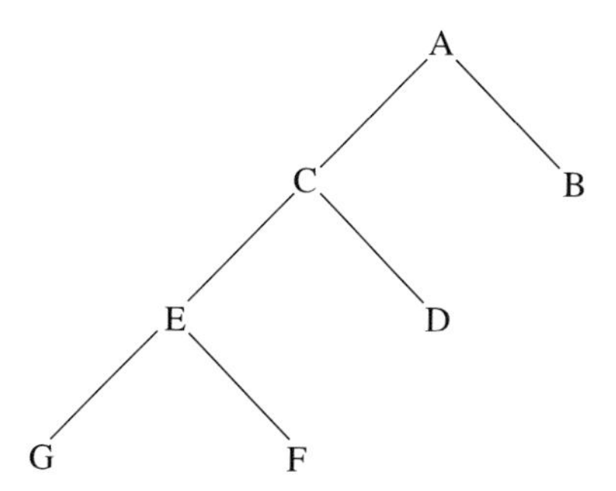

Figure 3.8
Re-ordered search tree.

If we go through the same steps again, the result is quite different.

1. Starting from A, examine the left branch (or child) from the parent node which leads to C.
2. If C is the goal, stop. It's not.
3. As C has branches, examine the left branch which leads to E.
4. If E is the goal, stop. It's not.
5. As E has branches, examine the left branch which leads to G.
6. If G is the goal, stop. It is.

Half the number of steps! The number of steps is very sensitive to the ordering, which can often lead to dramatic savings. Unfortunately it is almost impossible to predict in advance the ordering that will produce the most efficient search. On average the search time will be close to half the number of steps compared to an exhaustive search, particularly for large trees.

3.2.2 Breadth-first search

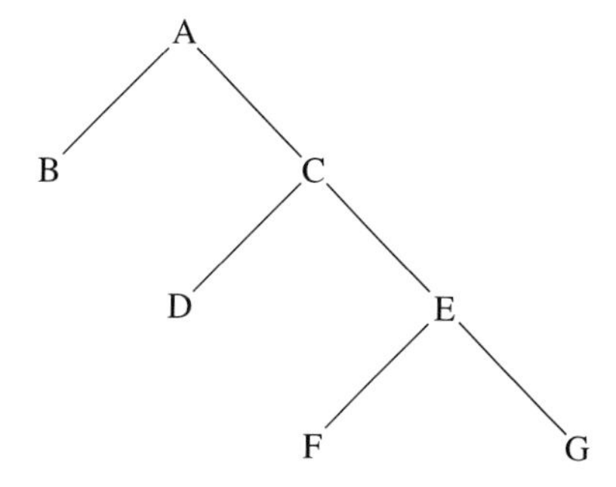

Figure 3.9

Consider the tree in Figure 3.7, shown again here as Figure 3.9. In a breadth-first search the tree is searched in horizontal layers. All of the nodes in each layer are examined before moving to the next layer.

Starting from node A, and assuming that the goal is G, a breadth-first search would proceed as follows:

1 Starting from A, examine the left branch (or child) from the parent node which leads to B.
2 If B is the goal, stop. It's not.
3 Now examine the right branch which leads to C.
4 If C is the goal, stop. It's not. This layer is done.
5 Go back to B. As the branch terminates at B, go back to C.
6 As C has branches, examine the left branch which leads to D.
7 If D is the goal, stop. It's not.
8 Examine the right branch of C which leads to E.
9 If E is the goal, stop. It's not. This layer is done.
10 Go back to D. As the branch terminates at D go back to E.
11 As E has branches, examine the left branch which leads to F.
12 If F is the goal, stop. It's not.
13 Examine the right branch that leads to G.
14 If G is the goal, stop. It is.

If the goal is at a relatively high level, it will be found more quickly than if it is at a lower level. Again, the search time is approximately half on average compared to an exhaustive search.

3.2.3 Best-first search

In this method looking for a termination is not enough. An ***evaluation*** at each node is required so that a decision can be made about which node should be explored next. The value at each node is calculated using a static evaluation function, which will vary from one situation to another.

The hill-climbing and gradient-descent methods described later in the chapter come under the heading of best-first search, and could be represented using a search tree. Each node would represent points in the search space with branches from each node to all the possible points in the space that the search could try next. The branch with the best evaluation would be chosen. It is unusual to see these methods shown as a tree search, however. Mostly one sees search trees being used to find goals and the methods employed aim to reduce the time needed to find the goals. In the gradient-descent methods the goal is to find the optimal solution, but there is no way of telling when that solution has been found, so it is difficult to use the tree search techniques described here.

The best-first method can be illustrated using an autonomous vehicle (AV) guidance system, where one example of an evaluation that can be made is the distance from the goal. Various schemes exist, several of which have been developed by Jarvis (1985). The methods are described as ***distance transforms***, and involve assigning a distance value to each square in the grid by scanning the grid first forwards then back, and repeating until the values no longer change. Figure 3.10 shows an example of an environment with a single object and a goal position.

0	0	0	0	0	0
0	[1]	[1]	0	**0**	0
0	[1]	[1]	0	0	0
0	0	0	0	0	0

0 = empty grid

[1] = object

0 = goal

Figure 3.10 AV environment.

The distance transform has several steps:

Step 1: Initialization

Initialize all the squares with a maximum value equal to the number of squares in the grid, in this example $6 \times 4 = 24$, except for the goal which should be set to 0.

24	24	24	24	24	24
24	[24]	[24]	24	**0**	24
24	[24]	[24]	24	24	24
24	24	24	24	24	24

24 = empty grid

[24] = object

0 = goal

Figure 3.11 Initialization.

Step 2: Forward scanning

Starting in the top-left corner, ***forward propagate*** to assign new values to the squares. This is done by scanning each square in the first column, then moving to the top of the second column and scanning down, and so on until the bottom of the right-hand column is reached.

At each square forward propagation is carried out as follows. For any particular square, look at the four neighbours which consist of the three in the row below and the one to the left, as shown in Figure 3.12.

1	current square	
2	3	4

Figure 3.12 Squares examined during forward scanning.

(When the square being examined is at the edge of the grid, such as in the left-hand column or at the bottom, just ignore the neighbours which are outside the grid.)

For each of the neighbouring squares marked 1 to 4 calculate the distance transform:

distance transform = (neighbour value + distance) × factor

where the factor in this equation is either 1 for *current* empty squares or a number greater than 1 for *current squares containing objects*, typically 3. The distance is the number of moves needed to get from the current square to the neighbouring square. If we use the AV where only N, S, E or W directions can be travelled, the distances to each of the neighbouring squares are shown in the following figure:

2	1	2
1	current square	1
2	1	2

Figure 3.13 Distances to each neighbouring square.

That is, only one step is needed to get to a neighbouring square if it is directly above, below, to the right or to the left. Two steps are needed to get to squares which are diagonally adjacent.

After calculating the distance transform for the four neighbours shown in Figure 3.12, the minimum value from those four distance transforms is found. If it is less than the value in the current square then it replaces it, otherwise the value in the current square doesn't change.

Applying this to every square in the grid, we get:

24	24	24	2	1	2	24 = empty grid
24	[24]	[24]	24	**0**	1	[24] = object
24	[24]	[24]	24	24	24	**0** = goal
24	24	24	24	24	24	

Figure 3.14 Grid after forward scanning.

Let's look at this in more detail, for example the 2 at the top of the fourth column from the left. After initialization it was 24, and the forward scan looked like this:

	current square ↓	
24	**24**	
[24]	24	**0**

Figure 3.15 Initial values of grid.

Using the equation for the distance transform

distance transform = (neighbour value + distance) × factor

the distance transforms of the four neighbours are:

1 (24 + 1) × 1 = 25
2 (24 + 2) × 1 = 26
3 (24 + 1) × 1 = 25
4 (0 + 2) × 1 = 2

The smallest value of the four is 2. Since 2 is smaller than 24, the new value for the square is 2.

Step 3: Backward scanning

This is essentially the same as forward scanning except that you start at the bottom right-hand corner of the grid and look at the three neighbours above and one to the right of the current square. Figure 3.16 shows the squares to be examined. Again, if any of the squares are off the grid then they are ignored.

1	2	3
	current square	4

Figure 3.16 Squares used in backward scanning.

As will be explained, the result of backward scanning on the entire grid is:

5	4	3	2	1	2	24 = empty grid
6	[15]	[6]	1	**0**	1	[24] = object
17	[24]	[9]	2	1	2	**0** = goal
6	5	4	3	4	24	

Figure 3.17 Grid after backward scanning.

Let's look at one of the squares in the object – the one that ends up with a value of 9 say. When the scan reaches this square, the grid looks like this:

24	24	24	2	1	2	24 = empty grid
24	[24]	[24]	1	**0**	1	[24] = object
24	[24]	[**24**]	2	1	2	**0** = goal
24	24	4	3	4	3	

Figure 3.18 The grid mid-way through backward scanning.

The four neighbours therefore look like this:

[24]	[24]	1
	[**24**]	2

Figure 3.19 Initial values of neighbouring squares.

The distance transforms are:

1 $(24 + 2) \times 3 = 78$
2 $(24 + 1) \times 3 = 75$
3 $(1 + 2) \times 3 = 9$
4 $(2 + 1) \times 3 = 9$

The minimum is 9, which is less than the present value of the square, so it is replaced.

Step 4: Repeat

Repeat steps 2 and 3 until there is no further change in the values. In this example, only one more forward and backward scan are needed, resulting in the following grid:

5	4	3	2	1	2	24 = empty grid
6	[15]	[6]	1	**0**	1	[24] = object
7	[18]	[9]	2	1	2	**0** = goal
6	5	4	3	2	3	

Figure 3.20 The final grid.

Now we can construct a search tree from any point on the grid, where each branch will have an evaluation equal to the distance to the goal. Searching the tree consists of only selecting the branch with the lowest value, so it is a best-first search. This means that at any point during the search, if there are two (or more) branches that can be selected, the best one, which is the one with the lowest score, is chosen. If the score on two branches is the same, one of them is arbitrarily chosen.

For example, start from the top-left corner. The first branch gives two possibilities – to travel S with a value of 6 or travel E with a value of 4. The choice would be to go E, so the S branch is effectively pruned since no more searching will take place along it. Figure 3.21 shows the search tree, where each node shows the score associated with the position of the AV.

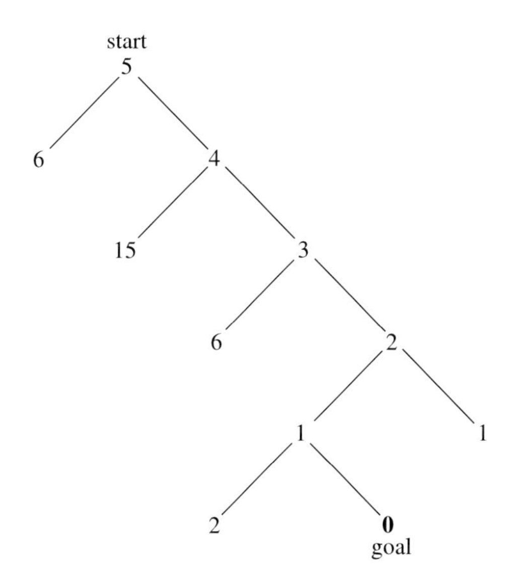

Figure 3.21
Best-first search tree.

This method effectively 'prunes' large sections of the tree. It doesn't necessarily give the optimal solution, but it usually gives a good solution in a short time.

3.2.4 The A* search algorithm

In the previous example only one branch was selected at each node, the other branch or branches being pruned, so that large parts of the tree were left unexplored. When there are many paths to the same goal it may be desirable to select the best path. In order to do this, the search algorithm has to be able to go back to unexplored parts of the tree if its current exploration proves to be more difficult than first anticipated. The ***A* algorithm*** developed by Winston does just this.

To illustrate the A* algorithm, let's assume that an autonomous vehicle (AV) has to travel from A to G in Figure 3.22, and that it can take several routes which go via sites B, C, D, E and F. The distances between sites are known, and the whole environment can be described with a diagram as in Figure 3.22(a). At each stage of the algorithm all of the existing paths that have been found so far are evaluated and the shortest one is taken up and advanced.

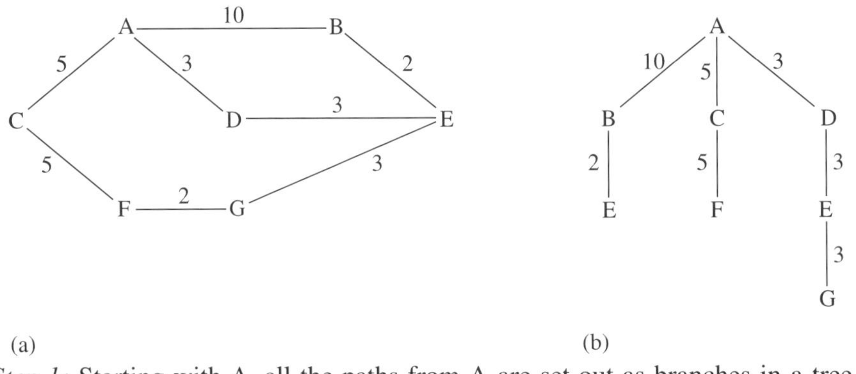

Figure 3.22
(a) Diagram showing the distances between all the sites; (b) search tree for the A* algorithm.

Step 1: Starting with A, all the paths from A are set out as branches in a tree structure shown in Figure 3.22(b) and the shortest path to another site is found. A to D (A–D) is the shortest path, with a length of 3.

Step 2: From A–D, the next path is A–D–E which has a length of 6. The three existing paths so far are A–D–E (6), A–B (10) and A–C (5), so A–C is chosen.

Step 3: From A–C, the next path is A–C–F which has a length of 10. The existing paths at this stage are A–C–F (10), A–D–E (6) and A–B (10), so A–D–E is chosen.

Step 4: From A–D–E the next path is A–D–E–G which has a length of 9. The existing paths are A–D–E–G (9), A–B (10), A–C–F (10), so A–D–E–G is chosen. As no shorter path can be found the algorithm terminates.

In this way some of the tree never gets explored, and so the search time is reduced. Also, because many alternative paths are explored and are never totally abandoned the solution will always be the optimum. This is therefore a very powerful breadth-first tree-search method which is useful when some evaluation is available at each node.

3.3 Calculus-based search

3.3.1 Mathematical models

In this section some of the basic mathematical tools that are often used in mechatronics will be discussed. It is assumed that basic calculus, namely differentiation and integration, doesn't have to be explained, and various formulae will be derived based on that assumption.

In many applications a well-defined mathematical model of a solution space exists as a function or formula. An answer can be found by 'solving' the mathematical function. In general, solving a mathematical formula means finding values for the variables in the formula such that the equation is satisfied.

For example, a mechatronic system such as an autonomous vehicle (AV) can move about a factory floor without any external guidance. One of its goals is to travel in a straight line from A to B, which is a short distance, x, in a time, t. Assuming that initially the AV is stationary, one way of doing this would be to accelerate at a constant rate up to a particular velocity, stay at that velocity for a certain length of time, and then slow down at a constant rate to a halt, having covered the distance, x. Figure 3.23 shows this motion as a graph of v against t. The distance travelled is found as the area under the graph.

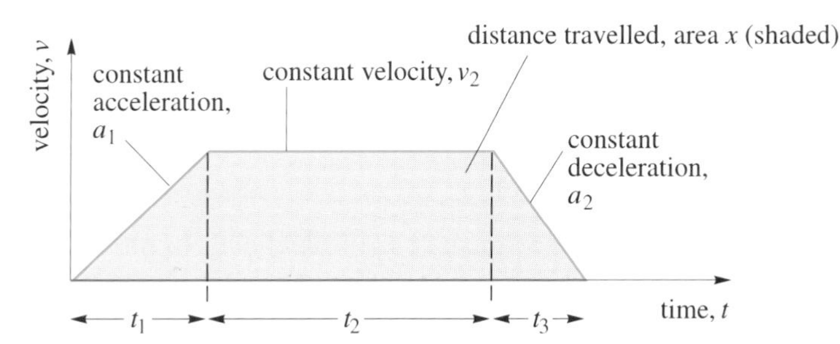

***Figure 3.23* Graph of velocity against time for the autonomous vehicle.**

This is a very simple model, and clearly in a practical situation you couldn't expect a vehicle to travel in such a perfect way. However, the model can still be used to calculate the 'ideal' values for the acceleration, velocity and deceleration. When these values are tried on the actual system, the performance will almost certainly be worse than that predicted from the model, but should still be sufficiently close to the desired performance. So a model, even a simplified ideal one, can still be useful.

The formula for the distance travelled is

$$x = 0.5a_1t_1^2 + v_2t_2 + 0.5a_2t_3^2 \tag{3.1}$$

where a_1 is the acceleration and a_2 is the deceleration.

If we assume that the acceleration and deceleration are both equal to a (for no other reason than to make the problem less complicated) then $t_1 = t_3$ and the formula can be simplified to

$$x = 0.5at_1^2 + v_2t_2 + 0.5at_1^2 = at_1^2 + v_2t_2 \tag{3.2}$$

Equation (3.2) describes the motion of the vehicle. The details of how this is derived do not concern us here. What is important is that it is possible to find a relationship between v_2 (which in the subsequent discussion we will call simply v), and a, which turns out to be

$$v^2 - avt + ax = 0 \tag{3.3}$$

where $t = t_1 + t_2 + t_3$, and $at_1 = at_3 = v$.

This equation can be solved by finding values for v and a which satisfy the equation; that is, values which make the whole expression equal to zero. It is clear that there is not one solution but a whole family of solutions which satisfy this problem. If we assume that we are interested in the specific problem where, say, $x = 10$ m and $t = 5$ seconds, then the solution space of $a = v^2/(vt - x)$ can be drawn with axes as a and v, as shown in Figure 3.24. All the solutions lie on this curve.

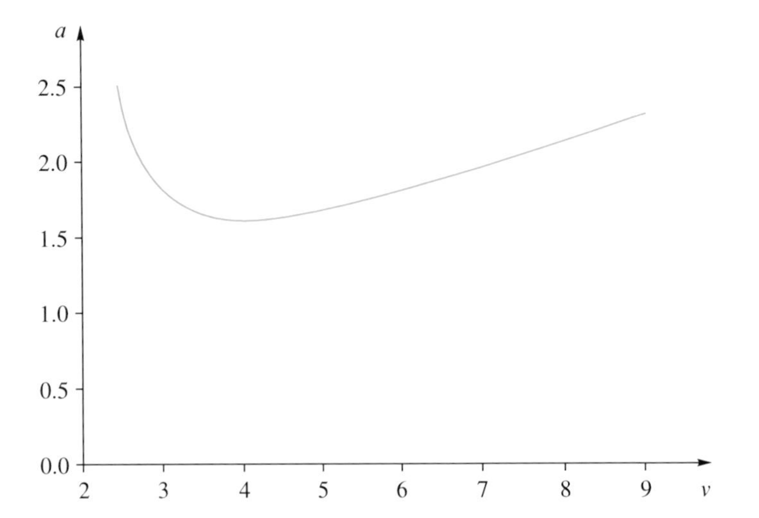

Figure 3.24
Graph of a against v, showing all the solutions, $a = v^2/(vt - x)$.

Faced with this problem the vehicle could search the solution space until it found a point which was on the curve, which would give suitable values for a and v. These values would be selected, and the vehicle could start to accelerate up to the required velocity.

In this example, the solution space had to be searched to find suitable values for the acceleration and velocity so that the vehicle travels the correct distance in the correct time. Sometimes solutions exist which can be calculated exactly given some additional data. For example, suppose one of the system objectives was to conserve power in the AV. This could be done by selecting the solution with the lowest value for the acceleration, a_{min}. This, as we shall see in the next section, turns out to be

$$a_{min} = \frac{4x}{t^2} \tag{3.4}$$

Since the solution can be found by calculation alone, there would be no need to search the solution space in any other way.

The lesson to be learned from all this is that if a mathematical model exists, it may be possible simply to calculate the solution, which will almost certainly be quicker than searching the solution space using the methods developed later in this chapter.

If the problem is more complex there may be no known methods of finding the solution directly. However, if the problem can still be expressed as a mathematical formula, methods may exist for approximating to the solution and maybe even finding the exact solution.

For example, suppose a problem can be defined by a polynomial equation in x, $P(x)$. If the polynomial is up to fourth order (has terms in x, x^2, x^3 and x^4) then it can be solved directly. Surprisingly, if there are higher order terms such as x^5, no method exists for solving the equation exactly. Even so, there are methods such as that in the next section which successfully search for good approximations.

3.3.2 Newton–Raphson method

The ***Newton–Raphson method*** is a popular algorithmic method for finding the solution to polynomial expressions. For illustration, we'll take the example of the autonomous vehicle again, so that the solution is known in advance. Let's assume, as before, that the required distance to be travelled is 10 m in a time of 5 seconds. The minimum acceleration solution given in equation (3.4) is

$$a_{\min} = \frac{4x}{t^2} = 1.6\ \mathrm{m\,s^{-2}}$$

Substituting $a = 1.6$, $x = 10$ and $t = 5$ into equation (3.3) for the vehicle gives

$$v^2 - 8v + 16 = 0$$

This equation is satisfied when the value of v is 4. Now let's assume that we don't know this solution. Figure 3.25 shows a plot of the equation

$$f(v) = v^2 - 8v + 16$$

Notice that it crosses (or in this case just touches) the v-axis when $v = 4$.

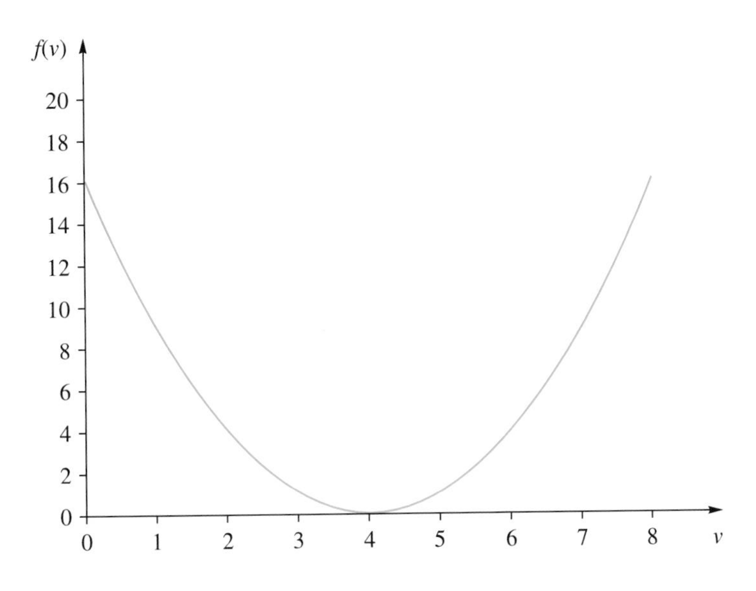

Figure 3.25

Recall that the derivative of $f(v)$ is an expression for the value of the slope of $f(v)$ at any point. In Figure 3.26, the slope is shown at the point $v = 6$, and it can be measured by drawing a tangent to the curve at $v = 6$ and measuring the sides of the right-angled triangle formed with the v-axis.

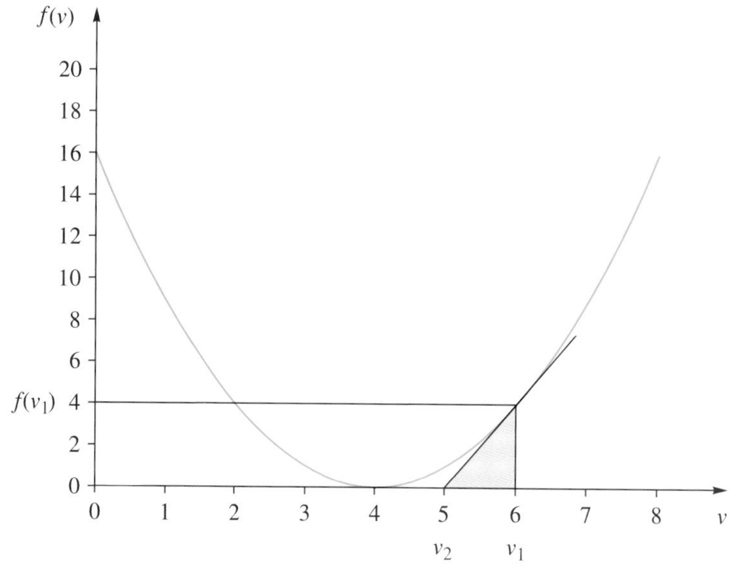

Figure 3.26

The Newton–Raphson method uses the fact that the slope of the curve gives an indication of the direction in which you have to travel to get closer to the point where the curve crosses the v-axis. It is at this point that one of the solutions to the expression $v^2 - 8v + 16 = 0$ exists. So the tangent at $v = 6$ could be extended and the value of v noted where the tangent crosses the v-axis, as in Figure 3.26, where the value is found to be $v = 5$. Now if $v = 6$ had been our first guess at the solution to the equation, we could say that 5 is our second guess. This could be continued, so that the slope at $v = 5$ could be found, and the tangent drawn and the point where it intercepts the v-axis would be our third guess.

This method can be expressed as follows:

Let our first guess be v_1. From this we'll make a new guess, v_2.

The slope at v_1 is $f'(v_1)$, where $f'(v)$ is the derivative of $f(v)$ with respect to time. The slope equals the height of the vertical side of the shaded triangle divided by the width of the base of the triangle. The height of the vertical side is equal to the value of $f(v_1)$. The value of the width of the base is the first guess v_1 minus the new second guess v_2. So,

$$f'(v_1) = \frac{f(v_1)}{(v_1 - v_2)}$$

$$v_1 - v_2 = \frac{f(v_1)}{f'(v_1)}$$

$$v_2 = v_1 - \frac{f(v_1)}{f'(v_1)}$$

In general, the new guess is obtained from the old guess using the formula

$$v_{k+1} = v_k - \frac{f(v_k)}{f'(v_k)} \tag{3.5}$$

Now, instead of measuring the slope at each point, the derivative can be obtained directly from the equation of the problem:

$$f(v) = v^2 - 8v + 16$$

$$f'(v) = 2v - 8$$

So,

$$v_{k+1} = v_k - \frac{(v_k^2 - 8v_k + 16)}{(2v_k - 8)}$$

To show how this works, let's continue with this example, starting with $v_1 = 6$.

Step 1:

$$f(v_1) = 4$$

$$f'(v_1) = 4$$

$$v_2 = 6 - \frac{4}{4} = 5$$

Step 2:

$$f(v_2) = 1$$

$$f'(v_2) = 2$$

$$v_3 = 5 - \frac{1}{2} = 4.5$$

Step 3:

$$f(v_3) = 0.25$$

$$f'(v_3) = 1$$

$$v_4 = 4.5 - \frac{0.25}{1} = 4.25$$

and so on.

The iterations continue while the value of v continues to get closer and closer to 4. The iterations stop if the value of $f(v) = 0$; that is, when the v-axis is reached. However, it is more usually the case that the point where $f(v) = 0$ is never quite reached. Eventually the method stops when the value of $f(v)$ is so close to 0 that the error can be neglected.

Thus, the Newton–Raphson method allows us to find a solution to a problem which can be expressed mathematically. This can be considered to be a search method: the solution exists in some search space, and the Newton–Raphson method searches that space by moving in a direction towards the solution. It can be used when the derivative is known. Later we shall look at a gradient descent method which can be applied when the mathematical form of the gradient is not known.

3.3.3 Minimization

The previous section showed how a polynomial expression of the form

$$P(x) = 0$$

could be solved using the Newton–Raphson method.

Many problems exist where the desired solution is the minimum or maximum value of a function. If the problem can be expressed as an equation, this can be translated into a problem of the sort just described by making use of the derivative.

In the example in the previous section the equation relating the velocity to the acceleration of the vehicle was given in equation (3.3) as

$$v^2 - avt + ax = 0$$

The graph in Figure 3.24 showed that there is a minimum value of a. This minimum is a ***turning point***, so called because the curve changes direction at that point. The value of the derivative at that point is 0, so we can calculate the value of v where the minimum occurs by finding the derivative and equating it to 0. In the above example, the derivative of $a = v^2/(vt - x)$ with respect to v is

$$\frac{da}{dv} = \frac{at - 2v}{x - vt}$$

Equating to 0 gives

$$at - 2v = 0$$

$$a = \frac{2v}{t} \text{ or } v = \frac{at}{2}$$

Substituting this expression for v into equation (3.3) gives the value of v and hence a in terms of x and t alone as

$$v = \frac{2x}{t} \qquad (3.6)$$

$$a = \frac{4x}{t^2}$$

The equation for a is the same as that used earlier in equation (3.4). Given a more complicated function, it may not be so easy to find the minimum, in which case the Newton–Raphson method can be employed again. Identical arguments apply to finding the maximum of a function where the derivative is also 0.

For example, if an object is thrown straight up into the air, it slows down to a halt, and then falls back to the ground. A graph of its height against time is shown in Figure 3.27. Notice that when it is at its maximum height the velocity, i.e. the derivative of the curve, is zero.

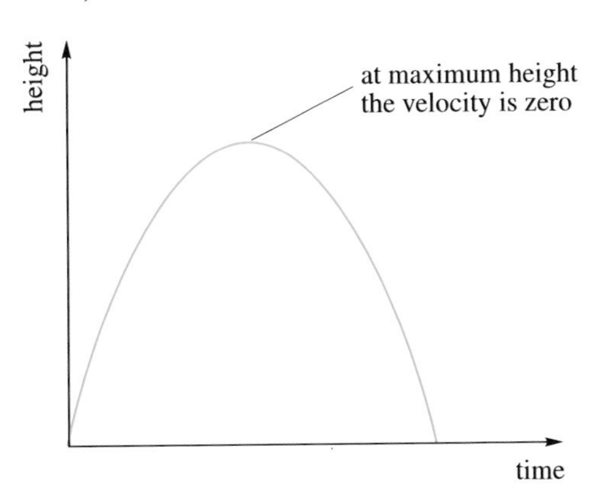

Figure 3.27
Trajectory of an object thrown into the air.

Care must always be taken to ensure that when finding a minimum or a maximum the correct turning point is found. This can be done by looking at the value of the derivative either side of the solution that has been found. Figure 3.28 shows three types of turning point and the sign of the slope either side of the turning point. Notice that a curve is also shown which has a point where the derivative is zero but it doesn't change direction. This is called a ***point of inflexion.***

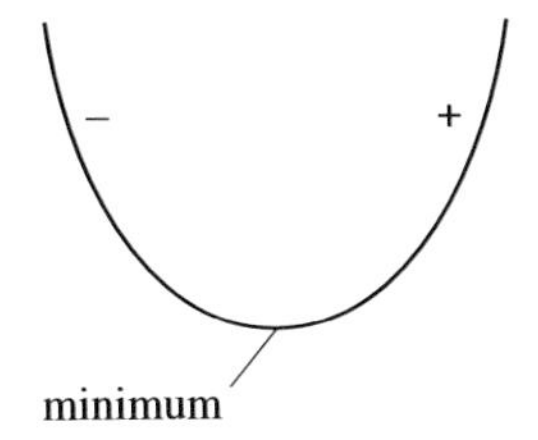

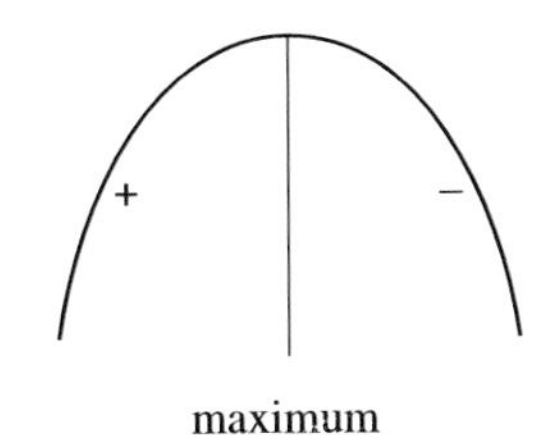

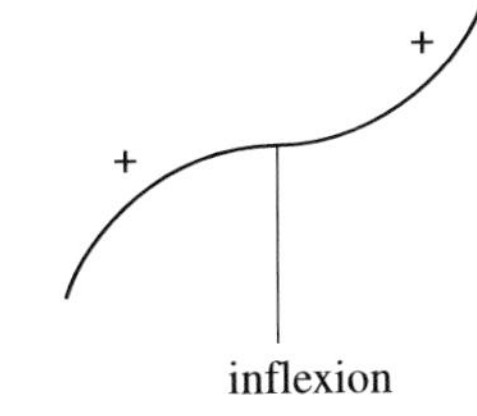

Figure 3.28
The three possible turning points.

Given a differentiable mathematical formula you can therefore find a minimum by searching the space defined by the derivative of the expression for the point where it has a value of zero.

3.3.4 Gradient descent

The Newton–Raphson method can sometimes become very complex. In examples such as the autonomous vehicle we have been considering, the process can be simplified. The expression for the Newton–Raphson algorithm derived earlier as equation (3.5) was

$$v_{k+1} = v_k - \frac{f(v_k)}{f'(v_k)}$$

and for the specific case where $x = 10$ m and $t = 5$ seconds,

$$v_{k+1} = v_k - \frac{(v_k^2 - 8v_k + 16)}{(2v_k - 8)}$$

This can be rationalized as follows:

$$v_{k+1} = v_k - \frac{(v_k - 4)^2}{2(v_k - 4)}$$

$$v_{k+1} = v_k - \frac{(v_k - 4)}{2}$$

The term $(v_k - 4)$ is proportional to the derivative of the function. So in general, the algorithm can be expressed as

$$v_{k+1} = v_k - \alpha f'(v_k) \tag{3.7}$$

where α is a constant.

This is the equation for a ***gradient-descent method***. It is called gradient descent because the change to the variable, v in this case, is proportional to the size of the gradient, $f'(v)$. It approximates the Newton–Raphson method in this example because the minimum of the function

$$f(v) = v^2 - 8v + 16$$

lies on the v-axis. It applies more generally to any function, but instead of terminating when it reaches the v-axis, it terminates when it reaches a minimum where $f'(v) = 0$.

For the particular problem that we've been considering, the formula for the gradient descent is

$$v_{k+1} = v_k - \alpha(v_k - 4)$$

If we let its value of α be 0.1, the formula becomes

$$v_{k+1} = v_k - 0.1v_k + 0.4 = 0.9v_k + 0.4$$

Let $v_1 = 6$.

$$v_2 = 0.9 \times 6 + 0.4 = 5.8$$

$$v_3 = 0.9 \times 5.8 + 0.4 = 5.62$$

$$v_4 = 0.9 \times 5.62 + 0.4 = 5.458$$

$$v_5 = 0.9 \times 5.458 + 0.4 = 5.3122$$

and so on. We find that when n is very large,

$$v_n = 4$$

So the method converges to the solution, which means that it gets closer and closer to the correct value with each iteration. The number of iterations is selected to give the appropriate accuracy for the finally selected value.

Gradient descent, therefore, gives a useful means of finding minima when the actual derivative is not known exactly, but the slope can be estimated from local information. This is particularly true when the solution space is not a smooth continuous differentiable space but a discrete one, which is discussed in the next section.

An extension of gradient descent is ***steepest descent***. In a multidimensional search space there will be gradients in many different directions, and the choice is made to follow the steepest gradient. This will also be discussed in the next section in the context of discrete search spaces.

3.3.5 Discrete search spaces and hill climbing

So far the mathematical models described have assumed a continuous differentiable search space. Very often the search space is not continuous but discrete, i.e. made up of individual points. Sometimes the discrete search space arises from sampling a continuous space, or sometimes the search space is discrete by nature, and sometimes just a set. An example of the latter is the set of pairs of dominoes that add up to 20. The search space consists of all possible pairs of dominoes which can be thought of as points in a space with nothing in between. Searching consists of 'hopping' from one point to another.

In order to perform an equivalent of gradient descent in a discrete space the nature of a gradient has to be considered in a different way from a continuous space. Let there be a measure associated with each point x, written $f(x)$. If the current position in the space is x_k and a neighbouring position is x_{k+1}, then the gradient is approximated by $(f(x_{k+1})-f(x_k))$. Secondly, since the step size has to be fixed so that it is the distance between points in the space, there cannot be a constant like α which would produce variable step sizes. So, a mechanism is needed to jump from the currently examined point to another. For example, the points could be arbitrarily laid out on a grid (e.g. Table 3.1), and all the neighbours examined to see if any were better. This illustrates ***hill climbing***, which simply ensures that the value selected at each iteration is less than (or greater than) the previous value.

Let the best solution so far be at the point $x = x_k$ and the measure associated with it be $f(x_k)$. A neighbouring point in the solution space is x_{k+1} and has a measure $f(x_{k+1})$. The search moves to x_{k+1} according to the following criterion:

$$\begin{array}{llll} \textit{If} & f(x_{k+1}) < f(x_k) & \textit{then} & x = x_{k+1} \\ \textit{else if} & f(x_{k+1}) \geqslant f(x_k) & \textit{then} & x = x_k \end{array} \qquad (3.8)$$

In this way, a solution is finally found which has a measure that cannot be reduced by any further moves. Just like gradient descent, it moves down the slope to a minimum. What hasn't been mentioned is what happens when a point has several neighbours, as would be the case in a multidimensional search space. The distinction is made that in hill climbing, an arbitrary neighbour can be selected and examined, and if its value is less than the present value the search moves to that neighbouring point. The alternative is that all the neighbours are examined and the one chosen is the one which has the lowest value and therefore makes the biggest change. This method is called ***steepest descent***.

In the following section we discuss some of the limitations of these gradient descent methods and also some attempts to overcome those limitations.

3.4 Probabilistic search

3.4.1 Limitations of gradient descent

To illustrate the concept of ***local optimum***, consider the following one-dimensional space of numbers:

6 5 5 4 4 **2** 3 3 4 5 6 7 **8** 6 3 **1** 3 4 4 5 7 8 **9** 8 7 6 **5** 6 6 6 **7**

Here the numbers **2**, **1** and **5** are ***local minima***, and of these **1** is the ***global minimum***. The numbers **6**, **8**, **9** and **7** are ***local maxima***, and of these **9** is the ***global maximum***.

1 How do you avoid ending up at a local minimum?
2 How do you know when the minimum that you have found is just a local minimum or the global minimum?

If you use gradient descent, the point at which the search starts is crucial in determining which minimum is found. The problem of finding the global minimum is that the search can get stuck in one of the local minima. In order to overcome this problem, the idea of a ***probabilistic search*** has been developed in which the search in general is still a descent, but occasionally the search is allowed to jump to a higher value. This allows the possibility of a search 'jumping out' of a local minimum.

This helps to overcome one difficulty, but in general whatever search method is used (except exhaustive search) there is no way of knowing with complete certainty that the minimum which is found is the global minimum. All we can do is to improve the probability of it being the global minimum using the techniques described in this section.

3.4.2 A two-dimensional problem

Table 3.1 shows part of a two-dimensional discrete solution space for a particular problem. This sort of space could arise as part of a perception subsystem which uses a neural network such as the Hopfield network discussed in the next chapter. It is a network that stores data at the minima of the search space. This gives it the ability to reconstruct data from partial or corrupted input patterns.

TABLE 3.1

4	5	7	6	4	4
4	**3**	6	5	5	4
5	4	6	4	4	3
5	5	5	4	3	3
5	6	5	4	**2**	3
4	5	5	5	4	4
4	4	3	4	5	5
5	4	**1**	3	4	5
6	6	4	4	6	5
6	6	5	6	5	5

By looking at the numbers you should be able to see that there are three minima (the bold numbers): that is, points where the values are smaller than all of their neighbours. The global minimum is the point with the smallest value, which in this example is the point with a value of 1. The two local minima have values of 2 and 3 respectively.

To find the global minimum you have to search the whole space. If there are N points in the space then you have to look at all N points to be absolutely certain that you have found the global minimum. This is an exhaustive search, and if N is very large this could be very time-consuming and therefore impractical. It is also possible that the search space is not finite, in which case an exhaustive search is impossible. This is why methods have been developed for searching the space more efficiently. To measure the efficiency, the probability of finding the global minimum will be used (probability is explained in detail in Chapter 6). If the probability of finding the global minimum is 1, then you can be totally confident that it can be found. If the probability is 0, then the global minimum cannot be found. The probability will always lie somewhere between 0 and 1 inclusively.

In an exhaustive search the probability of finding the global minimum increases linearly with the number of points examined. So initially the probability is 0, and after looking at all of the points the probability is 1. When half the points have been examined, the probability is 0.5. If the number of points examined is n, the probability of finding the global minimum, $p(\mathrm{g})$, can be expressed as

$$p(\mathrm{g}) = \frac{n}{N} \tag{3.9}$$

Now let's see if other methods exist that can do better than this, starting with ***random search***. Here, points are selected at random, and the value at that point is examined. If the value is the smallest value that has been seen so far it will be stored as the minimum value. What is the probability that after selecting n points the global minimum has been found?

The probability of *not* finding the global minimum in one trial is $(N-1)/N$. The probability of *not* finding the global minimum in n trials is $((N-1)/N)^n$, so the probability of finding the global minimum in n trials is:

$$p(\mathrm{g}) = 1 - \left(\frac{N-1}{N}\right)^n \tag{3.10}$$

For the space in Table 3.1, $N = 60$. Figure 3.29 shows the graphs of this probability and of the previous exhaustive search.

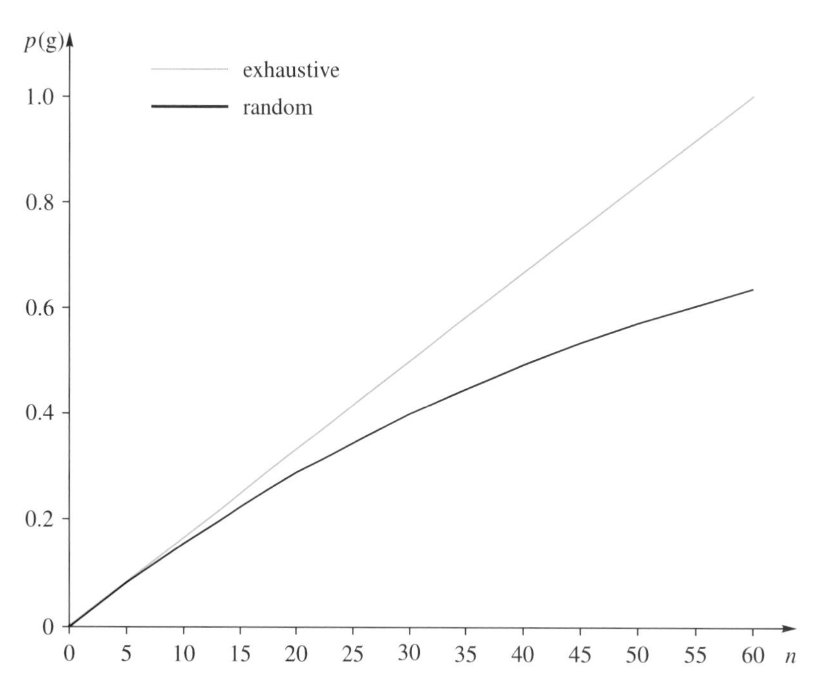

Figure 3.29 Probability of exhaustive and random search finding the global minimum.

The graph shows that random search performs worse than exhaustive search. This is because the assumption has been made that points in the solution space can be visited more than once. If this is changed so that a record is kept of which points have been seen, then the probability becomes the same as the exhaustive search because you would know for certain when all the points had been visited.

3.4.3 Hill climbing

As you've seen, hill climbing is a sort of gradient descent (or ascent) which can be used when a gradient cannot be defined, such as in a discrete search space. The method simply ensures that the value selected at each iteration is less than (or greater than) the previous value.

In the problem of finding a minimum in Table 3.1, hill climbing is applied by choosing a point at random, and then selecting a neighbouring point also at random. If the neighbouring point has a value which is less than or equal to the value at the current position, move to the neighbouring point. This was tried, and the results are shown in Figure 3.30, which also shows the exhaustive and random search.

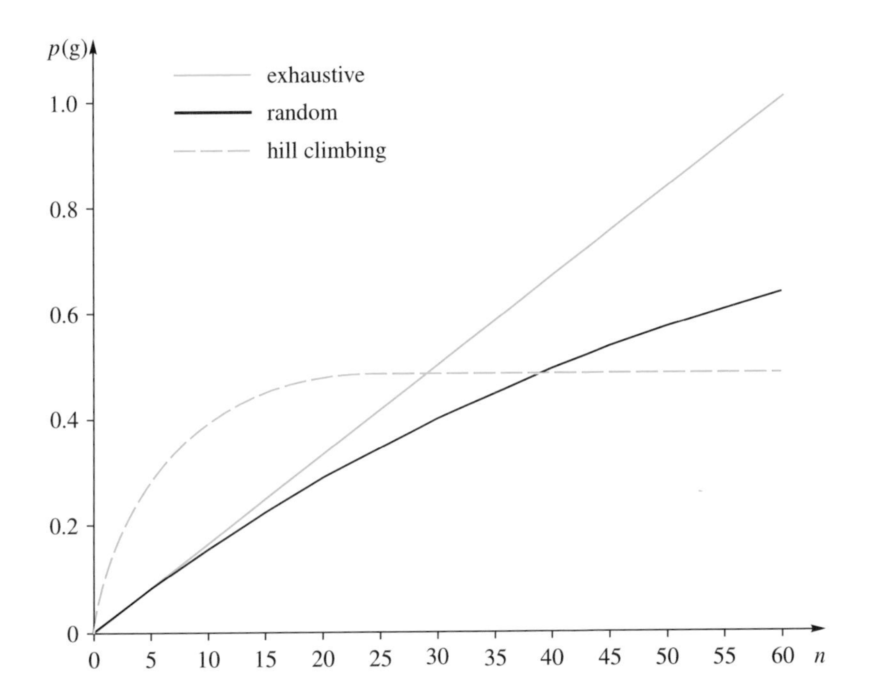

Figure 3.30 Probability of hill climbing finding the global minimum.

From this graph it is clear that hill climbing gives a better probability of finding the global minimum in fewer steps, but that it never gets better than a probability of about 0.5. This is because hill climbing finds local minima. In this example, the search gets stuck in the two other local minima about half of the time.

Next, a steepest descent algorithm is tried, where a point is selected at random and the eight neighbours examined. The search moves to the neighbour with the lowest value. The results of an experiment with the same data as before are shown in Figure 3.31.

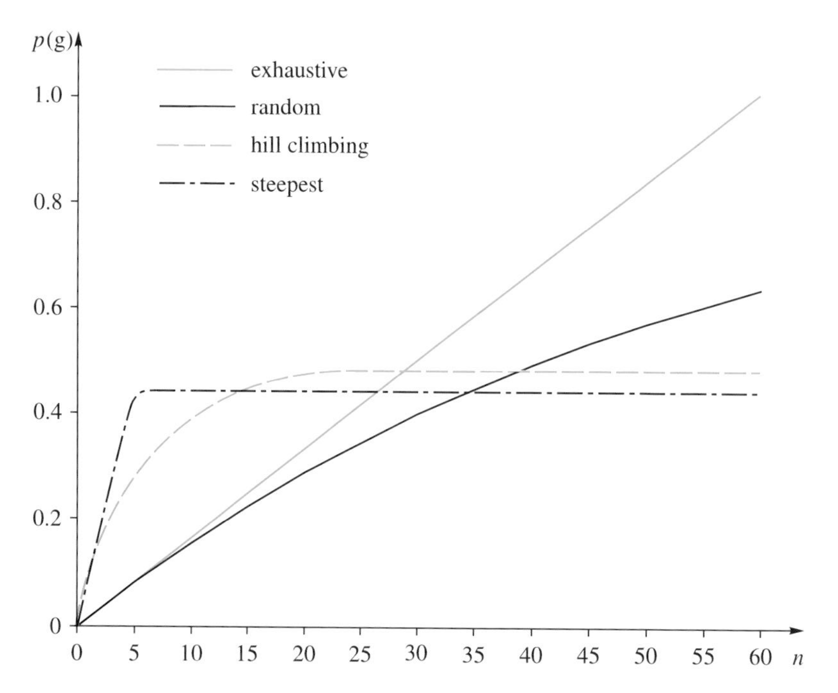

Figure 3.31 Probability of gradient descent finding the global minimum.

This figure shows that the steepest descent method gets to the minimum very quickly, but suffers from the same handicap as hill climbing which is the problem of ending up at a local minimum. However, this does show that within five steps the search will arrive at a minimum, and that about 50% of the time this will be the global minimum for this example.

3.4.4 Simulated annealing

A popular method of probabilistic search has been developed by physicists based on their understanding of some of the processes that take place when substances cool – in particular, the method of annealing a metal, where it is heated up and then cooled very slowly. The molecules of the metal form crystals which are in the minimum energy state for the metal. The metal has therefore settled at a global minimum for this example.

Simulated annealing, as the name suggests, mimics this process. The energy of a system, E, has to be defined, and this becomes the search space for the problem, which is to find a point of minimum energy.

It is similar to the hill-climbing method except that the decision about whether to keep the new solution or throw it away is probabilistic. This means that sometimes the new solution will be kept *even though the measure associated with it is worse than the best solution so far*. This allows the search to jump out of a local minimum.

The probabilities are such that if the new solution, at x_{k+1}, has a measure of energy, E_{k+1}, that is less than the best solution so far at x_k, then there is a high probability (between 0.5 and 1) that the new solution becomes the best solution. Similarly, if E_{k+1} is greater than E_k, there is a low probability (between 0 and 0.5) that the new solution becomes the current best solution.

Let's say that in a particular example the probability of accepting a new solution turns out to be 0.8. This means that in eight cases out of ten the new solution becomes the best solution so far, but in two cases out of ten it doesn't. The values of probability always lie between 0 and 1.

The probabilities, P_k, are calculated as follows:

If	$E_{k+1} < E_k$	$x = x_{k+1}$ with a probability of P_k	$1 \geqslant P_k > 0.5$
else	$E_{k+1} \geqslant E_k$	$x = x_{k+1}$ with a probability of P_k	$0.5 \geqslant P_k > 0$

rearranging:

If	$(E_k - E_{k+1}) > 0$	$x = x_{k+1}$ with a probability of P_k	$1 \geqslant P_k > 0.5$
else	$(E_k - E_{k+1}) \leqslant 0$	$x = x_{k+1}$ with a probability of P_k	$0.5 \geqslant P_k > 0$

What is needed therefore is a function that produces a value for the probability between 0.5 and 1 when $(E_k - E_{k+1}) > 0$, and a value between 0 and 0.5 when $(E_k - E_{k+1}) < 0$. A function which has this property is the ***sigmoid function***, described by the equation

$$y = \frac{1}{1 + e^{-x}}$$

When $x = 0$, $y = 0.5$.

When $x > 0$, $1 > y > 0.5$.

When $x < 0$, $0.5 > y > 0$

As this method simulates annealing, a factor equivalent to temperature has to be included in the model. This is done by dividing $(E_k - E_{k+1})$ by a notional 'temperature' T, and then substituting for x in the equation of the sigmoid. The probability is therefore

$$P_k = \frac{1}{1 + e^{-(E_k - E_{k+1})/T}} \tag{3.11}$$

When T is very large, P_k approaches 0.5, which means that the decision about keeping the new solution or throwing it away is purely random. When $T = 0$, $P_k = 1$ and the decision is not probabilistic but is equivalent to the hill-climbing method described earlier. So if the 'temperature' starts out high, the decisions seem arbitrary. As the temperature drops, the decision to make the new solution the current best solution or not becomes more deterministic. The effect is that the search can jump out of local minima, and should end up when $T = 0$ at the global minimum or a relatively good minimum.

This method can be understood in terms of jumping around in an 'energy landscape'. Hill climbing gets the search to a lower energy but can be caught in a local minimum, and simulated annealing allows jumps to higher energies, so escape from local minima becomes possible. Inevitably, there is more to this method when it comes to practical implementation. Firstly, cooling has to follow a schedule, and secondly decisions have to be made about the limits of the notional temperature.

For simulated annealing to work, a ***cooling schedule*** has to be constructed. This means that the 'temperature' has to be set to an initial value, and held at this value for a length of time while the search continues. How long is a difficult question to answer. The originators of this method say that the system has to reach ***thermal equilibrium*** before the temperature can be lowered. However, no way is given to determine when thermal equilibrium is reached.

To understand the importance of thermal equilibrium, you have to know about the probability of finding yourself at a particular point in the search space. The way that the problem has been configured, this probability is proportional to the energy at that point. In other words, the search will spend more time at a point if it is lower than any other point. This means that the search will spend more time at the global minimum than at any other point. However, these probabilities apply only when the system is in thermal equilibrium.

Imagine starting the search at a particular point. Shortly afterwards the search is halted and the statistics about which part of the space have been searched are obtained. Inevitably, the area immediately around the starting point will have been visited, and points remote from the starting point will not have been visited. It is important, therefore, to ensure that the search will have had enough time to cross the entire search space. This needs to be estimated for any given problem.

Returning to the problem defined earlier in Table 3.1, simulated annealing can be applied. The numbers in the table are interpreted as the values of the energy at any point in the space. The following cooling schedule was chosen:

For 10 steps: $T = 1.0$

For 10 steps: $T = 0.8$

For 10 steps: $T = 0.6$

For 10 steps: $T = 0.4$

For 10 steps: $T = 0.2$

For 10 steps: $T = 0.1$

In this particular problem, the maximum value of $E_k - E_{k+1} = 7 - 1 = 6$, so with $T = 1.0$ the probability is about 0.1, so that 10% of the time the search will be able to jump out of any minimum, which seems reasonable. With this cooling schedule, statistics were gathered and are presented in Figure 3.32.

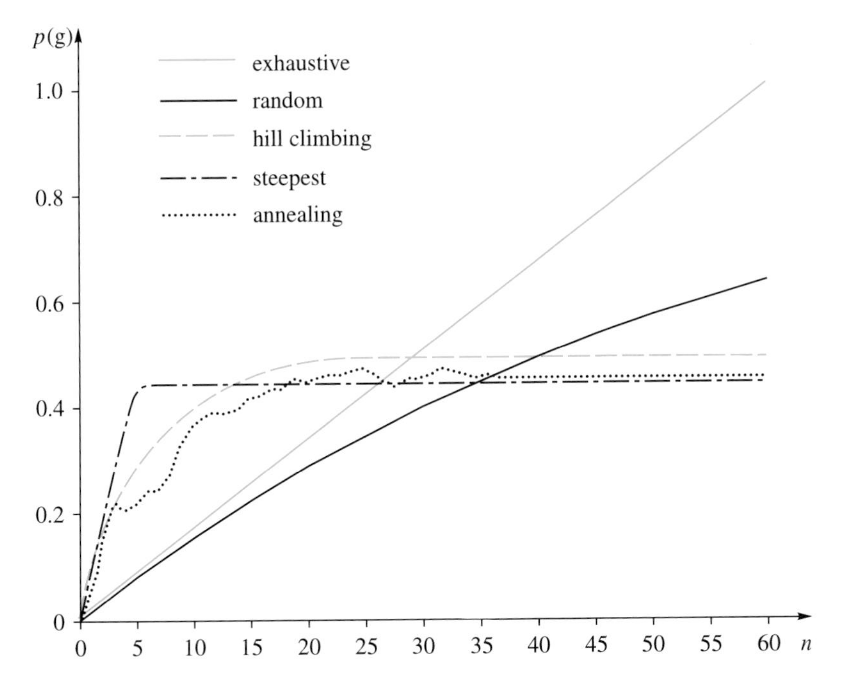

Figure 3.32 Probability of simulated annealing finding the global minimum.

The curve shows that annealing performs about as well as hill climbing. In fact, if the data are examined in detail, the performance is just slightly better than hill climbing since the global minimum is eventually reached a few times more often. The choice of a cooling schedule is critical to the performance. Several cooling schedules were tried and the result given here was the best that was obtained.

The very simple nature of the search space in Table 3.1 does not demonstrate the power of simulated annealing very well. In other experiments in which hill climbing was compared with simulated annealing in a 50-city travelling salesman problem, hill climbing was consistently out-performed by simulated annealing.

All of the gradient-descent methods described so far are limited to searching local regions of the search space. This means that the solution found is highly dependent on the starting point, since the gradient descent will move from this solution to the nearest minimum. Simulated annealing attempts to overcome this by allowing the search to reach 'thermal equilibrium', which means that when simulated annealing starts the temperature is high, so that the search can move about freely. This ensures that a large area of the search space is covered before cooling takes place, and so frees the search from the constraint of finding only solutions near the initial solution. In the next section another method is introduced which tries to combine the power of local gradient descent searching with the ability to cover large parts of the search space.

3.4.5 Genetic algorithms

Genetic algorithms were invented specifically to avoid getting stuck in local minima and to cover as much of the solution space as possible. They are a very efficient means of searching a solution space. Their inspiration came from nature, where it is believed that evolution has provided solutions to the difficult task of adapting life forms to suit particular niches.

The essential features of a genetic algorithm are the ***chromosomes*** that contain the genetic information. These are strings of data that define a particular solution. For example, a chromosome representing six genes might be specified as a six-digit binary number, say 110011. A population of these chromosomes, corresponding to a number of individual solutions to the problem, is created. The population of chromosomes at any one time will represent only a small number of the possible 'good' solutions. The population is initially created randomly, although it is possible to 'seed' the initial population with individuals which are known to be good solutions.

Next we need some way of measuring the ***fitness*** of the chromosomes so that the good solutions are selected to be parents more often than the not-so-good solutions. This is analogous to natural selection, where 'survival of the fittest' is said to occur. The ***fitness function*** selected is specific to the particular application, but generally it has a positive value which is large when a solution is good, and small when the solution is bad.

The mechanism for producing a new population from the current one is called ***breeding***. Parents are chosen in proportion to their fitness using a mechanism called ***roulette-wheel selection***. Each chromosome has a fitness, which can be regarded as a portion of the total fitness of the population. If this is drawn as a pie chart where the total area of the pie corresponds to the total fitness of the population, then an individual has a slice of the pie with a size that is proportional to its own fitness. This is shown in Figure 3.33.

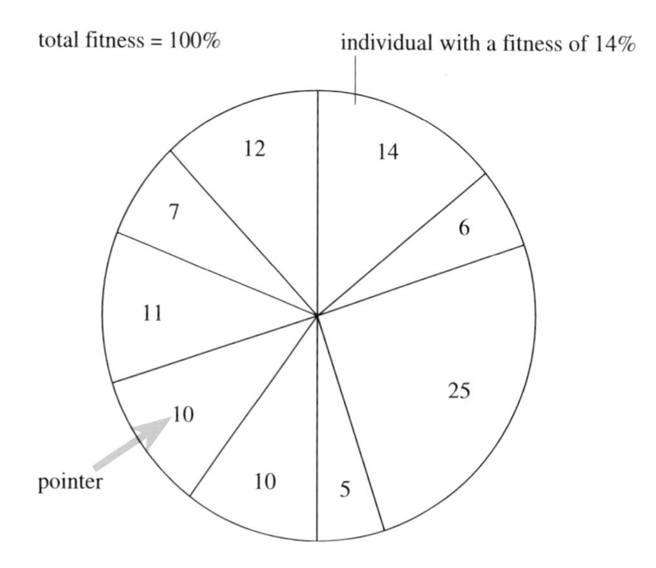

Figure 3.33
The fitness of a population of nine individuals.

Imagine that the pie is spun like a roulette wheel with a pointer at a fixed position. When the wheel stops spinning, the pointer indicates which individual is selected to be a parent. The probability of the pointer pointing at any individual is proportional to the size of the slice allocated to that individual. In other words, the number of times that an individual will be selected to be a parent is proportional to its fitness. If there are N individuals in the population, then the wheel is spun N times to select new parents.

Let's look at a simple example. Suppose there are 6 individuals in a population. Initially they are randomly generated and their fitness calculated, as in Table 3.2.

TABLE 3.2

Individual	Fitness	Running total
A	12	12
B	5	12 + 5 = 17
C	23	17 + 23 = 40
D	13	40 + 13 = 53
E	1	53 + 1 = 54
F	16	54 + 16 = 70

The total fitness is 70, so the fitness of A as a proportion is 12/70, which is about 17%. Figure 3.34 shows the roulette wheel.

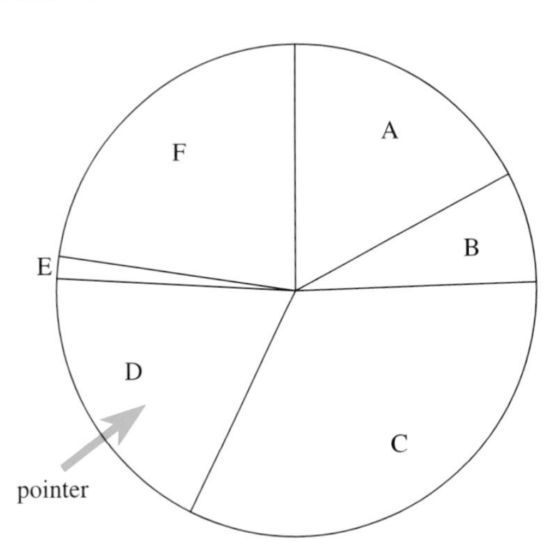

Figure 3.34 Roulette wheel of the six members of the population.

The roulette wheel is a good way of imagining what is going on in selection. What actually happens is that a random number is generated between 0 and the total fitness, in this example 70. The number generated is compared with the running total, shown in Table 3.2. The first individual with a running total greater than the random number is then selected. For example, if the random number generated is 45, D would be chosen because D is the first individual found when scanning down the table which has a running total, 53, which is greater than the random number.

In this example there is a population of 6, so a random number is generated six times. Let's assume that the six numbers generated are 45, 23, 31, 57, 4 and 55. The corresponding individuals would be D, C, C, F, A and F. These are the parents of the new population.

Offspring are produced by selecting pairs of parent chromosomes and ***crossing over*** some of the genetic material. In the example just given, the parents would be paired D and C, C and F, and finally A and F. Each pair of parents then produces two offspring. It is permissible for two parents to be the same, such as A and A, even though the offspring are also A and A.

Figure 3.35 illustrates a crossover for two binary-valued, six-digit parent chromosomes. The result is two offspring chromosomes, combining the digits of the parents according to the crossover point chosen.

parent 1	**1**	**1**	**0**	**0**	**1**	**1**		offspring 1	**1**	**1**	**0**	0	1	0
parent 2	1	0	1	0	1	0		offspring 2	1	0	1	**0**	**1**	**1**

crossover point

Figure 3.35 Breeding using single-point crossover.

This form of crossover is called ***single-point crossover***. The actual point at which crossover takes place is randomly chosen. Other forms of crossover exist such as two-point crossovers, but we will only use single-point in this book.

Again, let's take our six chromosomes, and let's assume that they have the following binary structure:

A	1	1	0	0	1	1
B	1	0	1	0	1	1
C	0	1	0	1	0	1
D	1	1	1	0	1	0
E	0	0	0	0	0	1
F	0	1	0	0	1	0

Crossover takes place by generating a random number that corresponds to the position along the chromosome. The genetic material is then swapped over between the parents at that point to create two new individuals. The random numbers generated for these examples would be in the range 0 to 5, corresponding to the six points along each chromosome where there is a 0 or a 1, starting from the right. If the random numbers are 3, 1 and 2, then crossover takes place after bit 3 in the first pair of parents, after bit 1 in the second pair and after bit 2 in the third pair. Crossover would then look like this (one parent in each pair is shown in bold to show where the genetic material comes from):

	parents						offspring					
	5	4	3	2	1	0	5	4	3	2	1	0
D	1	1 |	1	0	1	0	1	1	**0**	**1**	**0**	**1**
C	**0**	**1** |	**0**	**1**	**0**	**1**	**0**	**1**	1	0	1	0
C	0	1	0	1 |	0	1	0	1	0	1	**1**	**0**
F	**0**	**1**	**0**	**0** |	**1**	**0**	**0**	**1**	**0**	**0**	0	1
A	1	1	0 |	0	1	1	1	1	0	**0**	**1**	**0**
F	**0**	**1**	**0** |	**0**	**1**	**0**	**0**	**1**	**0**	0	1	1
	before crossover						after crossover					

Figure 3.36 ***Breeding using single-point crossover.***

In addition to crossover, ***mutation*** is allowed. This happens when some of the genetic material changes randomly, as shown in Figure 3.37.

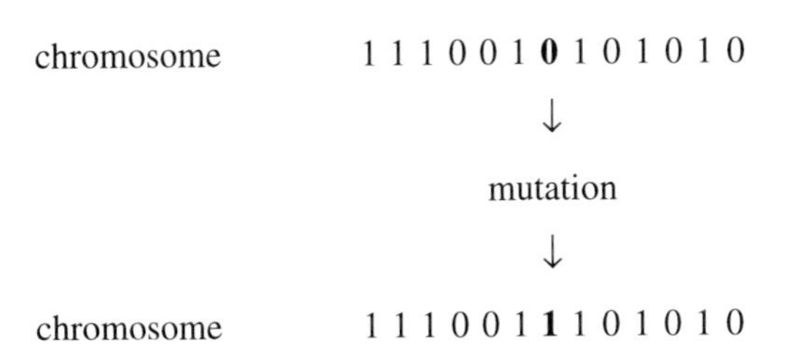

Figure 3.37 Mutation.

Mutation is usually defined by the ***mutation rate***, which is normally set to quite a low value, 0.001 say. This corresponds to one change in a thousand bits of data. The bits that are actually mutated are randomly selected. After a bit has been selected for mutation it is inverted, so a 0 becomes a 1 and vice versa.

In our simple example, the new population, after mutation, might look something like Figure 3.38, where two bits have been mutated (shown in bold):

A′	1	1	0	1	0	1
B′	0	1	**0**	0	1	0
C′	0	1	0	1	1	0
D′	0	1	0	0	0	1
E′	1	1	0	0	1	**1**
F′	0	1	0	0	1	1

Figure 3.38 Example of population after mutation.

When a genetic algorithm is applied to a problem an initial population is created with randomly generated chromosomes. Each chromosome is tested and an evaluation of its fitness is made. Having evaluated the whole population, breeding can take place. Breeding continues until a new population is created, at which point the old population is replaced by the new one. (In the ***elitist strategy***, a proportion of the fittest parents would also be carried over into the next generation, so there would be correspondingly fewer offspring in the next generation.)

This breeding and evaluation process continues, with the average fitness of the population being monitored together with the fitness of the fittest individuals. In a typical situation the maximum individual fitness will rise quickly and then at some point it will 'flatten off'. The average fitness of the population will rise more slowly, and if left to run for a long time would equal the value of the fittest individual. This latter fact arises because, if left for a long time, the population would eventually consist of replicas of the fittest individual and no others. Usually the search is terminated when the maximum individual fitness flattens off but the average fitness of the population is still rising.

It is not immediately obvious why genetic algorithms should be so good at searching for solutions. The answer is that the mutation operation tends to move the chromosome to a neighbouring position in the search space, and so could be considered as a local search. If the mutated offspring is fitter than its parent then its chances of breeding are improved. This is rather like the hill-climbing methods that have already been described. In addition, the crossover operation allows large jumps to be made in the solution space. This ensures that large areas of the space are searched and that solutions do not get stuck in local minima.

Genetic algorithms were also tried on the example problem that has been used throughout this section, originally given in Table 3.1. To try to get some sort of comparison with the other methods described, a population of 10 was selected and the algorithms used to obtain six generations. Each chromosome contained a binary representation of the x and y coordinates of the two-dimensional surface – four bits for the y-coordinate and three bits for the x-coordinate. In order that all of the codes that could be generated by the chromosomes would be meaningful, the two-dimensional surface was extended so that the whole pattern was a grid of 8 points by 16 points – the additional points being given a high value of 10. Single-point crossover was used and the mutation rate was set at 0.28.

The values given in Table 3.1 were used to calculate the fitness. The actual fitness function used was

$$\text{fitness} = \frac{60}{x} - 6$$

where x is one of the values given in Table 3.1. This fitness function ensures that when the value of x is 10, the fitness is 0, and that the largest values of fitness occur at the minima.

After running the algorithm for 2000 trials, the statistics for the number of times the local and global minima were found for each generation were as shown in Table 3.3.

TABLE 3.3

Generation	Found global minimum	Found local minima	Failed to find any minimum
1	10%	20%	70%
2	15%	30%	55%
3	17%	31%	52%
4	21%	31%	48%
5	21%	35%	44%
6	21%	35%	44%

Since each generation contains 10 individuals, roughly speaking the genetic algorithm found the global optimum 21% of the time after 6 generations or about 60 iterations. More often it found one of the other local minima, and just under half of the time it failed to find any of the minima.

Just as with simulated annealing, the example of Table 3.1 is too simple to demonstrate the power of genetic algorithms properly. On more complex problems they have been found to give good results, and there are many examples in the literature.

3.4.6 Summary of the optimization techniques described

The techniques described in this chapter all try to follow some kind of gradient to continually improve some measure of performance, with the aim of finding the solution that gives the best performance. Most of the techniques only ever search the immediate neighbourhood, and so can never break away from the locality in the search space in which the search is started. Simulated annealing tries to overcome this by allowing jumps to intermediate solutions that may perform worse than the present solution so that more of the search space can be examined. Genetic algorithms also use crossover to create new solutions in unexplored areas of the space.

Different methods suit different search spaces. The example problem was a relatively simple space, with the result that simple gradient-descent algorithms worked well. If the space was more complex – for example, if it was much bigger with much more diversity – then the gradient-descent methods would usually perform less well than simulated annealing and genetic algorithms. There exist some search spaces for which, until simulated annealing and genetic algorithms came along, there was no way of finding a solution apart from random or exhaustive search. These new tools at least allow machines to find solutions, possibly sub-optimal, faster than before. They do not find solutions easily, but they are a first step in overcoming the barrier of finding solutions to problems with a computational complexity which is NP-indeterminate. They will be demonstrated later in the book: simulated annealing will be shown applied to the travelling salesman problem in Chapter 5 on Scheduling, and genetic algorithms will be shown being applied to neural networks in control applications in Chapter 9 on Intelligent control.

3.5 Conclusion

The principles of applying a search method can be abstracted from this chapter. They can be found by posing the following questions:

- What sort of search problem is this?

  combinatorial search of finitely-generated set, e.g. the domino problem

  quantified optimization problem, e.g. the calculus-based methods.

- How big is the search space?

  small – a few thousand to a million points

  large – many millions

  infinite.

- What is known about the search space?

  is it continuous?

  is it differentiable?

  is there a formula(e) to represent it?

- What are the information sources?

  databases

  generative calculations, e.g. deduced knowledge (see Chapter 6 on Reasoning)

  sensors (continuous and discrete).

- What methods are available for this kind of search?

  exhaustive search

  random search

  breadth-first, depth-first, best-first

  hill climbing

  gradient or steepest descent

  simulated annealing

  genetic algorithms.

- How well do each of these methods work for each kind of problem?

  use exhaustive search for small finite spaces when it is essential that the global minimum is found;

  use random search for large evenly distributed homogeneous spaces;

  use hill climbing for discrete spaces where a sub-optimal solution is acceptable;

  use gradient descent for continuous or discrete spaces when a fast but probably sub-optimal solution is acceptable;

  use simulated annealing for large continuous or discrete spaces where a better solution than gradient descent is required, possibly the optimum solution, but with the cost of longer times needed for calculation;

  use genetic algorithms for large or infinite search spaces with sparse and diverse data;

  use tree search when a lot is known about the search space which is usually discrete, when a decision can be made at each step as to which direction to search and when there is a distinct goal. Sometimes this can be exhaustive, and therefore not fast except when the space is relatively small. Depth-first, breadth-first and best-first can speed up the search, each method being appropriate to different problems.

- Can the search problem be converted to another search problem ?

  using a different representation

  using new information (data fusion).

Answering these questions would help to pin down which search methods are applicable. However, it is not such a simple task and there are no absolutely clear guidelines to help. We hope this chapter has provided you with a set of tools that can be applied to specific problems, and an insight into how they work and when they work best.

Search lies at the heart of artificial intelligence, since almost all problems require a search to find a solution. Conventional mathematics can be used to find solutions to problems which have a mathematical representation, such as a formula or set of equations. Searching for solutions to the many other problems which do not have such a representation requires some understanding of the nature of the search space and how it is structured. This knowledge then guides the selection of an appropriate search technique and becomes the basis of heuristics aimed at giving acceptable solutions most of the time within acceptable costs.

Many of the techniques described in this book involve search in one guise or another. In pattern recognition we search for an interpretation of a given representation. In neural networks we have to search for a set of network weights which minimize the error of the system. In scheduling, we search for the best schedule of activities and places in time. In reasoning, rule-based systems and learning, we search for relevant new knowledge, given a knowledge base and new information. In intelligent control we search for a control strategy that will keep ill-defined and complex systems within specifications, and in computer vision we seek an interpretation of images. For these reasons the concepts appearing in this chapter will recur throughout the book.

References and further reading

Davis, L. (ed.) (1991) *Handbook of Genetic Algorithms*, Van Nostrand Reinhold.

Jarvis, R.A. (1985) 'Collision free trajectory planning using distance transforms', *Mech. Eng. Trans. of the I. E. Aust.*, ME10, 3, pp.187–191.

Winston, P.H. (1984) *Artificial Intelligence*, Addison-Wesley.

CHAPTER 4
NEURAL NETWORKS

4.1 Introduction

Artificial neural networks are emerging as an exciting new information-processing paradigm for intelligent systems. They differ in many respects from conventional sequential computers, and it is claimed that neural networks have the following advantages.

Potential advantages of neural networks

- They do not need to be programmed, as they can *learn* from examples.
- They can generalize from their training data to other data.
- They are *fault tolerant:* they can produce correct outputs from noisy and incomplete data, whereas conventional computers usually require correct data.
- On being damaged, they *degrade 'gracefully'* (that is, in a progressive manner), unlike sequential computers which can fail catastrophically after isolated failures.
- They are *fast*: their many interconnected processing units work in parallel.
- They are *relatively inexpensive* to build and to train.

These potential advantages have created considerable interest in the possibilities for applying neural networks in engineering, and have resulted in a great deal of research over the last ten years. Some of the claimed advantages are exaggerated, but others are certainly proven, and neural networks are becoming a standard technology for engineers.

Some of the many applications of neural networks include:

systems which detect explosives at airport gates;

character recognition and document reading systems;

robot vision systems;

speech understanding systems, e.g. telephone systems which can recognize and distinguish between words such as *yes*, *no*, *one*, *two*, *three*, etc.;

financial investment systems.

The fundamental feature of any neural network is that it is composed of a large number of interconnected processing units. These units are often relatively simple, and the network gets its computational power from the many units being connected, with outputs from the units being inputs to others. The way the units are connected is called the ***network topology*** (Figure 4.1).

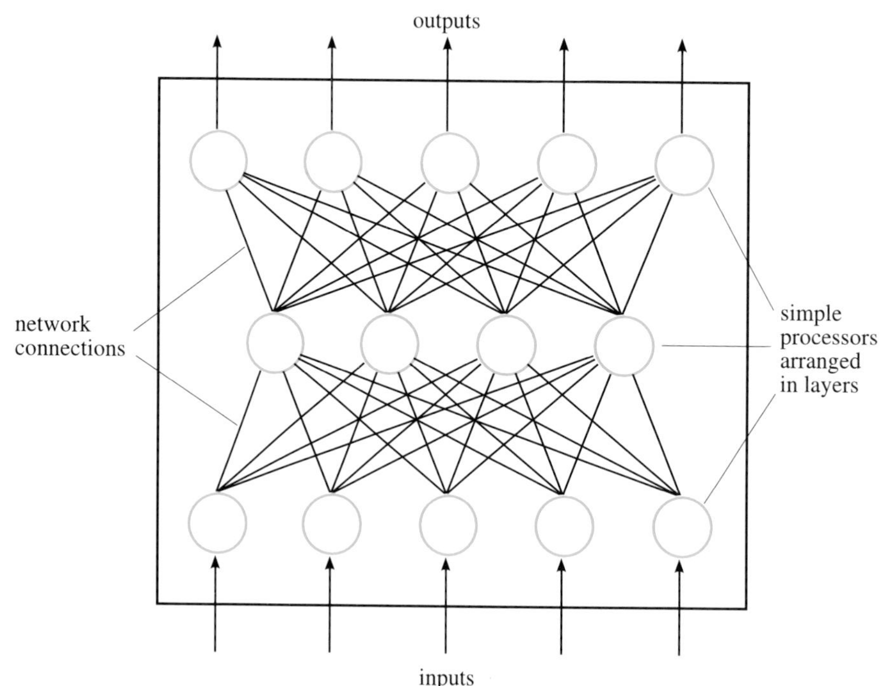

Figure 4.1
General architecture of a neural network.

Neural networks excel at ***classification***; they are pattern recognizers *par excellence*. When used in this way they are presented with information about objects or cases to be recognized, and their output signifies the class to which the object belongs. For example, later in the chapter we will see how a neural network can recognize characters on the basis of black and white information. We ***train*** the network by showing it examples of 'ideal' characters. On seeing new cases the network can tell us which class or character it best fits. The network ***generalizes*** from the characters it has seen to be able to recognize other characters in that class.

In this book we will present neural networks as powerful ***black-box classifiers***. By this we mean that they take input data, process them, and give output data. In general, we do not know precisely what is going on inside the network, which is why we say it is a black-box system. In general, the output is of the yes/no binary type: 'yes' the object belongs to this class, or 'no' the object does not belong to this class (Figure 4.2). This is why we call them classifiers.

Figure 4.2
Character classification.

To understand why classifiers are such powerful information processors, consider the following questions that an intelligent machine might have to resolve:

> What should I do next?
> Which way should I go?
> Is there an obstacle in my way?
> Are explosives present in this suitcase?
> Is this atmosphere poisonous?
> Should I invest in this currency?
> Is the system I am monitoring in a 'normal' state?
> Is this character a 1 or a 2?
> Is the camera aperture correct for this light level?

These can all be considered to be classification problems. If we assume that each question has a finite number of answers, and we can find appropriate training data, then answering these questions amounts to classifying the outputs as 'correct' or 'incorrect'.

Defining appropriate inputs and outputs, and finding appropriate training data, lie at the heart of successful engineering applications of neural networks. This can require considerable knowledge and ingenuity on the part of the engineer. In particular, it is essential that the input data are in the right form for a network to operate on, and they must contain sufficient information for the classification to be made. It is also essential, of course, that the outputs are relevant.

There are many examples in the literature of networks that can do wonderful things, such as the applications listed above. Almost all of them work because the system designers understood the overall nature of the problem they were trying to solve, and created appropriate ***pre-processing and post-processing subsystems***, as illustrated in Figure 4.3.

For example, neural networks are commonly used for classifying objects in image data. A very simple example of this is shown in Figure 4.4, where the characters 0, 1, + and $\times$ are formed on a 3×3 square grid. For consistency with generally accepted terminology, the squares will be called *pixels* (picture elements).

How can such graphic data be input into a neural network?

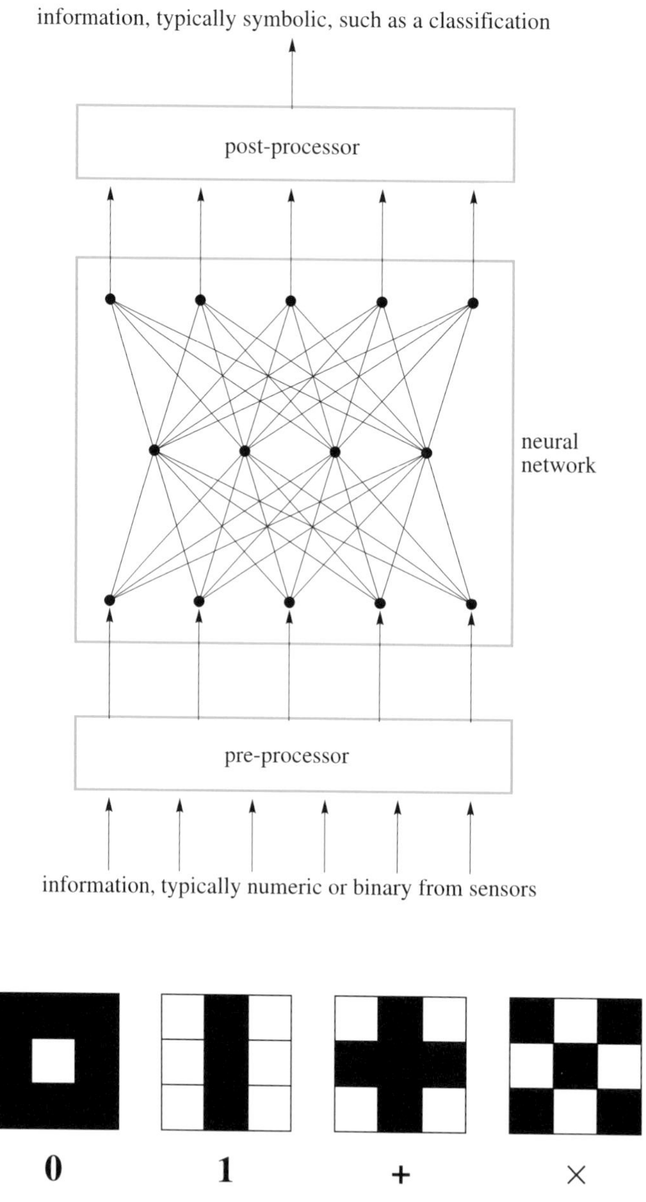

Figure 4.3
A typical neural network architecture sandwiches the network between a pre-processor and a post-processor.

Figure 4.4
Characters in a 3 × 3 image

The answer to this question is that the graphical data must be transformed into a sequence of numbers in order to be input into a neural network. In other words, a ***pre-processor*** is required. For a given application, satisfactory pre-processing may be achieved in a number of different ways. In some cases the design of the pre-processor is an essential feature in building a useful neural system.

Let us construct a pre-processor as follows. First, let us number the pixels from 0 to 8 as shown in Figure 4.5(a). (In computing it is usual to begin counting from 0 rather than 1.)

0	1	2
3	4	5
6	7	8

(a) Assigning numbers to each pixel of the 3×3 grid

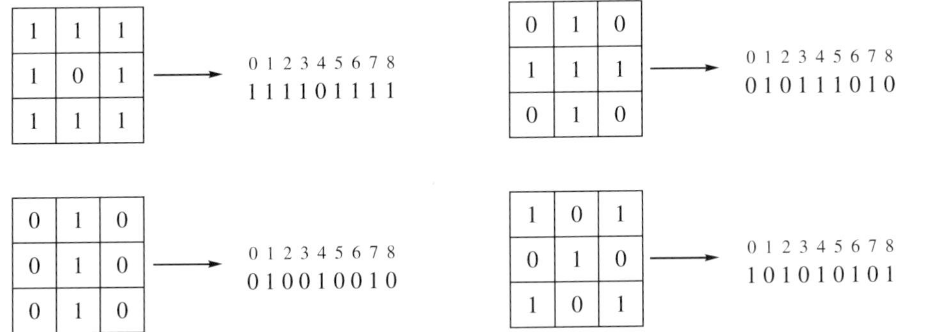

(b) Converting the grid of pixels (1 = black pixel, 0 = white pixel) into a sequence of numbers

Figure 4.5
The action of the image-to-numbers pre-processor.

The pre-processor specifies the inputs for the networks. Let the outputs be defined as follows:

1 0 0 0   0
0 1 0 0   1
0 0 1 0   +
0 0 0 1   ×

The characters assigned to each class give the information required by the post-processor to make a classification.

The training data for the network are therefore:

Inputs	**Outputs**	**Class**
1 1 1 1 0 1 1 1 1	1 0 0 0	0
0 1 0 0 1 0 0 1 0	0 1 0 0	1
0 1 0 1 1 1 0 1 0	0 0 1 0	+
1 0 1 0 1 0 1 0 1	0 0 0 1	×

The pre-processor converts the raw data available to the system into a form that can be input to a neural network, i.e. it *encodes* the input data as a list of numbers. The network then does the essential classification work to give one of the desired outputs. However, the outputs are presented as a list of numbers which may require *decoding* by the post-processor.

In this case the post-processor might be a module of a computer program which accepts outputs of the network such as (0.01, 0.04, 0.98, 0.11) and *thresholds* these numbers to obtain the binary string (0, 0, 1, 0), matches this binary string against data in memory and passes the symbol + to the cognition/execution subsystems. For example, it could give a message on a computer screen such as 'the character + was recognized'.

Thus the pre-processor converts image data into a sequence of nine numbers which can be input to a network, while the post-processor interprets as a character the four output numbers which come out of the network.

We began this section by listing some of the attributes of neural computers which people think are advantageous. We will end it by listing some of the features that are sometimes considered less advantageous. As always, when choosing between different alternatives such as information-processing paradigms, the engineer must be aware of the pros and cons.

Potential disadvantages of neural networks

- Neural networks have *no model* of the universe in which they work.
- Whereas neural networks work well for inputs reasonably similar to their training data, they may give completely unpredictable outputs outside this region.
- Although they require no programming, a considerable effort may go into the pre-processing and post-processing subsystems for a neural network.
- Much of the knowledge about neural networks is empirical.

Most of the remainder of this chapter will be devoted to explaining the technicalities of neural networks. Our objective is to give you sufficient information on this subject for you to be able to design and build your own neural processing systems.

4.2 The artificial neural unit

The fundamental feature of any ***neural network*** is that it is composed of many interconnected ***units***, each of which performs a ***weighted sum*** of its inputs. Figure 4.6 shows an example of one of these units.

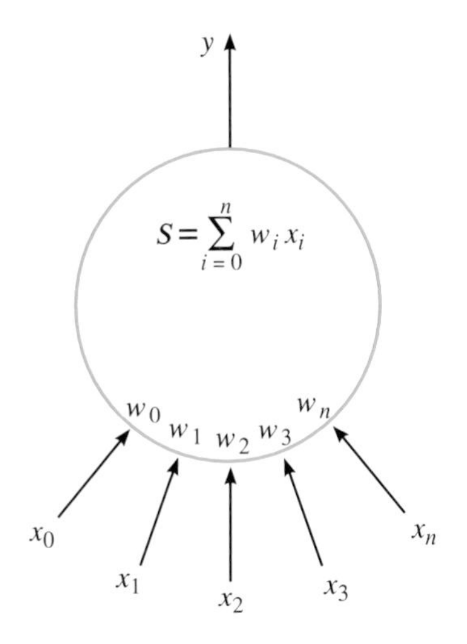

Figure 4.6
A unit in a neural network.

The unit has $n+1$ inputs, x_0 to x_n, and a single output, y. Associated with each input is a ***weight***, which is a real number. The value of each weight, w_0 to w_n, can be either positive or negative (or zero), very large or very small. When a particular set of input values arrives at the unit, each of the inputs is multiplied by its associated weight value and the sum of all the weighted inputs is found. Mathematically, this can be summarized by the following expression:

$$S = w_0 x_0 + w_1 x_1 + \ldots + w_n x_n = \sum_{i=0}^{n} w_i x_i$$

This is the weighted sum of the inputs, S. The value of this weighted sum determines the output of the unit, y. Exactly what that output is depends on the ***output function***, which in turn depends on the particular type of neural network. A typical output function would produce either an output of 1 if the weighted sum is positive, or an output of 0 if the weighted sum is negative. What happens when the weighted sum is zero? Again this can vary but typically this is treated the same as a positive weighted sum, so the output would be 1. This sort of output function is shown graphically in Figure 4.7, and is described as ***hard-limiting***.

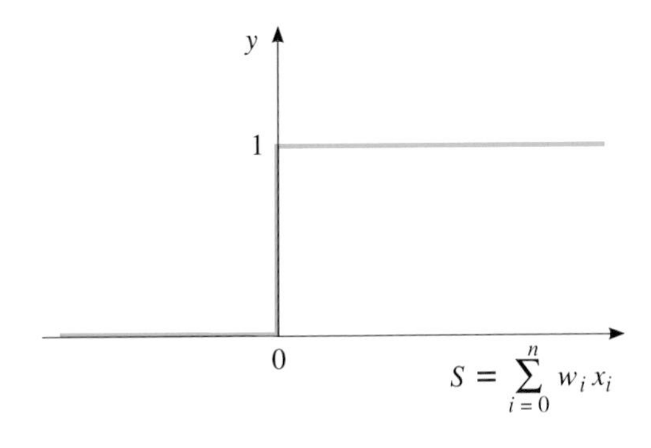

Figure 4.7
Hard-limiting output function.

You will notice that the expression for the weighted sum shows that the sum is taken for all the inputs x_1 to x_n, which are variable, but that it also includes a term x_0 with a corresponding weight w_0. This extra input x_0 has a constant value of 1 and provides an ***offset*** of w_0 to the weighted sum. This is essential for the correct working of the unit, as we shall see later.

The way that the units in an artificial neural network function has often been compared to the way that biological ***neurons*** work. It is known that neurons are connected together via ***synapses***, as shown in Figure 4.8.

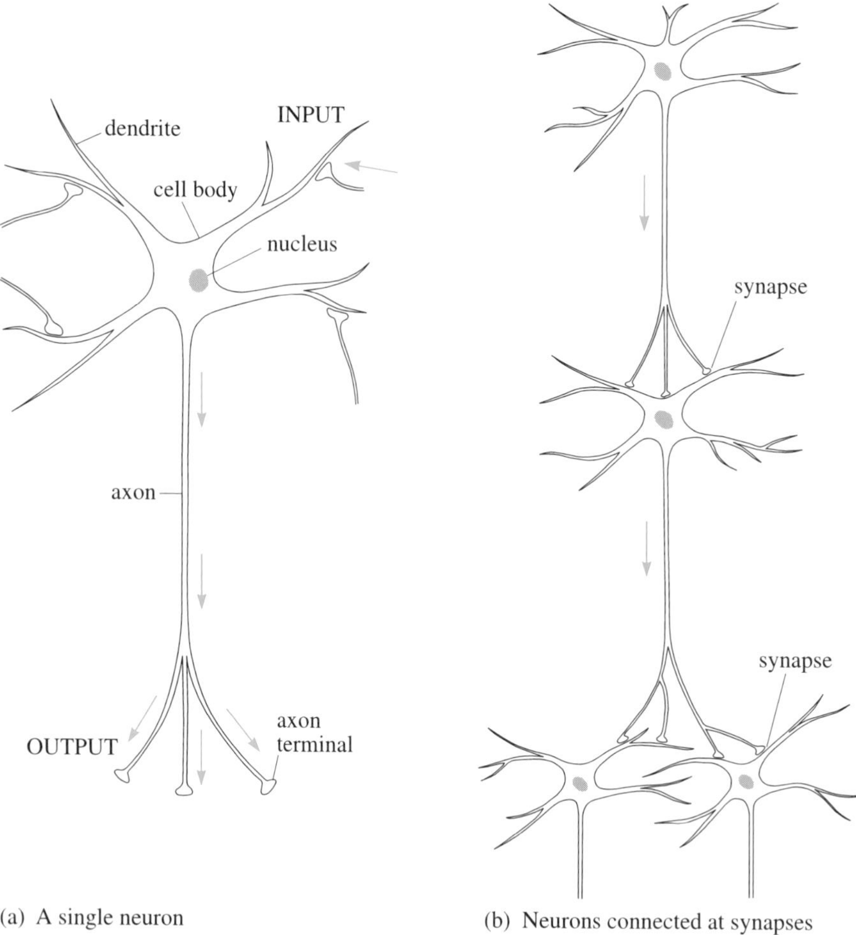

(a) A single neuron

(b) Neurons connected at synapses

Figure 4.8 Typical biological neuron.

A synapse produces a chemical response to an input. The size of the response can vary, and the mechanism is analogous to the weights in the units of an artificial neural network. The biological neuron 'fires' if the sum of all the reactions from the synapses is sufficiently large, so there is a similarity in their behaviour to the 'units' that have been described. In fact, the units were originally invented as an attempt at modelling biological neurons, hence the use of the term 'neural networks'. However, an element of caution is needed. Although there are some similarities between the functioning of these units and biological neurons, it would be untrue to say that an artificial neural network is like a brain. Biological neurons are far more complex than these simple models.

Artificial neural networks are composed of large numbers of these units connected together. However, it is worth looking at the properties of just one of these units to start with. First of all we want to show that a single unit is capable of performing the Boolean logic functions AND, OR and NOT when the inputs are binary with the values of either 0 or 1.

Figure 4.9 shows a single unit with two inputs $x_1$and x_2. For technical reasons explained in Section 4.3, there is a another fixed input, x_0, called the *offset*, which has a fixed value of 1.0. The weights are set to $w_0 = -1.5$, $w_1 = 1.0$ and $w_2 = 1.0$. It is assumed that the unit is hard-limited as explained previously, so that the output y is 0.0 if the sum, S, of the input times the weights is less than zero, and $y = 1.0$ otherwise.

When both the inputs x_1 and x_2 are 0.0,

$$S = -1.5 \times 1.0 + 1.0 \times 0.0 + 1.0 \times 0.0 = -1.5$$

and the output of the unit is 0.0. When x_1 is 0.0 and x_2 is 1.0,

$$S = -1.5 \times 1.0 + 1.0 \times 0.0 + 1.0 \times 1.0 = -0.5$$

and the output of the unit is 0.0. When x_1 is 1.0 and x_2 is 0.0

$$S = -1.5 \times 1.0 + 1.0 \times 1.0 + 1.0 \times 0.0 = -0.5$$

and the output y is again 0.0. Finally, when x_1 is 1.0 and x_2 is 1.0,

$$S = -1.5 \times 1.0 + 1.0 \times 1.0 + 1.0 \times 1.0 = +0.5$$

and in this case, after hard-limiting, the output of the unit is $y = 1.0$. Table 4.1 summarizes these results.

TABLE 4.1 THE AND FUNCTION

x_0 (fixed)	x_1	x_2	S	y
1	0	0	−1.5	0
1	0	1	−0.5	0
1	1	0	−0.5	0
1	1	1	+0.5	1

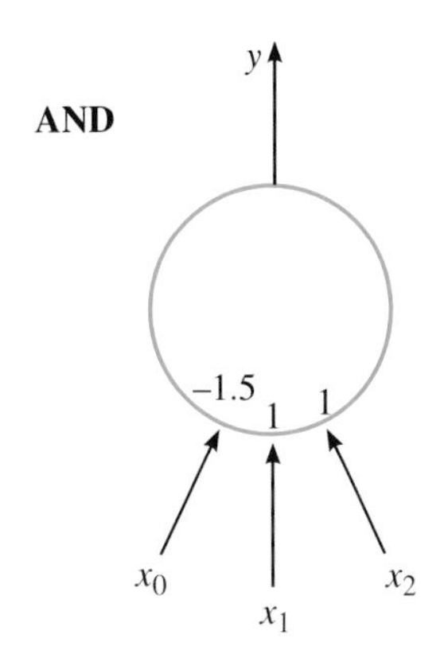

Figure 4.9
The AND function.

The output in Table 4.1 is only 1 when *both* inputs x_1 AND x_2 are 1, and therefore the unit performs the logical AND function.

Figure 4.10 shows a unit with two inputs which is capable of performing the logical OR function. The output response is summarized in Table 4.2. You can see that the output is 1 when *either* x_1 is 1 OR x_2 is 1.

TABLE 4.2 THE OR FUNCTION

x_0 (fixed)	x_1	x_2	S	y
1	0	0	−0.5	0
1	0	1	+0.5	1
1	1	0	+0.5	1
1	1	1	+1.5	1

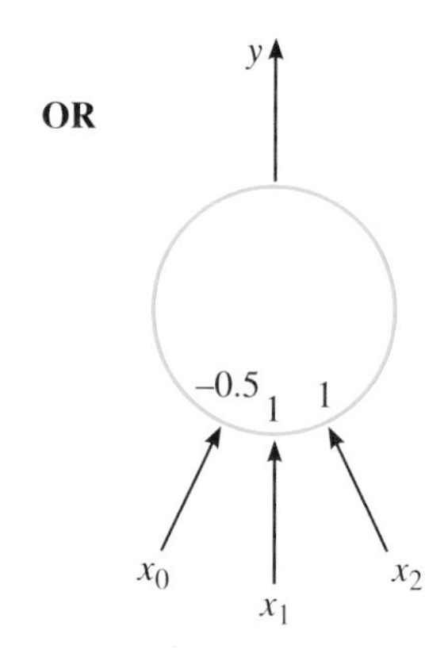

Figure 4.10* *The OR function.

Finally, Figure 4.11 shows a unit that can perform the logical NOT function. If the input x_1 is 0 then the output is 1, and vice versa. This is summarized in Table 4.3.

TABLE 4.3 THE NOT FUNCTION

x_0 (fixed)	x_1	S	y
1	0	+0.5	1
1	1	−0.5	0

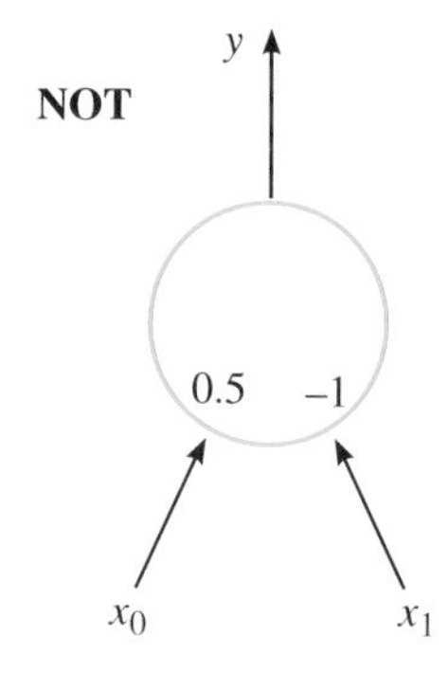

Figure 4.11* *The NOT function.

We have therefore demonstrated how these simple units can perform the logical functions AND and OR for the case where the number of inputs is two (not counting the constant input x_0) and NOT where there is one input and x_0. It was stated in Chapter 3 of Volume 1 of this book that any logic function can be constructed from gates that perform these three basic logical functions. This means that conventional computers *could* be built using these artificial neural units alone instead of the usual transistor-based logic circuits.

We are not suggesting that computers can or ever will be built using neural networks, but their potential was the reason for the initial excitement over neural networks. Researchers felt that they had shown that, since brains are made of neurons, and neurons behave like logic gates, and computers are made from logic gates, it follows that the brain is like a computer. Therefore it should be possible to mimic or simulate the functions of the brain on a computer. Unfortunately this was too optimistic, because neurons are much more complex than this simple model. The quest for producing an artificial brain is still highly elusive and is likely to remain so for many years to come.

In engineering terms, on the other hand, these units provide a medium in which certain operations can be carried out with more success than conventional algorithmic methods. In particular, pattern classification, which was introduced earlier in the book, can be implemented relatively successfully using neural networks.

4.3 Pattern classification

Objects are said to belong to a particular class if they have properties which are similar to other objects in that class. If we wanted to make a two-way classification of fruit into either apples or bananas, for example, we could select a set of measurements such as size and weight. When presented with an unknown example of a fruit at a later stage we could classify it as either belonging to the class 'apple' or the class 'banana' according to which class has the most features in common with the previously learnt examples.

What makes the 'apples and bananas' example difficult for machines to carry out is that the objects to be classified don't match up exactly with any of the previously recorded examples. One way to resolve this is to store some 'ideal' object that is representative of each class – the perfect apple and the perfect banana. This ideal object is then the ***model*** or ***template*** against which we compare new objects. In a similar way, new discoveries of fossils are classified by comparing them with the large collection of previously identified fossil specimens in the British Museum and elsewhere.

Neural networks provide an alternative approach in which there is no model. Somehow the general characteristics of the class have to be inferred from the examples that have been seen. It is sometimes said that the neural network has a ***distributed model***, which means that the model is not stored in one place but is distributed throughout the network in the values of the weights.

A single unit can sometimes be enough to be able to carry out a pattern classification. As an example, we will carry out a two-way classification into classes A and B, using only two measurements x_1 and x_2. Table 4.4 lists the data that we collected from ten samples. Notice that the data here are not binary numbers: the values are decimals. In this example we are going to leave the data in this form.

TABLE 4.4 MEASUREMENTS OF TEN SAMPLES, CLASSIFIED AS OBJECTS A AND B

Sample	x_1	x_2	Classification
1	2.0	3.5	A
2	3.0	1.5	B
3	4.5	1.5	B
4	1.5	2.0	B
5	3.0	4.5	A
6	2.0	5.0	A
7	4.0	3.0	B
8	3.0	3.0	B
9	3.0	5.5	A
10	4.0	4.5	A

It would not be too difficult to carry out the classification of any new sample given the two measurements. For example, what class does the object with the following measurements belong to?

$$x_1 = 2.0, \quad x_2 = 4.5$$

It isn't immediately obvious, but yes, it's A. Don't worry if you can't see why – it will become clearer as we go along.

We could display the same data on a graph using x_1 and x_2 as the two axes, and mark the position of each sample on the graph with a symbol representing either 'A' or 'B'. This is shown in Figure 4.12. Chapter 2 gave various methods for separating the clusters of points, and neural networks provide another method with the great advantage of *learning from examples*.

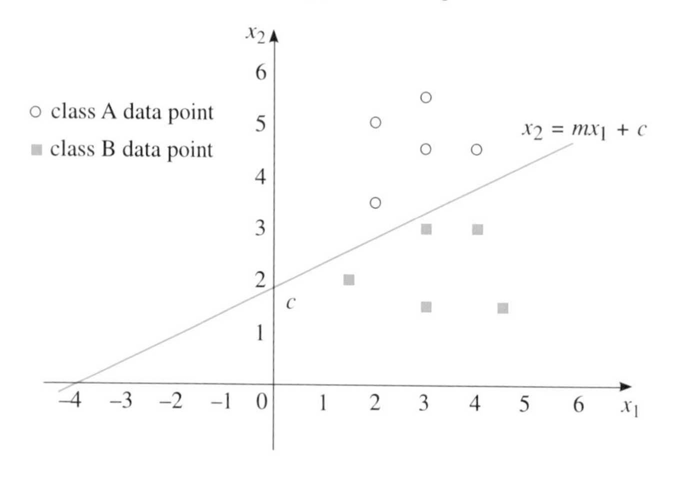

Figure 4.12 Graph of the two-object data, for the samples of Table 4.4.

The graph, called the ***pattern space***, shows that the two types of object lie in quite distinct clumps. We can separate the clumps quite easily by drawing a line between them, as shown in Figure 4.12. Given data for some new object, we would only have to test to see if the data correspond to a point in the pattern space which lies on one side of the line or the other in order to classify the object. These two classes can therefore be described as ***linearly separable***.

How do we test whether a point is on one side of a line or the other? A line has the general form

$$y = mx + c$$

where m is the slope or gradient of the line and c is the point on the y-axis where the line intersects.

On our graph, y corresponds to x_2 and x corresponds to x_1, so the equation for a general line in our pattern space is

$$x_2 = mx_1 + c \tag{4.1}$$

This can be rearranged to give

$$x_2 - mx_1 - c = 0 \tag{4.2}$$

We can find the value of c for the straight line in Figure 4.12 by noting where the line passes through the x_2 axis, i.e. at $x_2 = 2.0$. So, when $x_1 = 0$, $x_2 = 2.0$. Substituting into equation (4.2), we get

$$2.0 - m \times 0 - c = 0$$

$$c = 2.0$$

Similarly, by looking at the point where the line goes through the x_1 axis we can find m. In this instance, the line crosses the x_1 axis at $x_1 = -4.0$, so when $x_2 = 0$, $x_1 = -4.0$. Substituting again, we get

$$0 - m(-4.0) - 2.0 = 0$$

$$m = 0.5$$

So the equation for the straight line in Figure 4.12 is

$$x_2 - 0.5x_1 - 2.0 = 0 \tag{4.3}$$

When a pair of coordinates x_1 and x_2 are substituted into this equation, if the result is 0 then the coordinates correspond to a point that lies on the line.

It turns out that if x_1 and x_2 are substituted into this equation and the result is greater than 0, the point lies *above* the line. Similarly if the result is less than 0 the point lies *below* the line. So for classification, when the two measurements are substituted into equation (4.3) for the line separating the two classes, if the result is greater than 0 the object is of type A. Alternatively, if the result is less than 0 the object is of type B.

For example, consider $x_1 = 2.0$ and $x_2 = 4.5$. Substituting into the left-hand side of equation (4.3) we get

$$4.5 - (0.5 \times 2.0) - 2.0 = 1.5$$

The result is greater than 0, so the object is of type A, confirming our earlier intuitive conclusion.

Now, let's return to neural networks. Recall that a single unit gives an output of 1 if the weighted sum is greater than or equal to 0, and an output of 0 if the weighted sum is less than 0. If we code the objects such that 1 corresponds to A and 0 corresponds to B then we have to find a set of weights that will produce 0 and 1 when appropriate. The equation for the weighted sum of a two-input unit is

$$\sum_{i=0}^{2} w_i x_i = w_0 x_0 + w_1 x_1 + w_2 x_2$$

The output of the unit, y, is 1 when the weighted sum is greater than 0, i.e.

$$y = 1 \quad \text{when } w_0 x_0 + w_1 x_1 + w_2 x_2 > 0$$

Similarly, the output is 0 when the weighted sum is less than 0, i.e.

$$y = 0 \quad \text{when } w_0 x_0 + w_1 x_1 + w_2 x_2 < 0$$

The dividing line between them corresponds to the weighted sum being equal to 0:

$$w_0 x_0 + w_1 x_1 + w_2 x_2 = 0 \tag{4.4}$$

This can be rearranged to give

$$w_2 x_2 = -w_1 x_1 - w_0 x_0 \tag{4.5}$$

$$x_2 = -\frac{w_1}{w_2} x_1 - \frac{w_0}{w_2} x_0 \tag{4.6}$$

Comparing equation (4.6) with equation (4.1) and putting $x_0 = 1$ gives

$$m = -\frac{w_1}{w_2}, \quad c = -\frac{w_0}{w_2}$$

There is no unique solution to these two equations. Just as an example, let us start by assuming that $w_2 = 1.0$. Then equation (4.3) is obtained from the values

$$w_0 = -2.0 \quad \text{and} \quad w_1 = -0.5$$

With these weights the unit can discriminate between the two classes of objects. A single unit with the weights shown in Figure 4.13 could classify objects A and B on the basis of the data in Table 4.4.

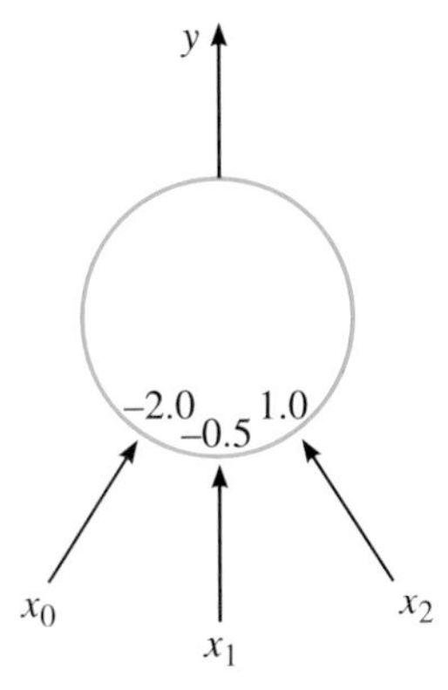

Figure 4.13
Two-way classifying unit.

We can test this by presenting the unit with the two values shown earlier, and see if the output corresponds to the correct classification. The values used were

$$x_1 = 2.0, \quad x_2 = 4.5$$

When these values are presented to the unit the weighted sum is

$$w_0x_0 + w_1x_1 + w_2x_2 = (-2.0 \times 1) + (-0.5 \times 2.0) + (4.5 \times 1.0) = 1.5$$

The weighted sum is greater than 0 so the output is 1, which, by our design, corresponds to object A. What has been achieved is described as ***generalization***, which means that although the weights were selected on the basis of a set of known input–output data, the unit can correctly classify new data that it has not seen before.

This shows some of the capabilities of a single unit, but clearly it would be far better if the weights could be determined automatically. In this example it was not difficult to find values for the weights because the pattern space was two-dimensional. Problems with more inputs would produce a pattern space with more than two dimensions, which is difficult to visualize, and so it is more difficult to find simple lines or equations that separate the data.

A further problem is the fact that it is not always possible to separate data using straight lines. Figure 4.14 shows two examples of a pattern space with two classes of data. Neither of these sets of data could be separated using just a straight line, and they are therefore described as ***non-linearly separable***. Fortunately this can be overcome if you use several units rather than just one.

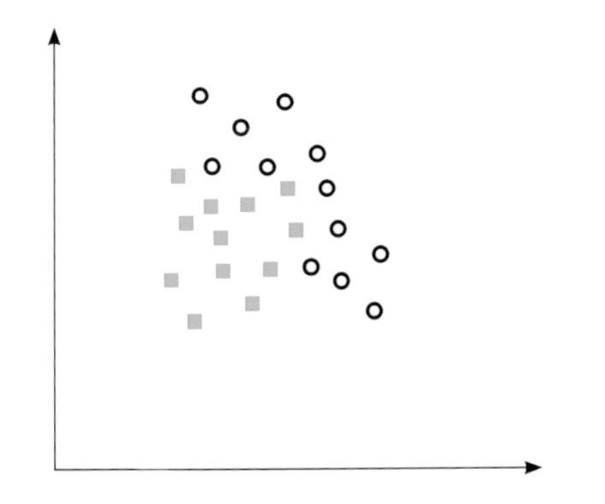

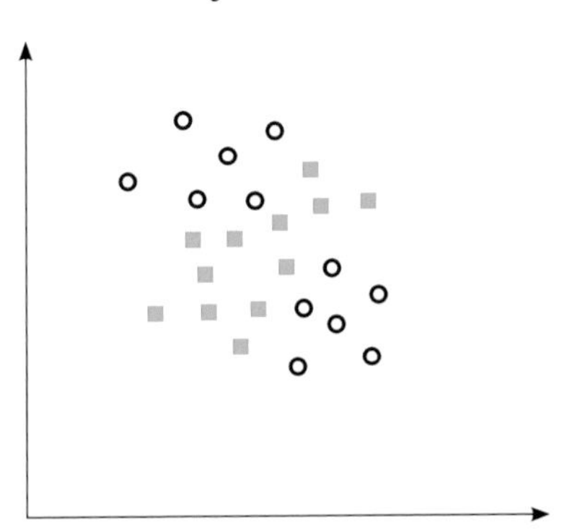

Figure 4.14
Examples of non-linearly separable problems.

The way that the units are connected determines whether a network is a feedforward or a feedback network. In the following sections these two types of network will be examined. Also, you will be shown how the weights can be determined by the network itself.

4.4 Feedforward networks

The single unit in the previous section was found to have the ability to perform a pattern classification only when the data are linearly separable. For our simple example of a two-input problem the line can be drawn in the pattern space, and the weights calculated by hand. Even so, the unit we considered had the ability to generalize from the initial set of data, and therefore provides a very powerful method for pattern classification.

Two problems remain:

1. How can these units be used to classify non-linearly separable data?
2. How can the weights be determined in cases where there are more than two inputs?

The first problem can be overcome by connecting several units together to form a feedforward network, as shown in Figure 4.15. In a ***feedforward network*** the units are grouped into ***layers***. The reason that this type of network is called a feedforward network is that the outputs of units in one layer are only ever

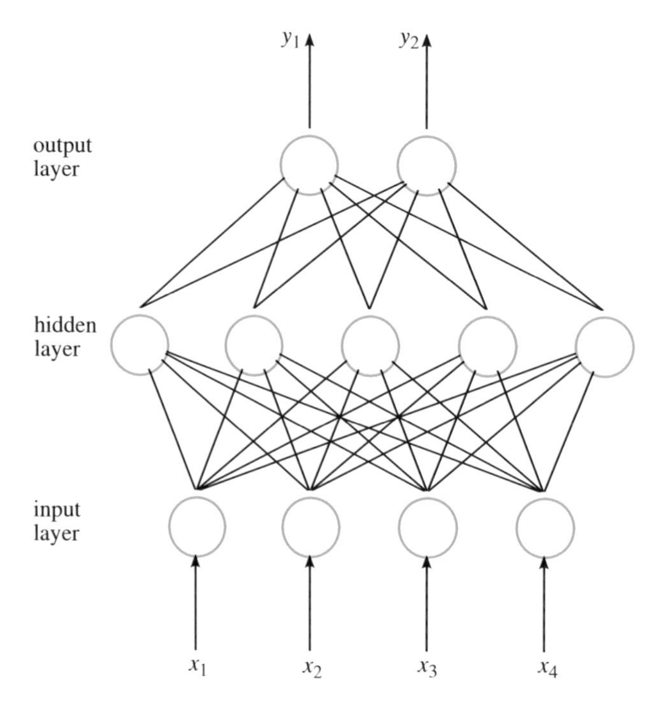

Figure 4.15
A three-layer feedforward network.

connected to the inputs of units in a later layer, usually the next layer. The information in the network is therefore always flowing from the inputs to the outputs.

The first layer is called the ***input layer***, and is usually a ***fan-out layer***. This means that a unit in this layer has one input and several outputs which all have a value equal to the input so that no actual processing takes place.

The next layer is called a ***hidden layer*** because its inputs and outputs are not connected to the outside world, so there is no direct access to the units in this layer. The example shown in Figure 4.15 has only one hidden layer, but in general there could be many hidden layers.

The final layer is called the ***output layer***. It has several inputs from the units in the previous hidden layer, and its outputs are the outputs of the network.

Units in the hidden layers and the output layer are of the type that was discussed in the previous section, with one difference. This difference is that the output isn't necessarily hard-limited. The exact nature of the output function is concerned with learning, and will be described in the next section.

The network in Figure 4.15 is a special case called a ***fully connected multilayer network***, in which *all* of the units in one layer are connected to *all* of the units in the next layer and *only* the next layer. Thus outputs from units in the input layer are connected to the inputs of units in the first hidden layer. Outputs from units in the first hidden layer are connected to the inputs of units in the second hidden layer, and so on, until finally the outputs from units in the last hidden layer are connected to the inputs of the units in the output layer.

We are not going to prove that a feedforward network is capable of classifying non-linearly separable data. We will, however, describe one classic problem and show how it can be implemented. This problem is called the EXCLUSIVE-OR, and has become a sort of test problem over the years to show the limitations of the single unit. Table 4.5 shows the EXCLUSIVE-OR function, which has two binary inputs. Figure 4.16 shows the same function in pattern space.

TABLE 4.5 THE EXCLUSIVE-OR FUNCTION

x_1	x_2	y
0	0	0
0	1	1
1	0	1
1	1	0

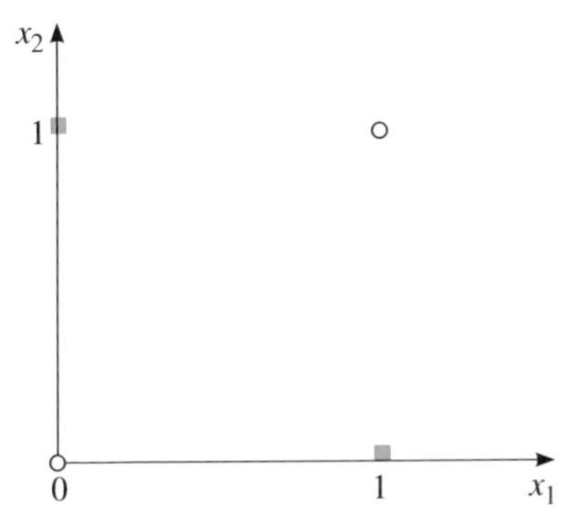

Figure 4.16
The EXCLUSIVE-OR function in pattern space.

It should be apparent from the pattern space that it is impossible to draw a single straight line that would separate the 0s from the 1s. A single unit would therefore be unable to solve this problem. The network in Figure 4.17, on the other hand, can implement the EXCLUSIVE-OR problem.

The network in Figure 4.17 can use hard-limiters because the solution can be found using the three types of logic gate (AND, OR and NOT) that were described at the start of this chapter rather than letting the network find the values for the weights itself. Later in this section we will describe the method for automatically finding a solution in which hard-limiting cannot be used.

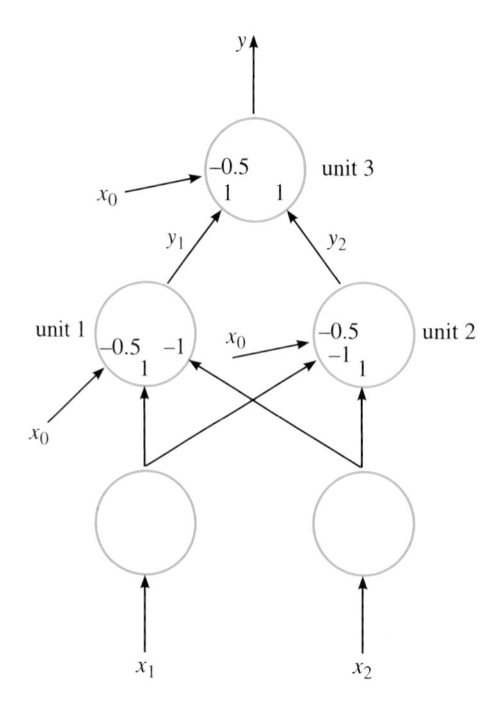

Figure 4.17
Feedforward solution to the EXCLUSIVE-OR problem.

The ***EXCLUSIVE-OR problem*** can be stated in words as:

> the output is 1 when either x_1 is 1 AND x_2 is NOT 1
> OR x_1 is NOT 1 AND x_2 is 1
> otherwise the output is 0.

The hard-limited units in Figure 4.17 are carrying out these logical functions:

> Unit 1: $y_1 = 1$ when x_1 is 1 AND x_2 is NOT 1
> Unit 2: $y_2 = 1$ when x_1 is NOT 1 AND x_2 is 1
> Unit 3: $y = 1$ when y_1 is 1 OR y_2 is 1 but not both

Note that the unnumbered units in the first (or input) layer do no more than receive the inputs and fan them out to the next layer – they do not have any weights and do no processing; they simply make each output the same as the input.

The processing can be summarized as follows:

x_0	x_1	x_2	y_1	y_2	y
1	0	0	0	0	0
1	0	1	0	1	1
1	1	0	1	0	1
1	1	1	0	0	0

Thus the inputs shown in Table 4.5, if applied to the inputs of this network, would give the correct output in each case. This shows that a feedforward network is capable of classifying non-linearly separable data.

4.5 Learning in neural networks

The next problem is that of finding the values for the weights automatically. Given a particular classification problem, if we assume that we have chosen a feedforward network with a sufficient number of layers and number of units in each layer, then there exists a set (or many sets) of weights which produce the correct responses. Finding a set of weights therefore requires us to *search* for an acceptable solution. When the values of the output are known, they can be used to find values for the weights by a process described as ***supervised learning***. (We will look more closely at what we mean by 'learning' in Chapter 8.)

4.5.1 Delta rule

Of the search techniques described in Chapter 3, the calculus-based search showed how a derivative could be used as an indication of the direction to the solution if the solution exists at a minimum (or maximum). The ***gradient-descent method***, described in Section 3.3.4, is also referred to as the ***delta rule*** when applied in feedforward networks. A feedforward network using the delta rule is called a ***multilayer perceptron***.

The search space is multidimensional, with the number of dimensions corresponding to the number of weights. The value used to measure the candidate solution is the ***mean squared error***, $\overline{E}$. The error is the difference between the desired output, d, and the actual output, y. This value, e, is squared and the average value found for all the examples in a training set. The ***training set*** consists of all the known input–output pairs. If there are P examples in the training set, then for a single processing unit the mean squared error is

$$\bar{E} = \frac{1}{P} \sum_{p=1}^{P} e_p^{\ 2} \tag{4.7}$$

where $e_p = d_p - y_p$ for each example, p.

It is possible to picture what is required for learning in problems with only two inputs (including the constant input x_0). An example of such a problem is the NOT function described earlier. The weighted sum is

$$S = \sum_{i=0}^{1} w_i x_i = w_0 x_0 + w_1 x_1$$

and, since $x_0 = 1$,

$$S = w_0 + w_1 x_1$$

The mean squared error for this case is

$$\bar{E} = \frac{1}{2} \sum_{p=1}^{2} (d_p - y_p)^2$$

In order to show how $\bar{E}$ changes with the values of w_0 and w_1, we are going to use a small 'sleight of hand'. We are going to remove the hard-limiting from the output and compare the desired value with the weighted sum, just for the moment. Since there are only two possible binary inputs (0 and 1), and the output in each case is the inverse (1 and 0), the mean squared error can be calculated as

$$\bar{E} = \frac{1}{2} \sum_{p=1}^{2} (d_p - S_p)^2$$

$$= \frac{1}{2} [(d_1 - S_1)^2 + (d_2 - S_2)^2]$$

When $x_1 = 0$, the desired output is $d_1 = 1$, and the actual output is $S_1 = w_0$. Similarly, when $x_1 = 1$, the desired output is $d_2 = 0$, and the actual output is $S_2 = w_0 + w_1$. The total error is

$$\bar{E} = \frac{(1 - w_0)^2 + (0 - w_0 - w_1)^2}{2}$$

When this function is plotted as in Figure 4.18 it can be seen that a surface is created called the ***error surface***. A valley or minimum exists at the point where $w_0 = 1$ and $w_1 = -1$, at which point the mean squared error is zero. So in this example, it turns out that the hard-limiter is not needed since the unit can produce the correct output without it. This is not generally the case, but serves to illustrate the error surface in this particular example.

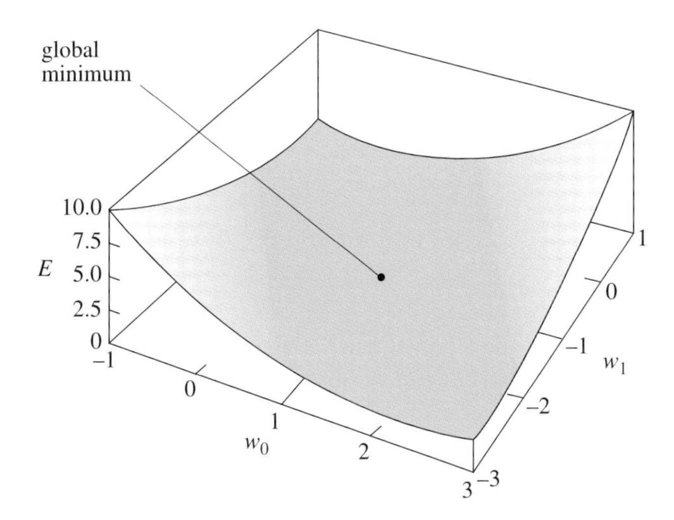

Figure 4.18 Error surface for the inverter problem.

In gradient descent, the aim is to move down a surface to a minimum. This is achieved by changing each of the variables (the weights in this case) by an amount that is proportional to the negative of the slope. That is,

$$\Delta w_i = -\alpha \frac{\partial \overline{E}}{\partial w_i} \tag{4.8}$$

where α is a constant, and $\overline{E}$ is the mean squared error. The symbol Δ is a delta, and the notation Δw_i means 'the change to w_i'. The derivative of the mean squared error with respect to a weight w_i is

$$\frac{\partial \overline{E}}{\partial w_i} = -\frac{2}{P} \sum_{p=1}^{P} e \frac{\partial y}{\partial w_i}$$

This is derived as follows. The mean squared error is defined as

$$\overline{E} = \frac{1}{P} \sum_{p=1}^{P} e^2$$

where $e = d - y$ for each example, and P is the number of examples.

Using gradient descent, the change to each weight w_i is proportional to the negative of the slope. That is,

$$\Delta w_i = -\alpha \frac{\partial \overline{E}}{\partial w_i}$$

where α is a constant. The derivative of the mean squared error with respect to a weight is

$$\frac{\partial \overline{E}}{\partial w_i} = \frac{1}{P}\sum_{p=1}^{P} \frac{\partial}{\partial w_i} e^2 = \frac{1}{P}\sum_{p=1}^{P} \frac{\partial}{\partial w_i}(d - y)^2$$

By the chain rule of calculus,

$$\frac{\partial \overline{E}}{\partial w_i} = \frac{1}{P}\sum \frac{\partial}{\partial y}(d - y)^2 \times \frac{\partial y}{\partial w_i} = \frac{1}{P}\sum_{p=1}^{P} -2(d - y)\frac{\partial y}{\partial w_i} = -\frac{2}{P}\sum_{p=1}^{P} e \frac{\partial y}{\partial w_i}$$

This shows that in order to find the derivative of the mean squared error with respect to the weights, the derivative of the output y with respect to the weights is needed. This is why the hard-limiter function used earlier in this chapter won't do, as it is not a differentiable function.

In order to overcome this problem a different output function is applied to the summed weights. This has to be differentiable and monotonic, which means that for every value of the weighted sum, there is only one value of output, and vice versa. A commonly used function is the ***sigmoid function***, shown in Figure 4.19.

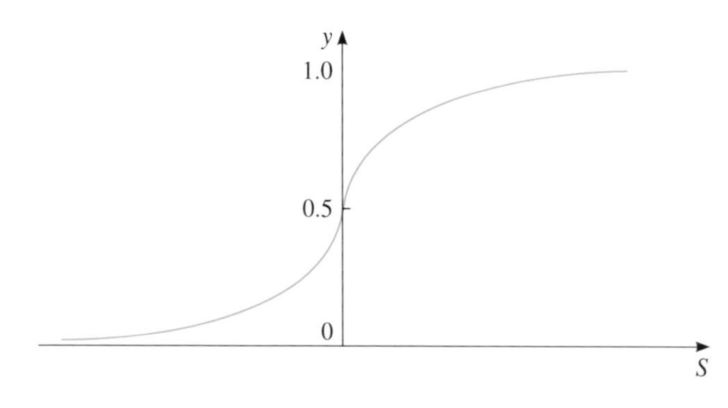

Figure 4.19 Sigmoid function.

The equation for this sigmoid output function is

$$y = \frac{1}{1 + e^{-S}} \qquad (4.9)$$

where $S = \sum_{i=0}^{n} w_i x_i$ and e is the base of natural logarithms.

As you can see, when the weighted sum is greater than 0 the value of the output rises to 1 as the value of the weighted sum increases, and similarly when the weighted sum is less than 0 the output falls to 0 as the value of the weighted sum decreases. When the weighted sum is 0 the output is 0.5.

The derivative of the sigmoid function with respect to the weighted sum, S, is:

$$\frac{\mathrm{d}y}{\mathrm{d}S} = \frac{\mathrm{e}^{-S}}{(1 + \mathrm{e}^{-S})^2} = y(1-y)$$

This gives us enough information to be able to evaluate the changes that have to be made to the weights to reduce the error and consequently to find the solution.

Suppose a sigmoid function is applied to give the output

$$y = \frac{1}{1 + \mathrm{e}^{-S}}$$

The derivative of the output with respect to a weight can be found using the chain rule,

$$\frac{\partial y}{\partial w_i} = \frac{\mathrm{d}y}{\mathrm{d}S} \times \frac{\partial S}{\partial w_i}$$

where $S = \sum_{i=0}^{n} w_i x_i$.

For a sigmoid function, $y = \frac{1}{1 + \mathrm{e}^{-S}}$, we have

$$\frac{\mathrm{d}y}{\mathrm{d}S} = \frac{\mathrm{e}^{-S}}{(1 + \mathrm{e}^{-S})^2} = y(1 - y)$$

Also

$$\frac{\partial S}{\partial w_i} = \frac{\partial}{\partial w_i} \sum_{j=0}^{n} w_j x_j = \sum_{j=0}^{n} \frac{\partial w_j}{\partial w_i} x_j = x_i$$

Substituting these, we get

$$\frac{\partial \bar{E}}{\partial w_i} = \frac{-2}{P} \sum_{p=1}^{P} e \frac{\partial y}{\partial w_i} = \frac{-2}{P} \sum_{p=1}^{P} e \frac{\mathrm{d}y}{\mathrm{d}S} \times \frac{\partial S}{\partial w_i} = -\frac{2}{P} \sum_{p=1}^{P} e\, y(1 - y)x_i$$

which can be written as

$$\frac{\partial \bar{E}}{\partial w_i} = -\frac{2}{P} \sum_{p=1}^{P} \delta x_i$$

where $\delta = ey(1 - y) = y(1 - y)(d - y)$.

Therefore, from (4.8),

$$\Delta w_i = -\alpha \frac{\partial \bar{E}}{\partial w_i} = \frac{2\alpha}{P} \sum_{p=1}^{P} \delta x_i$$

which can be written using the Greek symbol η (eta) as

$$\Delta w_i = \eta \sum_{p=1}^{P} \delta x_i$$

where $\eta = \frac{2\alpha}{P}$ is a constant, or

$$\Delta w_i = \sum_{p=1}^{P} \eta \delta x_i$$

This states that the adjustment to the weight w_i is the sum of $\eta\, \delta x_i$ taken over all of the examples in the training set. It is common practice to simplify this procedure by simply changing the value of w_i by an amount $\eta\, \delta x_i$ after each example in the training set.

The result is the delta rule formula:

$$\Delta w_i = \eta \delta x_i \tag{4.10}$$

where

$$\delta = y(1-y)(d-y) \tag{4.11}$$

and η is a small positive constant, usually between 0 and 1, called the ***learning rate***.

This formula shows us that the change to a weight is positive when δ and x_i are both positive or both negative. For example, if x_i is 0.3 and y is 0.2 when the desired output is 1, then

$$\delta = 0.2(1-0.2)(1-0.2) = 0.128$$

$$\Delta w_i = \eta \times 0.128 \times 0.3 = 0.0384\eta$$

Let us assume that $\eta = 0.5$ and that w_i is initially –0.8. The value of Δw_i is added to the old value of the weight, w_i, to produce the new value. The new value for w_i will be

$$\begin{aligned} w_i(\text{new}) &= w_i(\text{old}) + \Delta w_i \\ &= -0.8 + (0.0384 \times 0.5) \\ &= -0.7808 \end{aligned}$$

Here the value of the weight gets more positive so that the value of the weighted sum increases, and consequently the output will be closer to the desired value.

This gives us the formula for changing the weights for a single unit when the value of the error between the desired output and the actual output is available. In multilayer networks this formula has to be modified for the units in the hidden layer.

4.5.2 Back-propagation

In a multilayer perceptron, the value of the output of a unit in one of the hidden layers can be found, but in the first instance there is no 'desired' value, so an error cannot be formed. The error only exists at the output layer where we know what value we want the output to be.

The derivation of the formula for the weight changes in the hidden layers proceeds as follows.

Assume that the network is a three-layer network with a single perceptron in the output layer, as shown in Figure 4.20.

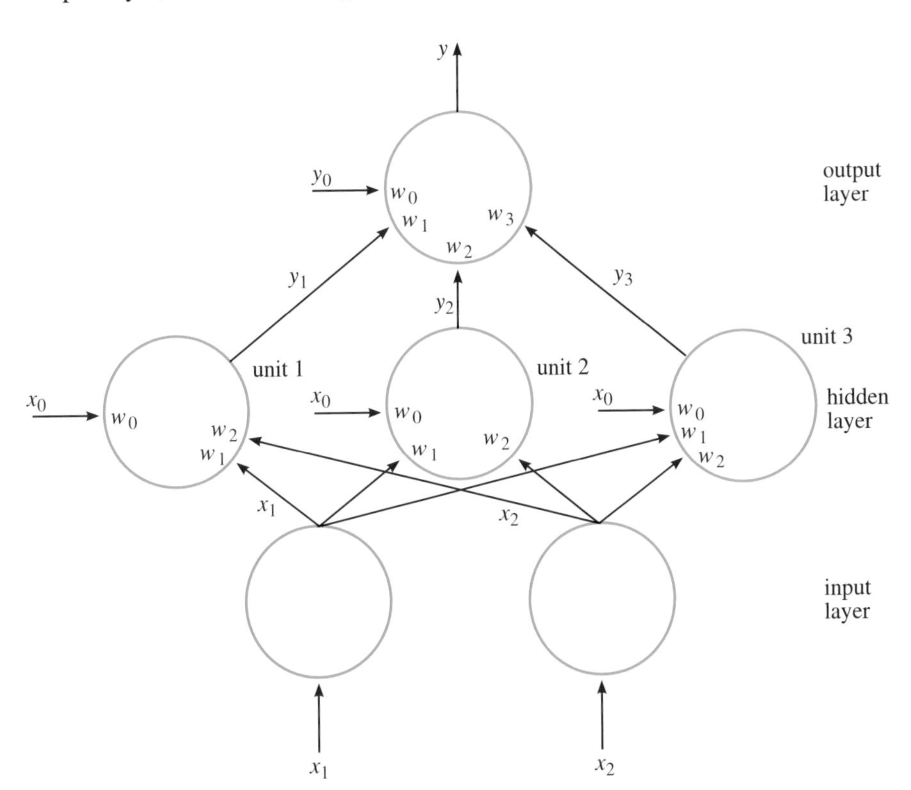

Figure 4.20 ***Three-layer, single-output perceptron.***

Consider a weight in the hidden layer. If we want to change w_1 in unit 2 in the hidden layer, we need to know the effect that it has on the final output, y. To do this we need the derivative of the error with respect to w_1.

The squared error at the output of the network is $E = (d - y)^2$.

We are making the assumption from the start that we are going to adjust the weight after every example in the training set, so the squared error is used and not the mean squared error; i.e. we assume

$$\Delta w_i = -\alpha \frac{\partial E}{\partial w_i}$$

where E is the current error. Then

$$y = f\left(\sum_{i=0}^{3} w_i y_i\right) \quad \text{for the unit in the output layer}$$

$$y_2 = f\left(\sum_{i=0}^{2} w_i x_i\right) \quad \text{for unit 2 in the hidden layer}$$

where $f(\ )$ is the output function, which we can assume is a sigmoid.

By the chain rule,

$$\frac{\partial E}{\partial w_1} = \frac{\mathrm{d}E}{\mathrm{d}y} \times \frac{\partial y}{\partial w_1}$$

$$= \frac{\mathrm{d}E}{\mathrm{d}y} \times \frac{\mathrm{d}y}{\mathrm{d}y_2} \times \frac{\partial y_2}{\partial w_1}$$

Now,

$$\frac{\mathrm{d}E}{\mathrm{d}y} = \frac{\mathrm{d}}{\mathrm{d}y}(d - y)^2 = \frac{\mathrm{d}}{\mathrm{d}y}(d^2 - 2dy + y^2) = -2(d - y)$$

Furthermore

$$\frac{\partial y}{\partial y_2} = \frac{\partial}{\partial y_2}\left(\frac{1}{1 + \mathrm{e}^{-\Sigma w_i y_i}}\right) = \frac{w_2\, \mathrm{e}^{-\Sigma w_i y_i}}{(1 + \mathrm{e}^{-\Sigma w_i y_i})^2} = w_2 y(1 - y)$$

and

$$\frac{\partial y_2}{\partial w_1} = \frac{\partial}{\partial w_1}\left(\frac{1}{1 + \mathrm{e}^{-\Sigma w_i x_i}}\right) = \frac{x_1\, \mathrm{e}^{-\Sigma w_i x_i}}{(1 + \mathrm{e}^{-\Sigma w_i x_i})^2} = x_1 y_2(1 - y_2)$$

So

$$\frac{\partial E}{\partial w_1} = -2x_1 y_2(1 - y_2) w_2 y(1 - y)(d - y)$$

and since, by (4.8), $\Delta w_i = -\alpha \dfrac{\partial E}{\partial w_i}$

$$\Delta w_1 = \eta x_1 y_2(1 - y_2)w_2 y(1 - y)(d - y)$$

or

$$\Delta w_1 = \eta x_1 y_2(1 - y_2)w_2 \delta$$

where $\delta = y(1 - y)(d - y)$, and in this case the constant, η, is $\eta = 2\alpha$.

And then

$$\Delta w_1 = \eta x_1 \delta_2$$

where $\delta_2 = y_2(1 - y_2)w_2\delta$.

So the change Δw_1 for the intermediate layer has been found from the parameter δ calculated for the output layer.

If there were more output neurons, the error E would be the sum

$$E = [d_1 - y_1]^2 + [d_2 - y_2]^2 + \ldots + [d_m - y_m]^2$$

So the derivative of the error with respect to the outputs is

$$\sum \frac{\partial E}{\partial y_i} = -2 \sum [d_i - y_i]$$

Consequently, the general form of the equation for δ_i for unit i in hidden layer k is

$$\delta_i = y(1 - y) \sum_{j=1}^{m} w_{i(j)}\delta_j$$

where y is the output of the hidden unit, m is the number of units in layer $k + 1$ that unit i is connected to via weight $w_{i(j)}$, and δ_j is the value of δ for each of the m units in layer $k + 1$.

The result is the same as equation (4.10), repeated here as equation (4.12):

$$\Delta w_i = \eta \delta x_i \tag{4.12}$$

The expression for δ, however, is quite different in this case:

$$\delta_i = y(1-y) \sum_{j=1}^{m} w_{i(j)}\delta_j \tag{4.13}$$

This is a rather fierce-looking formula but is actually not too difficult to interpret. Instead of the error $(d-y)$, the weighted sum of the δ values from the units one layer ahead are used. Layer $k+1$ could be the output layer or another hidden layer. It is assumed that the unit in layer k is connected to m units in layer $k+1$. The subscript i indicates which unit in the current layer, k, is being adjusted, and the subscript j indicates the unit in the next layer, $k+1$. The weight, $w_{i(j)}$, is therefore the weight in unit j of layer $k+1$ that is associated with the input from unit i in layer k.

In ***back-propagation***, the calculations start at the output layer, where the first values of δ are calculated using equation (4.11) and the weights adjusted. Next the values of δ in the hidden layer immediately preceding the output layer are calculated and the weights adjusted. This time equation (4.13) is used to calculate the δ values. This continues back through the network until all of the weights have been adjusted.

Figure 4.21 shows a specific case of a unit in a hidden layer, k, connected to two units in the next layer, $k+1$. In this case $m=2$.

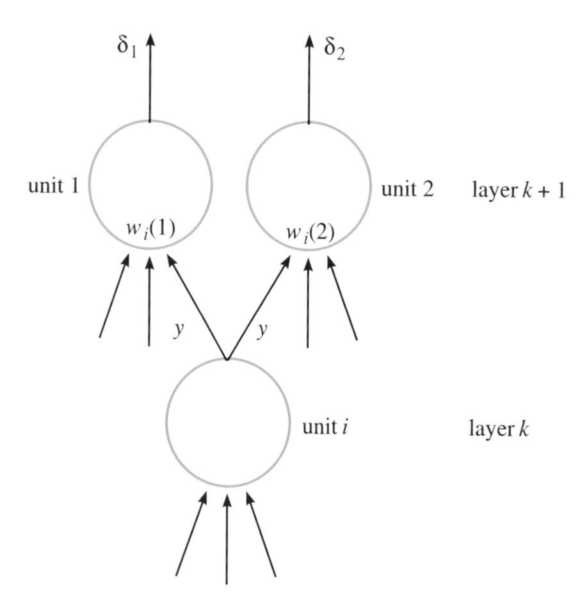

Figure 4.21 *Calculating δ for a unit in a hidden layer.*

To make life easier, let us assume that layer $k+1$ is the output layer. The two values, δ_1 and δ_2 are calculated using equation (4.11) and all the weights in units 1 and 2 in layer $k+1$ are adjusted according to the formula in equation (4.10). Next, the weighted sum of these δ values is found. The weights used are those in units 1 and 2 in layer $k+1$ that receive an input from unit i in layer k. Thus the weighted sum is

$$w_{i(1)}\delta_1 + w_{i(2)}\delta_2$$

This is then used to calculate the value of δ_i so that the weights in unit i can be adjusted.

4.6 Feedback networks

The main difference between feedback networks and feedforward networks is that the connections in feedback networks allow information to flow in either direction, including from the output to the input. As a result, back-propagation cannot be used as a means of learning. Solutions are therefore found not by minimizing the error function but by minimizing a function that is analogous to the 'energy' in the network. In the feedforward case the error was seen to form a surface, and gradient descent was used to move down the surface to the lowest point, called a minimum. In feedback networks the 'energy' also forms a surface.

In one example, the ***Hopfield network*** shown in Figure 4.22, a set of inputs and corresponding outputs are used to calculate weights which position the network at minima in the energy surface. Training is therefore much simpler than in back-propagation as it is achieved in one calculation. When the network is operating as a pattern recognizer, it is initialized by holding the inputs, x_i, at their desired values. Since the x inputs are connected to the outputs of the units, the values of x_i become the initial outputs of the network. The x inputs are then removed and the network is allowed to iterate until it settles to a stable solution. During the iterations, a unit is arbitrarily selected and the weighted sum of its inputs is calculated. If the weighted sum is greater than 0 the output is 1, otherwise the output is 0. The network is performing gradient descent, with the result that the network is directed towards the nearest local minimum, which hopefully corresponds to the correct output value.

Problems arise in the Hopfield network because the creation of the many local minima corresponding to input–output pairs also creates other spurious minima. Thus it is possible for one of the original input values taken from the training set to drive the network to one of these spurious local minima and produce the wrong output.

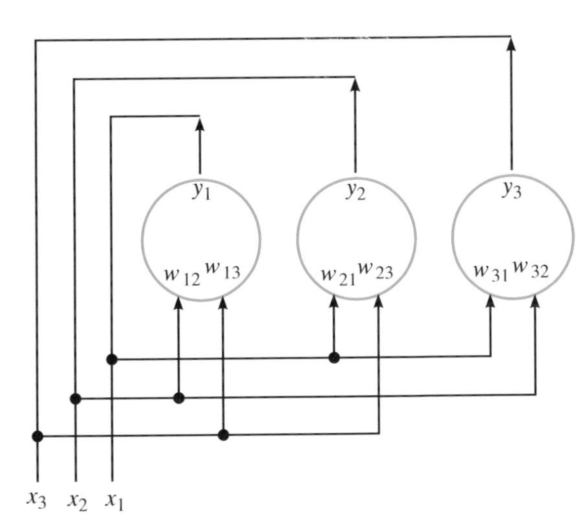

Figure 4.22
The Hopfield network.

One way of overcoming this problem is to ensure that the desired input–output relationship is placed at the global minimum in the energy surface while the system is learning. An example of such a network is the ***Boltzmann machine***, shown in Figure 4.23. It uses simulated annealing (Section 3.2.4) to set the values of the weights when learning input–output relationships so that the network is at a global minimum. Then when an input arrives at the network, simulated annealing again allows the network to settle into a global minimum, producing the correct output response.

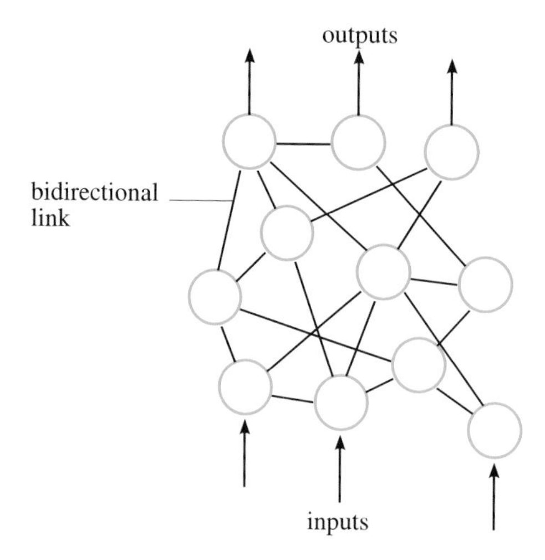

***Figure 4.23* The Boltzmann machine.**

Although feedback networks have very important properties, due to limitations on space we will focus exclusively on feedforward networks in the remainder of this chapter. At the time of writing, feedback networks were being used much less than feedforward networks partly due to the limits of the available technology to construct these networks and partly because of the length of time that these networks take to arrive at solutions. Interested readers can find more detailed descriptions of the Hopfield network and Boltzmann machine in other books, for example Rumelhart and McClelland (1986) and Picton (1994).

4.7 Uses of the multilayer perceptron

The multilayer perceptron consists of units which have a sigmoid as the output function and the weights are adjusted using back-propagation. Any problem that requires a particular input–output relationship which is known in advance, or where examples exist of correct input and output values, can be implemented using a multilayer perceptron.

In a mechatronic system, multilayer perceptrons can be used in all of the three subsystems that were identified in Volume 1 of this book.

(a) Perception

Multilayer perceptrons can be used to classify visual or other sensory inputs – an important function in pattern recognition.

(b) Execution

Multilayer perceptrons can be used to transform the desired target coordinates to, say, appropriate joint angles in a robot arm. They can also be used to control linear systems, but have found an important role in controlling complex non-linear systems where models are difficult to obtain.

(c) Cognition

Finally, the functions of the cognition subsystem are:

pattern recognition

searching

reasoning

learning.

It is clear that neural networks can recognize patterns and that they learn, but what about searching and reasoning? The way that neural networks learn makes use of search techniques such as gradient descent or simulated annealing, for example. Learning is equivalent to finding a solution in a multidimensional space, so a neural network can be thought of as a physical embodiment of a search technique. If a problem can be set out in such a way that it can be defined by a set of input and output pairs, then a neural network should be able to find a solution. A suitable choice of those pairs ensures that generalization is meaningful.

Although this sounds easy, it is in fact the most difficult aspect of any neural network implementation. If we can assume that when a problem is defined by a set of input–output pairs the neural network can find the solution, just how the network does it is no longer of interest. Getting a problem into that form, however, is usually non-trivial and may even be impossible.

Exact reasoning uses deductive logic to produce an output that can be used to make a decision from a set of inputs. At the beginning of this chapter you were shown how the neural units could behave like logical elements. It follows that a neural network is capable of logical reasoning. However, because of the network's ability to produce outputs other than just 0 or 1 when using sigmoid output functions, it is possible to have some form of inexact reasoning.

The ability of the network to generalize means that missing input data or noisy data can still sometimes produce an output which has a value other than 0 or 1, but which is still distinguishable from other outputs. For example, Figure 4.24 shows a network in which unit 1 has been set up to perform the AND function of three inputs, x_1 AND x_2 AND x_3, while unit 2 performs the function x_1 AND NOT x_2 AND x_3.

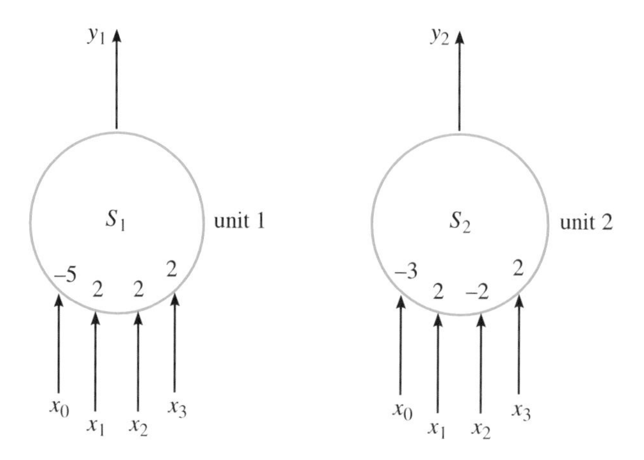

Figure 4.24 ***Two different logic functions.***

The value of S_1 is calculated as the weighted sum of the inputs when the weights w_0 to w_3 are –5, 2, 2, 2, and the value of S_2 is the weighted sum of the inputs when the weights are –3, 2, –2, 2. The outputs (Table 4.6) are calculated by substituting the weighted sum into the sigmoid function.

TABLE 4.6 INPUTS, OUTPUTS AND WEIGHTED SUMS FOR THE UNITS IN FIGURE 4.21

x_0	x_1	x_2	x_3	S_1	y_1	S_2	y_2
1	0	0	0	–5	0.01	–3	0.05
1	0	0	1	–3	0.05	–1	0.27
1	0	1	0	–3	0.05	–5	0.01
1	0	1	1	–1	0.27	–3	0.05
1	1	0	0	–3	0.05	–1	0.27
1	1	0	1	–1	0.27	+1	0.73
1	1	1	0	–1	0.27	–3	0.05
1	1	1	1	+1	0.73	–1	0.27

Now if input values of say 1, 0.8, 0.1 and 1 appear at the inputs x_0 to x_3, the outputs of the two units would be

$$y_1 = 0.23, \quad y_2 = 0.6$$

The outputs are not clear 0s or 1s. But it is still possible to reason that y_2 is greater than y_1 and therefore one could have more confidence in y_2 being correct than y_1 being correct. In both cases one could make a decision, even though neither of these values are binary 0s or 1s.

4.7.1 An example: optical character recognition

An area where neural networks have had some success is in optical character recognition (OCR) of typed, printed or handwritten characters. At the time of writing, OCR is certainly a growth area: scanners that can read documents are becoming more readily available, and the software to read and interpret the images produced by these scanners is becoming more sophisticated. Most of the OCR software that is available has some form of learning ability, although most do not use a neural network approach. However, it is claimed that some OCR software using neural networks is not only able to read typed text, but also some handwritten text.

There are many problems involved in reading text. Typically, the process involves a number of stages before the pattern recognition properties of the software are put to use. These stages include image processing, where the image is 'cleaned up' so that the text is made clearer. In the simplest case this just involves a threshold which converts a grey-scale image into a binary image, so that in each picture cell the black text is represented with a binary 0, while the white background is represented with a binary 1.

Next, there has to be some fairly sophisticated software which can isolate all of the characters. Some researchers have said that this is the real problem in OCR. It may seem quite trivial, but examples where the text has 'run together', that is where two characters are touching, are difficult to separate. There are also problems of scaling for size and adjusting for rotation. Usually OCR packages can handle a slight degree of rotation, due to the document not being aligned properly in the scanner. However, if a document is scanned upside-down, or rotated through 90°, then it is unlikely that the OCR software could process the data successfully.

Finally, there is a stage of pattern recognition. One way would be to use templates to compare with each character in turn. The template that is nearest or most similar to the character would produce the highest score, and so the character would be identified. In packages which can learn, the human operator would be asked to identify any character which either produces a very low score because no template matches very well, or because more than one template matches the character and so the correct one has to be selected. In these cases the templates would be modified or new templates created to accommodate this new information.

A neural network solution would consist of training a network on examples of a particular character set, such as a particular typographic font. All the characters in that font would be shown to the network, and when the error reaches a suitably low value, the network would have been trained. Showing the network any character from that font again should produce the correct response which would identify the character.

As an example, let us consider the problem of representing and recognizing the ten characters 0 to 9.

Considering first the numeral 0 (zero), we can represent it as an 8×8 pattern of black and white picture cells, and as a pattern of binary 0s and 1s:

Input pattern

0 0 0 0 0 0 0 0
0 0 0 1 1 0 0 0
0 0 1 0 0 1 0 0
0 1 0 0 0 0 1 0
0 1 0 0 0 0 1 0
0 0 1 0 0 1 0 0
0 0 0 1 1 0 0 0
0 0 0 0 0 0 0 0

Our desired output pattern for the numeral 0 is the following string of ten binary digits (bits):

Desired output pattern

1 0 0 0 0 0 0 0 0 0

If the first bit of the output is 1 and all the other bits are 0s, the output pattern represents the numeral 0. Similarly, if the second bit is 1 and all the others are 0s, the output pattern represents the numeral 1, and so on for all ten numerals.

So, for each of the ten numerals we can draw up an input pattern and a desired output pattern, as shown below and on the next page. This can be used as our training set for a neural network.

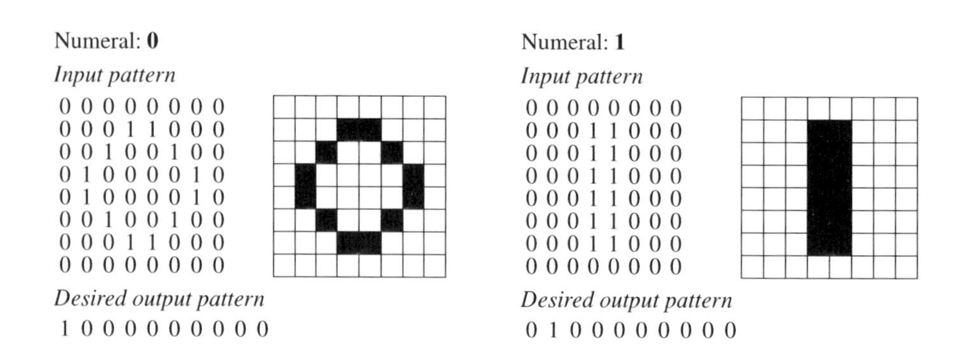

Numeral: **2**

Input pattern

0 0 0 0 0 0 0 0
0 0 1 1 1 1 0 0
0 0 0 0 0 1 0 0
0 0 0 0 0 1 0 0
0 0 1 1 1 1 0 0
0 0 1 0 0 0 0 0
0 0 1 1 1 1 0 0
0 0 0 0 0 0 0 0

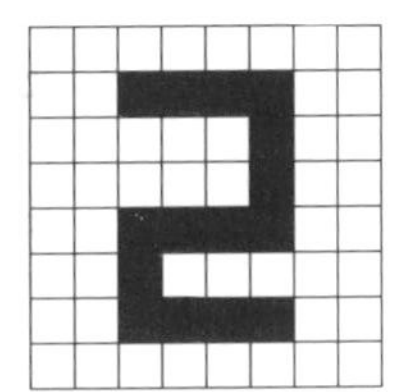

Desired output pattern

0 0 1 0 0 0 0 0 0 0

Numeral: **3**

Input pattern

0 0 0 0 0 0 0 0
0 0 1 1 1 1 0 0
0 0 0 0 0 1 0 0
0 0 0 1 1 1 0 0
0 0 0 0 0 1 0 0
0 0 0 0 0 1 0 0
0 0 1 1 1 1 0 0
0 0 0 0 0 0 0 0

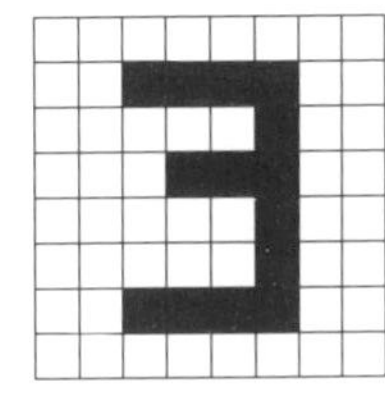

Desired output pattern

0 0 0 1 0 0 0 0 0 0

Numeral: **4**

Input pattern

0 0 0 0 0 0 0 0
0 0 1 0 0 0 0 0
0 0 1 0 0 0 0 0
0 0 1 0 0 0 0 0
0 0 1 0 1 0 0 0
0 0 1 1 1 1 0 0
0 0 0 0 1 0 0 0
0 0 0 0 0 0 0 0

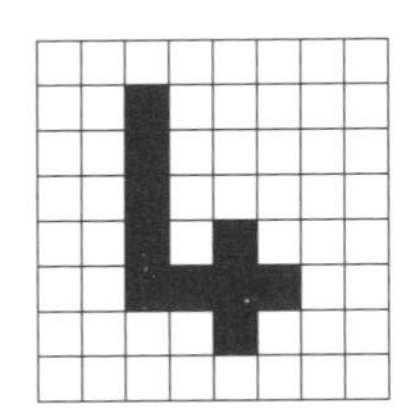

Desired output pattern

0 0 0 0 1 0 0 0 0 0

Numeral: **5**

Input pattern

0 0 0 0 0 0 0 0
0 0 1 1 1 1 0 0
0 0 1 0 0 0 0 0
0 0 1 1 1 1 0 0
0 0 0 0 0 1 0 0
0 0 0 0 0 1 0 0
0 0 1 1 1 1 0 0
0 0 0 0 0 0 0 0

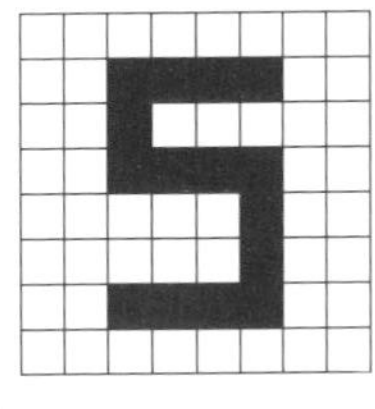

Desired output pattern

0 0 0 0 0 1 0 0 0 0

Numeral: **6**

Input pattern

0 0 0 0 0 0 0 0
0 0 1 0 0 0 0 0
0 0 1 0 0 0 0 0
0 0 1 0 0 0 0 0
0 0 1 1 1 1 0 0
0 0 1 0 0 1 0 0
0 0 1 1 1 1 0 0
0 0 0 0 0 0 0 0

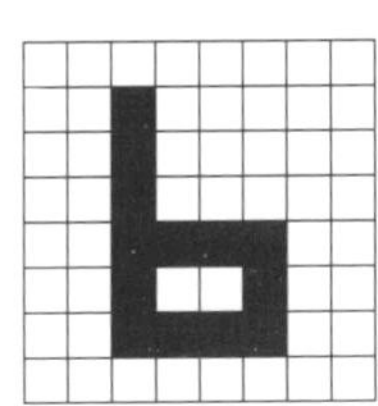

Desired output pattern

0 0 0 0 0 0 1 0 0 0

Numeral: **7**

Input pattern

0 0 0 0 0 0 0 0
0 0 1 1 1 1 0 0
0 0 0 0 0 1 0 0
0 0 0 0 1 1 0 0
0 0 0 0 1 0 0 0
0 0 0 1 1 0 0 0
0 0 0 1 0 0 0 0
0 0 0 0 0 0 0 0

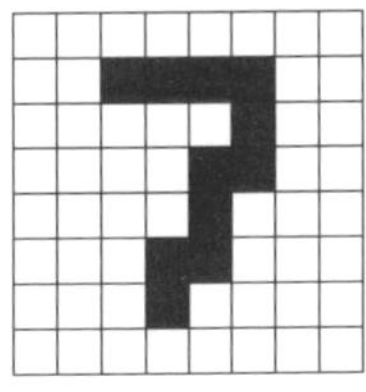

Desired output pattern

0 0 0 0 0 0 0 1 0 0

Numeral: **8**

Input pattern

0 0 0 0 0 0 0 0
0 0 1 1 1 1 0 0
0 0 1 0 0 1 0 0
0 0 1 1 1 1 0 0
0 0 1 0 0 1 0 0
0 0 1 0 0 1 0 0
0 0 1 1 1 1 0 0
0 0 0 0 0 0 0 0

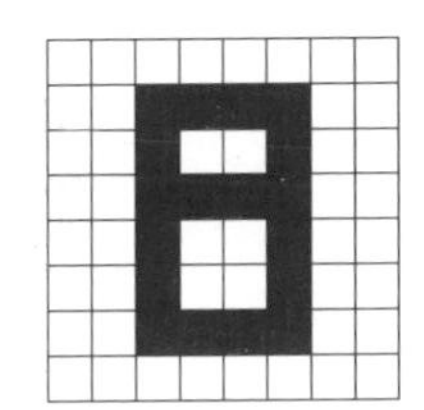

Desired output pattern

0 0 0 0 0 0 0 0 1 0

Numeral: **9**

Input pattern

0 0 0 0 0 0 0 0
0 0 1 1 1 1 0 0
0 0 1 0 0 1 0 0
0 0 1 1 1 1 0 0
0 0 0 0 0 1 0 0
0 0 0 0 0 1 0 0
0 0 0 0 0 1 0 0
0 0 0 0 0 0 0 0

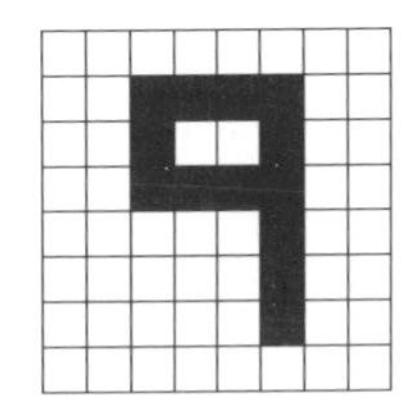

Desired output pattern

0 0 0 0 0 0 0 0 0 1

Now let us consider how these input and output pairs can be used to train a multilayered perceptron to recognize the ten numbers. As with all cases, a choice has to be made about the architecture of the network. In this example it was found, after trying a number of different configurations, that a network with the following parameters could be successfully trained to recognize the ten input patterns we have specified.

Number of inputs	64
Number of outputs	10
Number of hidden layers	1
Learning coefficient	0.5
Units in hidden layer	10
Number of training pairs	10

This describes a three-layer perceptron with 64 input units, 10 output units, and 10 units in the hidden layer, with the learning coefficient set at 0.5. Figure 4.25 shows the network.

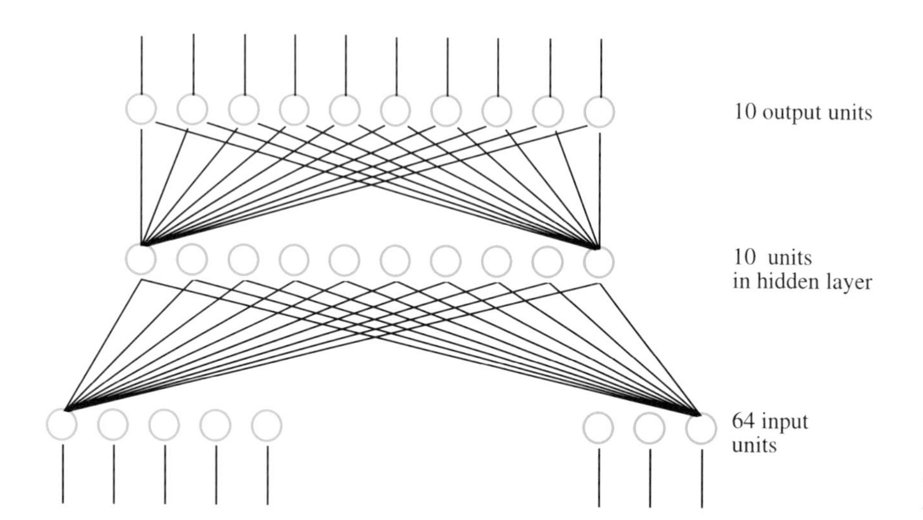

Figure 4.25
The numeral classifier network (for clarity, not all the connections are shown).

The network was initialized by setting all the weights to a random value between −1 and +1. After presenting the training data to the network 100 times, the outputs were as shown in Table 4.7.

TABLE 4.7

Numeral	Outputs									
0	**0.46**	0.02	0.17	0.13	0.08	0.09	0.02	0.13	0.19	0.15
1	0.06	**0.72**	0.07	0.10	0.12	0.10	0.00	0.31	0.06	0.18
2	0.12	0.05	**0.22**	0.15	0.04	0.10	0.10	0.05	0.14	0.07
3	0.09	0.06	0.12	**0.17**	0.02	0.15	0.02	0.09	0.13	0.14
4	0.11	0.11	0.09	0.06	**0.67**	0.07	0.14	0.18	0.09	0.12
5	0.08	0.04	0.10	0.14	0.03	**0.14**	0.02	0.10	0.13	0.16
6	0.07	0.03	0.17	0.11	0.23	0.12	**0.82**	0.05	0.14	0.07
7	0.13	0.17	0.05	0.08	0.11	0.10	0.00	**0.33**	0.09	0.27
8	0.15	0.02	0.13	0.12	0.03	0.12	0.04	0.07	**0.16**	0.15
9	0.11	0.04	0.06	0.09	0.04	0.12	0.01	0.19	0.12	**0.26**

The values highlighted in bold are the outputs that should be 1, while all the other outputs should be 0. You can see that there is little correspondence between most of the inputs and the outputs at this stage. However, after 200 iterations the outputs were as shown in Table 4.8.

TABLE 4.8

Numeral	Outputs									
0	**0.85**	0.00	0.06	0.05	0.07	0.00	0.00	0.11	0.13	0.08
1	0.01	**0.84**	0.05	0.07	0.03	0.04	0.00	0.11	0.00	0.04
2	0.07	0.08	**0.76**	0.18	0.01	0.08	0.03	0.01	0.18	0.01
3	0.05	0.08	0.13	**0.74**	0.00	0.25	0.00	0.06	0.06	0.04
4	0.06	0.04	0.02	0.00	**0.84**	0.01	0.08	0.04	0.06	0.04
5	0.00	0.02	0.03	0.14	0.02	**0.60**	0.02	0.01	0.16	0.18
6	0.01	0.02	0.09	0.02	0.11	0.16	**0.90**	0.00	0.18	0.05
7	0.09	0.10	0.01	0.09	0.06	0.03	0.00	**0.82**	0.02	0.17
8	0.07	0.00	0.10	0.02	0.04	0.11	0.05	0.01	**0.68**	0.20
9	0.01	0.01	0.00	0.01	0.06	0.16	0.00	0.10	0.14	**0.76**

At this stage all the outputs were greater than 0.5 where they are supposed to be 1, and all the outputs were less than 0.5 where they are supposed to be 0. So the network seems to be converging on a solution.

After 600 iterations the outputs were as shown in Table 4.9

TABLE 4.9

Numeral	Outputs									
0	**0.94**	0.00	0.01	0.03	0.04	0.00	0.00	0.03	0.04	0.03
1	0.00	**0.95**	0.03	0.01	0.01	0.01	0.00	0.04	0.00	0.01
2	0.01	0.03	**0.94**	0.04	0.00	0.01	0.01	0.01	0.04	0.00
3	0.03	0.01	0.03	**0.93**	0.00	0.06	0.00	0.02	0.01	0.01
4	0.03	0.03	0.01	0.00	**0.94**	0.00	0.03	0.01	0.01	0.00
5	0.00	0.01	0.01	0.04	0.00	**0.91**	0.02	0.00	0.04	0.04
6	0.00	0.01	0.03	0.00	0.05	0.06	**0.95**	0.00	0.04	0.01
7	0.03	0.03	0.01	0.03	0.01	0.00	0.00	**0.94**	0.00	0.03
8	0.03	0.00	0.03	0.00	0.01	0.02	0.02	0.00	**0.92**	0.03
9	0.01	0.00	0.00	0.00	0.02	0.04	0.00	0.03	0.03	**0.94**

Clearly, all the outputs are now within 0.1 of their correct values. We chose to stop the training here even though the outputs were not exactly correct. It is quite a common practice to stop the training when all the outputs are within a certain value of the desired value, since further training would reduce these errors but only at the expense of a lot more processing time.

To test the network we presented it with 12 input patterns. These consisted of the 10 numerals which had been corrupted by noise, shifted, or generally made slightly different (see over page), and two other test patterns which consisted of 'all 0s' and 'all 1s' respectively, as shown below.

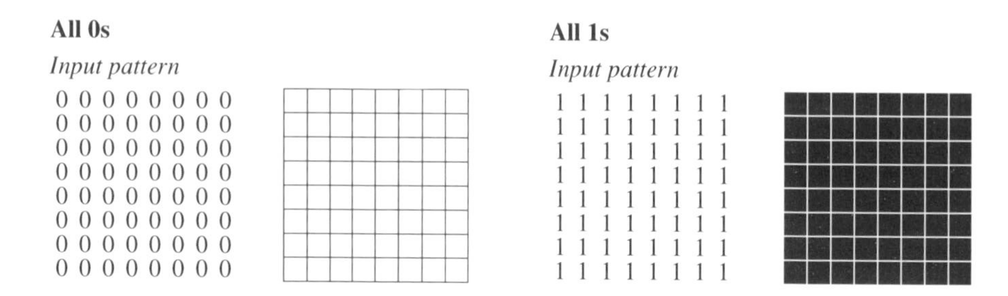

0 ?

Input pattern

```
0 0 0 0 0 0 0 0
0 1 0 1 1 0 0 0
0 0 1 0 0 1 0 0
0 1 0 0 0 0 1 0
0 1 0 0 0 0 1 0
0 1 1 0 0 1 0 0
0 0 0 1 1 0 0 1
0 0 0 0 0 0 0 0
```

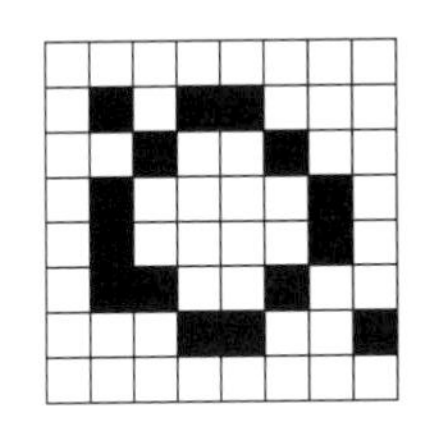

1 ?

Input pattern

```
0 0 0 0 0 0 0 0
0 0 0 0 1 1 0 0
0 0 0 0 1 1 0 0
0 0 0 0 1 1 0 0
0 0 0 0 1 1 0 0
0 0 0 0 1 1 0 0
0 0 0 0 1 1 0 0
0 0 0 0 0 0 0 0
```

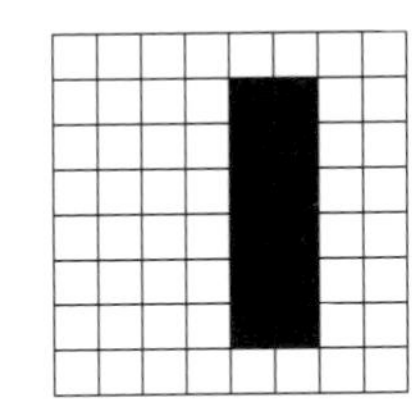

2 ?

Input pattern

```
0 0 0 0 0 0 0 0
0 0 1 1 1 1 0 0
0 0 0 0 0 1 0 0
0 0 1 1 1 1 0 0
0 0 1 0 0 0 0 0
0 0 1 0 0 0 0 0
0 0 1 1 1 1 0 0
0 0 0 0 0 0 0 0
```

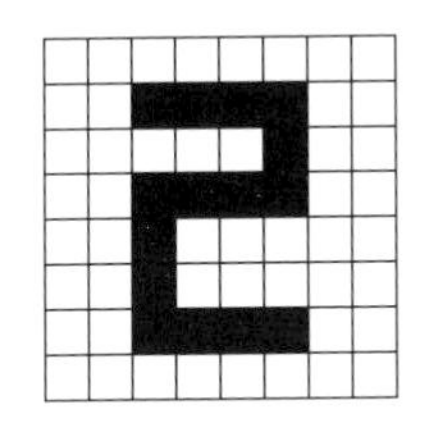

3 ?

Input pattern

```
0 0 0 0 0 0 0 0
0 0 0 1 1 1 0 0
0 0 0 0 0 1 0 0
0 0 0 0 0 1 0 0
0 0 0 1 1 1 0 0
0 0 0 0 0 1 0 0
0 0 1 1 1 1 0 0
0 0 0 0 0 0 0 0
```

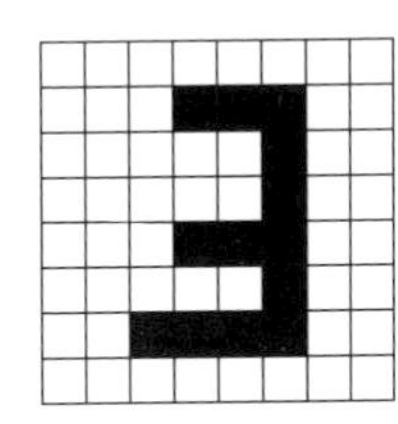

4 ?

Input pattern

```
0 0 0 0 0 0 0 0
0 0 0 0 0 1 0 0
0 0 0 0 1 1 0 0
0 0 0 1 0 1 0 0
0 0 1 0 0 1 0 0
0 0 1 1 1 1 1 0
0 0 0 0 0 1 0 0
0 0 0 0 0 0 0 0
```

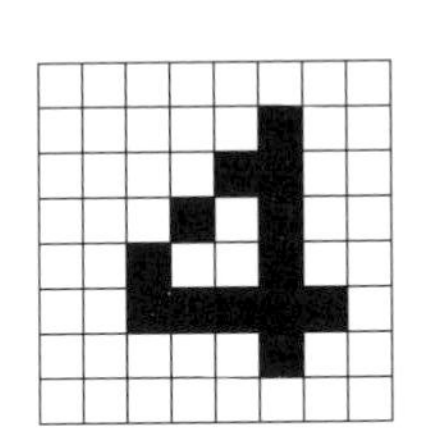

5 ?

Input pattern

```
0 0 0 0 0 0 0 0
0 0 1 1 1 1 0 0
0 0 1 0 0 0 0 0
0 0 1 1 1 1 0 0
0 1 0 0 0 1 0 0
0 0 0 0 0 1 0 0
0 0 1 1 1 0 0 0
0 0 0 0 0 0 0 0
```

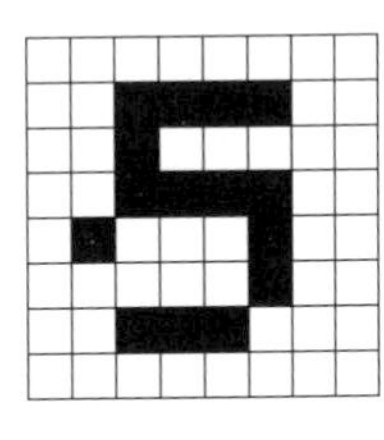

6 ?

Input pattern

```
0 0 0 0 0 0 0 0
0 0 0 1 1 1 0 0
0 0 1 0 0 0 0 0
0 0 1 0 0 0 0 0
0 0 1 1 1 1 0 0
0 0 1 0 0 1 0 0
0 0 1 1 1 1 0 0
0 0 0 0 0 0 0 0
```

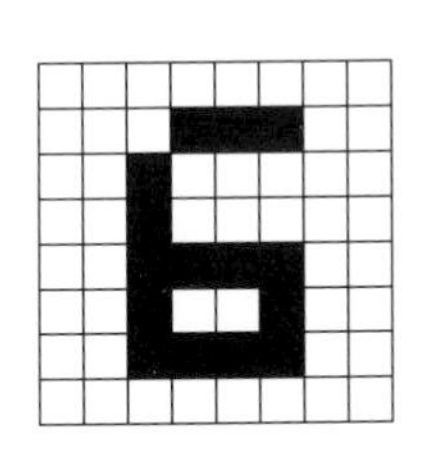

7 ?

Input pattern

```
0 0 0 0 0 0 0 0
0 0 1 1 1 1 0 0
0 0 1 0 0 1 0 0
0 0 0 0 0 1 0 0
0 0 0 0 1 0 0 0
0 0 0 1 0 0 0 0
0 1 0 1 0 0 0 0
0 1 0 0 0 0 0 0
```

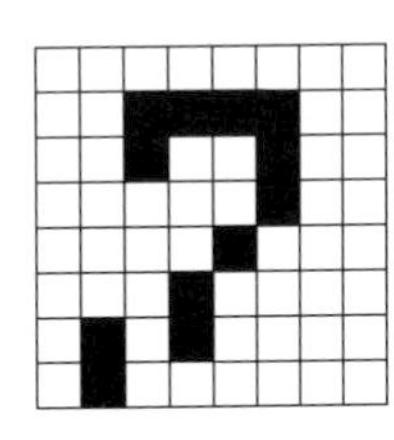

8 ?

Input pattern

```
0 0 0 0 0 0 0 0
0 0 1 1 1 0 0 0
0 0 1 0 0 1 0 0
0 0 1 1 1 1 0 0
0 0 1 0 0 1 1 0
0 0 1 1 0 1 0 0
0 0 1 1 1 1 0 0
0 0 0 0 0 0 0 0
```

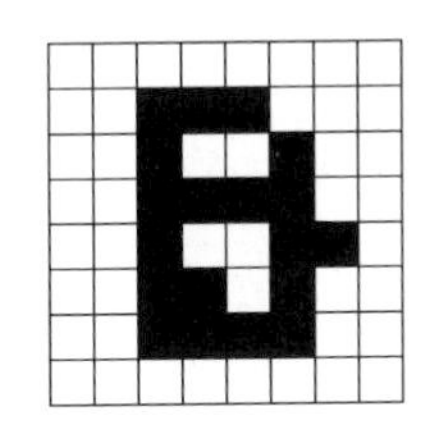

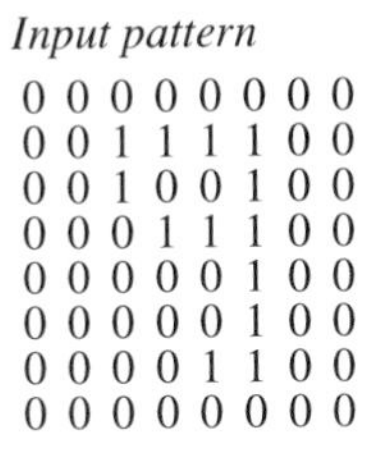

9 ?

Input pattern

```
0 0 0 0 0 0 0 0
0 0 1 1 1 1 0 0
0 0 1 0 0 1 0 0
0 0 0 1 1 1 0 0
0 0 0 0 0 1 0 0
0 0 0 0 0 1 0 0
0 0 0 0 1 1 0 0
0 0 0 0 0 0 0 0
```

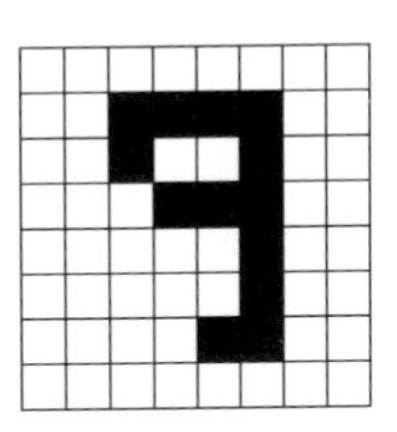

The responses were as shown in Table 4.10.

TABLE 4.10

Input pattern	Outputs									
All 0s	0.04	0.12	0.02	0.03	0.08	0.01	0.02	0.08	0.01	0.08
All 1s	0.00	0.01	0.00	0.00	0.00	0.00	0.00	0.03	0.00	0.22
0	**0.93**	0.00	0.01	0.02	0.04	0.00	0.00	0.03	0.03	0.02
1	0.04	**0.11**	0.01	0.32	0.01	0.00	0.00	0.18	0.00	0.01
2	0.05	0.00	**0.08**	0.01	0.00	0.01	0.01	0.01	0.82	0.01
3	0.06	0.06	0.89	**0.33**	0.00	0.01	0.01	0.01	0.01	0.00
4	0.25	0.00	0.01	0.00	**0.22**	0.00	0.00	0.06	0.16	0.04
5	0.00	0.01	0.01	0.04	0.00	**0.84**	0.01	0.00	0.04	0.03
6	0.00	0.02	0.17	0.00	0.03	0.12	**0.61**	0.00	0.08	0.00
7	0.01	0.01	0.00	0.01	0.02	0.00	0.00	**0.81**	0.01	0.05
8	0.09	0.00	0.03	0.00	0.05	0.01	0.01	0.01	**0.89**	0.01
9	0.04	0.00	0.00	0.17	0.00	0.01	0.00	0.04	0.02	**0.44**

First, the 'all 0s' and 'all 1s' patterns produced no definite high output. This is as we would expect, as they don't correspond to any of the training patterns. Next, the corrupted numerals 0, 5, 6, 7 and 8 produced good responses, even though the value of the output that should be 1 is lower. In the case of 9 the value is very low (below 0.5), but at least it is higher than any of the other outputs, so it would be correctly classified but with a low confidence.

The corrupted numerals 1, 2, 3 and 4 are all incorrectly classified. The network responds that 2 is an 8 and that 3 is a 2. In these cases the changes we made to the numerals were fairly major, just to see how the network copes. The number 1, for example, has been shifted to the right, which gives the network a problem. This emphasizes what we said earlier about an OCR system: shift and rotation would have to be dealt with by an earlier part of the system before pattern recognition takes place. A neural network is particularly sensitive to displacements, so that in many systems some form of transformation is used which produces an output that is invariant to these changes before attempting to classify the images.

In our test, the numerals 2, 3 and 4 have been altered so that, in effect, they are new symbols. It's not surprising that the neural network fails, because it hasn't seen these new symbols before. This shows the importance of selecting the training set. It must contain representative examples of all the sorts of patterns that it will encounter. If two different types of the number 4 are expected, for example, then it should be shown both types in the training set.

This shows that a neural network can generalize so that inputs that are similar but not exactly the same can produce the correct response. But in generalizing, some bizarre input patterns can produce outputs which appear to be correct but are in fact completely wrong. Care must always be taken to ensure that the training set contains examples of *all* the input patterns that the network is likely to encounter. Outside of the training set, the neural network can make catastrophic errors.

4.8 Conclusion

Artificial neural networks have become increasingly popular because of their ability to perform complex pattern classifications without having to be explicitly told how. Their ability to learn from examples and to be able to generalize from these examples makes them a very powerful tool. Inevitably, they have their supporters but also their critics. Critics of neural networks point out that although networks can be shown to work on small 'toy' problems there is a real difficulty in scaling them up to work on very large problems. Although they entered the field with an initial enthusiasm, many researchers into neural networks have found them difficult to apply successfully. However, it is becoming clearer that success in neural networks depends not only on a good knowledge of the networks themselves, but also a good understanding of their intended application. Too often people have tried to apply neural networks to problems which are themselves not well understood, in the hope that the network will be able to 'sort it out for itself'.

As we discussed at the beginning of this chapter, deciding on which information-processing paradigm to use is a decision in which the engineer must weigh up the pros and cons for the particular application. We can summarize these as follows:

- Neural networks are black-box classifiers. They are appropriate for applications in which matched input–output pairs are easy to define.
- Neural networks are well suited to applications in which the data are very noisy. In particular they are very good at transforming multiple sensor information into symbolic form for further processing by neural or conventional computers.
- Problems which may not appear appropriate for neural processing can be transformed by pre-processing into appropriate forms. Often this requires the system designer to have a good understanding of the system.
- Since they have no model, by themselves neural networks are not appropriate for knowledge-based processing involving reasoning.

- Neural networks cannot communicate their workings to humans, and so it may be difficult to see if they are going wrong (although computer graphics and other techniques are addressing this).
- Neural networks may be slow to train, and also unpredictable in their training. However, once trained they are inexpensive to copy for mass-production.
- Neural networks can be simulated in software on conventional sequential computers.
- Neural networks are very good at interpolating and very bad at extrapolating: they are well suited for applications in which the boundaries of the possible inputs are known but are less well suited for situations very different from their training data.
- Neural networks cannot detect inconsistent data which may result in unpredictable training behaviour or training which gives incorrect outputs.

When used intelligently, in problems which can be defined by a representative set of input and output pairs, neural networks are excellent. There is always a certain amount of experimentation needed to select the appropriate architecture with an appropriate set of parameters. Rules of thumb are starting to emerge to help in this selection process, and algorithms are being developed to speed up training. These aids will reduce the time taken to establish whether or not a proposed network will converge on a solution.

In this chapter we have only described the tip of the iceberg. There are many more neural network architectures which are proving to be successful in other areas. Networks with the ability to self-organize, for example, are becoming more common. But at the time of writing, the field is dominated by the multilayered perceptron, which we expect will be with us for some time to come. We strongly commend neural networks as an information-processing paradigm that really works for many difficult applications. We also urge caution in their use. Neural networks are not well suited to safety-critical applications, unless they are used in hybrid systems which also have some knowledge-based processing. As it happens, most applications of neural networks are indeed hybrid, combining the best of both the neural and sequential information-processing paradigms.

References

Rumelhart, D.E. and McClelland, J.L. (1986) *Parallel Distributed Processing*, MIT Press.

Picton, P.D. (1994) *Introduction to Neural Networks*, Macmillan Press, Basingstoke and London.

CHAPTER 5
SCHEDULING

5.1 Introduction

Scheduling in intelligent machines involves determining the order of activities for execution. This means developing a time sequence of actions, control heuristics, and positions for the machine to follow in order to achieve its goals.

Examples of scheduling in intelligent machines include deciding which activity takes priority to be performed next, and in what order activities will be performed after that. For example, a robot may decide that it must recharge its batteries before doing anything else. It may realize that it must pick up a part from one place before it can deliver it to another. In more complex situations a mechatronic system may have to reschedule 'on the fly' when there is a component failure, say. For example, a conveyor belt in a manufacturing system may break down and the system may need to decide how to transport parts until it is repaired. This might involve mobile robots being diverted from other, less urgent, tasks to shuttle the parts between the start and end of the conveyor belt. At the time of writing few industrial systems have anything like the intelligence needed for this.

The main elements of scheduling are:

- time
- position
- activities and their execution
- ordering activities in time
- ordering activities by position
- ordering positions in time (path planning).

An important class of scheduling algorithms relate to what is called the ***travelling salesman problem***. This problem involves finding the shortest circuit for a salesman between a set of cities, with the requirement that no city is to be visited twice. In other words, the salesman must find a *path* along which he can perform his activities in space and time. Furthermore the path must be *optimal* in some sense. Typically there are penalties or costs attached to both the distance travelled and the time it takes to traverse the path. The travelling salesman problem is important because its computational complexity means that it must be solved heuristically for any but the most simple problems. Even though the search space is finite, for as few as 100 cities it would take centuries to search exhaustively.

In an industrial context the travelling salesman problem might translate into the 'travelling robot problem' in which a robot has to plan a route in order to perform various tasks along the way while minimizing the energy load on its motors and minimizing its travel time.

The main problems related to scheduling that will be covered in this chapter, are:

- critical path analysis
- shortest path problems
- travelling salesman problems
- heuristic activity-path planning
- dynamic activity-path planning
- scheduling hierarchical systems.

5.2 Representation in scheduling

As always in AI, scheduling immediately raises the issue of *representation*. How should we represent the concepts of time, space and activity in order to design machines which can accomplish some stated purpose?

In physics and mathematics one often represents a point in three-dimensional space by Cartesian coordinates such as (x, y, z) where x, y and z are real numbers. In practice we can rarely represent the position of a machine in this way because we can only use finite decimal numbers. For example, a machine that theoretically stops after π seconds will probably stop after 3.142 seconds since it is impossible to represent exactly the irrational number π by a finite decimal expansion. In considering machines which move in space it is common to represent the space by an array of square cells. In this case we do not discriminate between the points within any particular cell.

Figure 5.1 shows an example of a two-dimensional grid. A point, **p**, in that space has Cartesian coordinates of $x = 2.135$ and $y = 3.076$, but the grid representation would simply say that the point is within the square with the coordinates $x = 2$ and $y = 3$.

Sometimes the cells are made to correspond to the grid of pixels on a computer screen so that an operator can see where the machine is. Within a Cartesian representation objects may be represented by geometric entities such as points, lines, polygons and polyhedra. Within the cellular representation objects may be represented as sets of cells, which may approximate to geometrical objects.

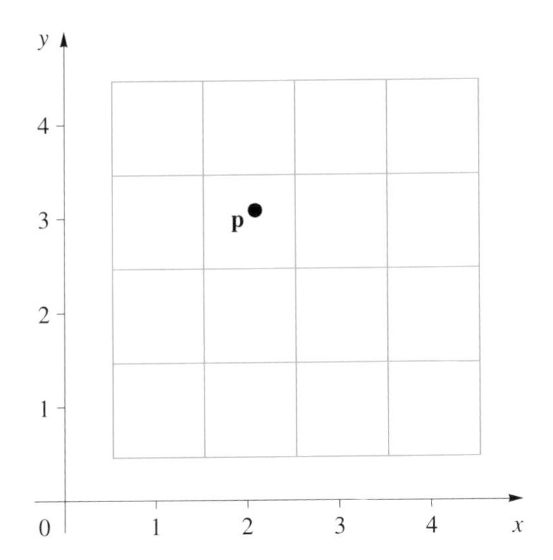

Figure 5.1
Grid representation of a two-dimensional space.

In Newtonian physics one represents time by real numbers such as t_1 and t_2, where the number $(t_1 - t_2)$ is called the *duration* of the interval between the time instants t_1 and t_2. In AI we must also divide time up into equal sampling intervals, and each action is considered to have taken place at the end of one of those intervals. This is rather like thinking of a clock ticking; we examine the machine and its surroundings at each tick but *not* in between.

Similarly, we must find an appropriate representation for the set of actions that the machine may take during execution. In simple cases the actions may just be a list, such as switching motors on and off in a washing machine. However the set of actions may be much more complex, as in the case of an autonomous vehicle. Here we may have atomic actions such as switching motors on and off, and we may have more complex actions such as picking up an object. This suggests that there can be a hierarchy of actions in complex machines in which composite actions result from sequences of simpler actions.

5.3 Graphs and networks for representing scheduling problems

The language of nodes, links and arrows in graph and network theory provides invaluable ways of representing many of the central ideas in scheduling.

Somewhat informally, we define a ***graph*** to be a set of objects called ***nodes*** (or *vertices*) and a set of objects called ***links*** (or *edges*), where every link is associated with two nodes. Intuitively the nodes are dots or points, and the links are lines between pairs of points. Thus in Figure 5.2 the link 'corridor' is associated with the node 'bin' containing parts and the node 'machine' which requires them.

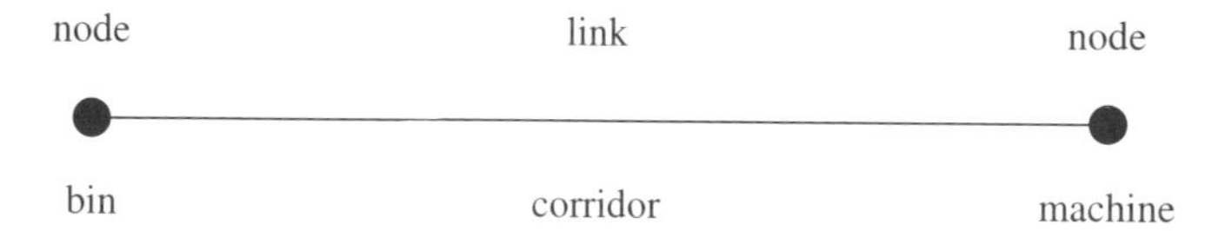

Figure 5.2
The concept of a link between two nodes.

A ***directed link*** is a link in which the order of its nodes matters. For example, it might be desirable to distinguish between the link from the bin to the machine on which parts are carried, and the return journey link from the machine to the bin. An arrowhead is placed on the link to denote its direction. For example, Figure 5.3 shows the case where the link is directed from the bin to the machine, so that the same link cannot be used to travel from the machine to the bin.

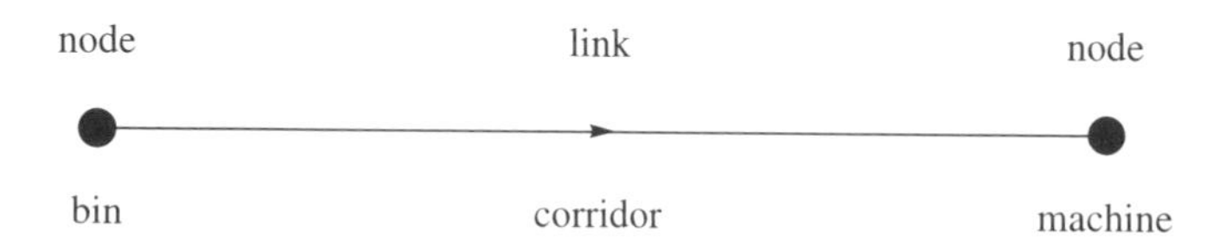

Figure 5.3
A directed link.

In general we may have many links or arrows between nodes, each one representing different things. For example, parts might be transported between two locations by robot or conveyor belt as shown in Figure 5.4.

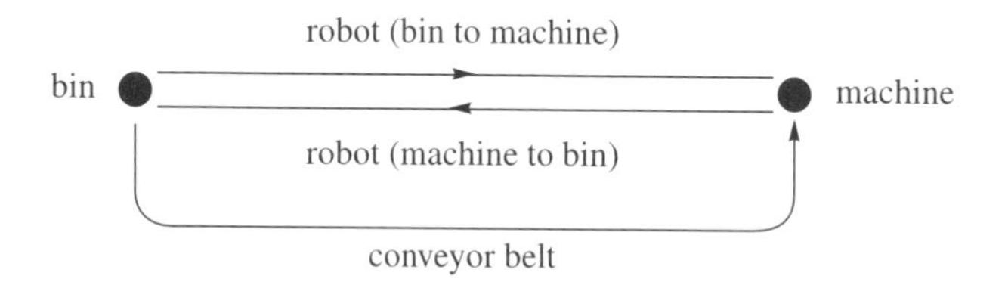

Figure 5.4
Many directed links between nodes.

This representation demonstrates that the robot is capable of more than the conveyor belt because it can move in two directions whereas this conveyor belt can only move in one.

As we've already said, a graph is any set of links with their nodes. A ***directed graph*** is a graph in which all the links are directed (represented by arrows). A ***network*** is a graph which has numbers attached to its links which represent some quantity. For example, the link may be weighted in terms of its length, the time it takes to travel it, the number of objects it can transport in unit time, and so on. Often one speaks of the ***weighting*** of the links.

A *path* through a graph is a contiguous set of links. The ***weight*** of a path in a network is the sum of the weights on its links. A ***minimum path*** between two nodes is one for which there is no other path with smaller path weight. When the weights refer to distance we speak of a ***shortest path***. A ***cycle*** in a network is a path in which the start and end nodes are the same.

A large class of problems in scheduling is concerned with finding shortest paths and cycles for the (usually very large) set of all possible paths or cycles between locations.

Figure 5.5 shows an example of a graph, a directed graph and a network, and a path or a cycle in each.

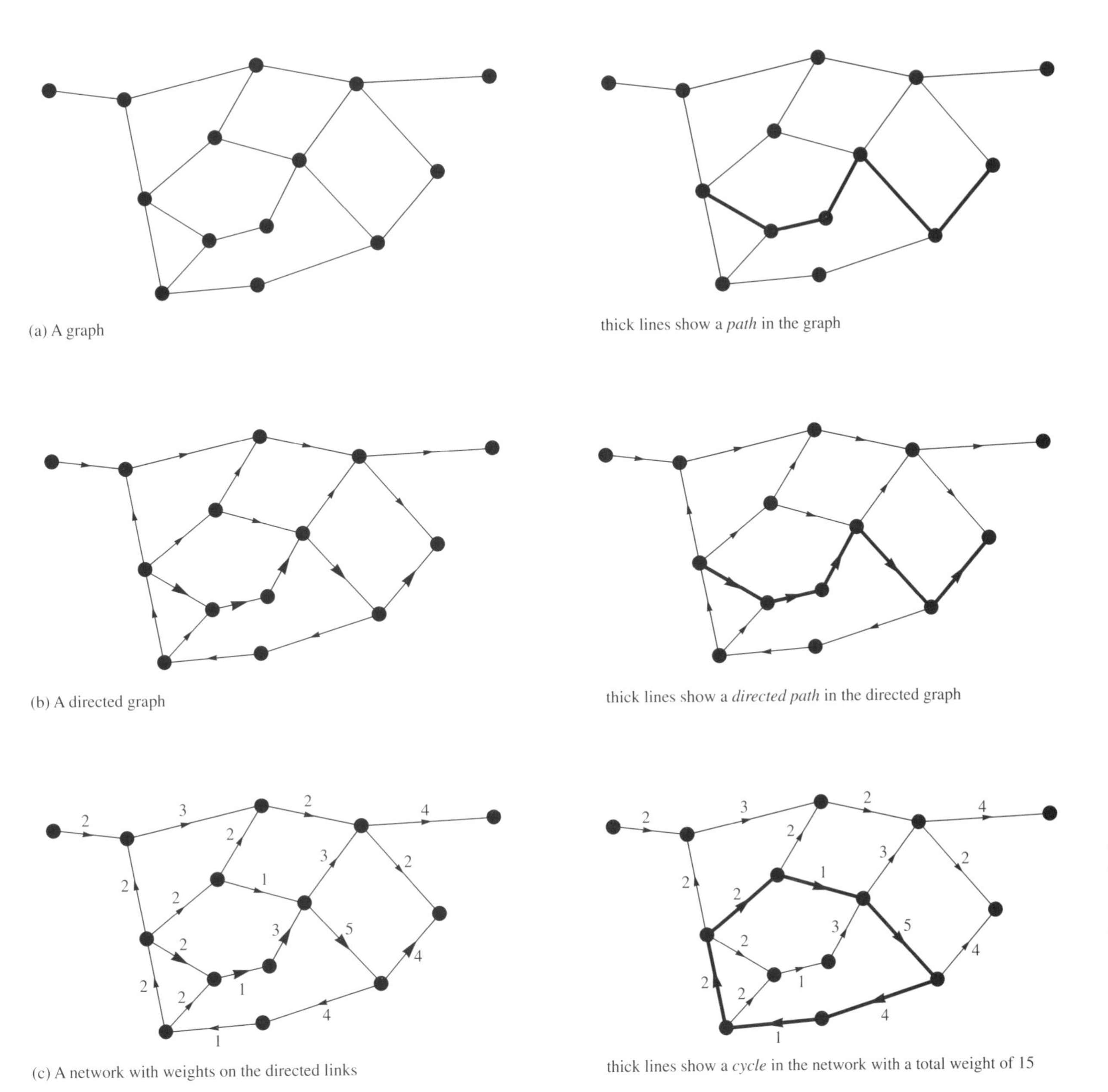

(a) A graph
thick lines show a *path* in the graph
(b) A directed graph
thick lines show a *directed path* in the directed graph
(c) A network with weights on the directed links
thick lines show a *cycle* in the network with a total weight of 15

▲ ***Figure 5.5***
Examples of graphs, directed graphs, networks, paths and a cycle.

Apart from representing paths in physical space, networks can be used to represent paths through what might be called *priority space*. For example, for a given set of activities, suppose we know that some must occur before others. Then we can represent this by placing an arrow from each of the activities to each activity which follows it. This then leads to the idea of *critical paths* through the activity network.

5.4 Shortest paths

In general, a machine that is moving from one position to another will attempt to find the 'shortest' path between them. Here the term *shortest* means the path with the smallest weighting in the network of possible movement links. When the link weightings represent actual distances, the term 'shortest' corresponds to its usual meaning of having the least length. However, the weightings may take into account other things such as travel times, gradients, reliability, danger or cost. In such cases the term 'shortest' is used rather loosely to mean the path with minimum weighting.

The problem of computing shortest paths has attracted a great deal of attention over the last forty years. This has been driven by the many commercial situations in which finding the shortest path will save money. It has also been driven to a great extent by the problem of trying to resolve urban, and increasingly rural, traffic congestion. Road traffic theory begins from a fundamental assumption stated by J. G. Wardrop in 1952: 'The journey times on all the routes actually used are equal, and less than those which would be experienced by a single vehicle on any other route'; that is, drivers choose what they expect to be the *shortest* time path between their origin and destination.

Increasingly, road traffic systems are becoming computer controlled, and they can be considered to be very large and complex mechatronic systems. Each vehicle on a road system has an autonomous intelligent controller (the human driver) which plans its route and determines its local behaviour at any given time. The vehicle is subject to external controls such as the existence of roads and intersections, traffic lights, speed restrictions, and road signpostings. The existence of widespread traffic jams suggests that the theory and control strategies for road systems are currently rather poor. It is believed that introducing more machine intelligence into the control of road systems will improve their overall behaviour.

Chapter 1 of Volume 1 distinguishes two strategies for the control of complex systems. The first is hierarchical top-down control, where everything is ultimately decided by a master controller. Some people believe that this approach to control cannot be viable for very complex systems such as those increasingly encountered. An alternative allows that the parts of the system can make autonomous control decisions out of which system behaviour will emerge. If all the control decisions taken by drivers were taken by computers, road systems would exemplify this kind of control. For this reason, mechatronic engineers can learn a lot from the research that has been conducted into road systems over many decades.

A number of algorithms have been proposed for finding shortest paths; these algorithms have different characteristics depending on the nature of the problem. In general, the amount of time it takes a computer to find a shortest path between an origin and a destination increases with the number of links and nodes in the network. A network has n nodes, so there are $n^2 - n$ ordered pairs of nodes. If it takes an average of t seconds to find a shortest path between one pair of nodes, and the shortest path were computed for each pair of nodes, the computation time would be $(n^2 - n)t$ seconds. Suppose $t = 0.0001$ s. Then for 100 nodes the computation time is about a second, while for 1000 nodes it is over a minute, and for 10000 nodes it is nearly three hours. To understand the practical significance of this, we can note that London has many more than 10000 road intersections.

Dijkstra's algorithm is more efficient at finding shortest paths between every origin and every destination in a network, having complexity $O(n \log n)$. It achieves this by doing all the calculations together, rather than doing them in pairwise sequence. Using Dijkstra's algorithm, a personal computer might take some 20 seconds to calculate a path between John O'Groats and Land's End through the 25000 nodes of the Ordnance Survey road network data. (In practice it also finds the shortest path to the other 25000 nodes.) The details of this algorithm are beyond the scope of this book, but they can be found in many standard texts.

These computation times are important because they determine whether or not a machine can calculate them 'on the fly' in real time. In order to reduce computation times, the environment is usually structured in some way. For example, only a subset of all possible links and nodes may be considered. In general, this means that 'shortest' paths obtained may not be the shortest possible, and the solutions obtained using them may be sub-optimal.

The failure of road traffic planners to design congestion out of road systems is partly due to a flaw in the representation. Although it is simple to represent roads in terms of their static features such as length, number of lanes, and gradients, the quantity optimized by most drivers is ***travel time***. This is not a linear function of any of these static measures; indeed it is not even a continuous function. The time taken to travel a given piece of road depends critically on the number of other vehicles travelling on that road. When the concentration of vehicles on the road reaches a certain level the *dynamics* become unstable, and shock waves may be experienced as drivers have to reduce speed from free flow to a crawl. In fact the representation is even more complex than this. If a road link in front is blocked then flow on the current link will also be blocked: the travel time on a given link may depend on what is happening on other links.

The lesson to be learnt from this is that simplistic representations may not allow the engineer to address the reality of the system, and so make it uncontrollable. Finding shortest paths through large complex mechatronic systems will undoubtedly be an important element in their control. These shortest paths must allow for the possibility of unexpected events such as links becoming blocked or other parts of the system behaving in unexpected ways.

5.5 Critical path analysis

Intelligent machines will often have to perform complex sequences of operations in order to achieve their goals. Each operation will take a certain time, and some will have to be completed before others can begin. The machine will have to plan the order in which to perform the various tasks, and estimate the time it will take.

For example, the process of assembling a bicycle can be split into the activities listed in Table 5.1.

TABLE 5.1

	Activity	Duration (minutes)
A	frame preparation, including front forks	7
B	mounting and aligning front wheel	7
C	mounting and aligning back wheel	7
D	attaching the derailleur gears to the frame	2
E	installing the gear cluster	3
F	attaching the chain-wheel to the crank	2
G	attaching the crank and chain-wheel to the frame	2
H	mounting the right pedal and toe-clip	8
I	mounting the left pedal and toe-clip	8
J	final attachments (handle-bars, seat, brakes, etc.)	18

Source: Dolan and Aldous, 1993

The ***precedence*** relations of these activities, i.e. what activities must be completed before these activities can begin, are shown in Table 5.2. (Activities with no preceding activities are omitted.)

TABLE 5.2

Activity	Preceding activities
C	D, E
E	D
F	D
G	F
H	E, F, G
I	E, F, G
J	A, B, C, D, E

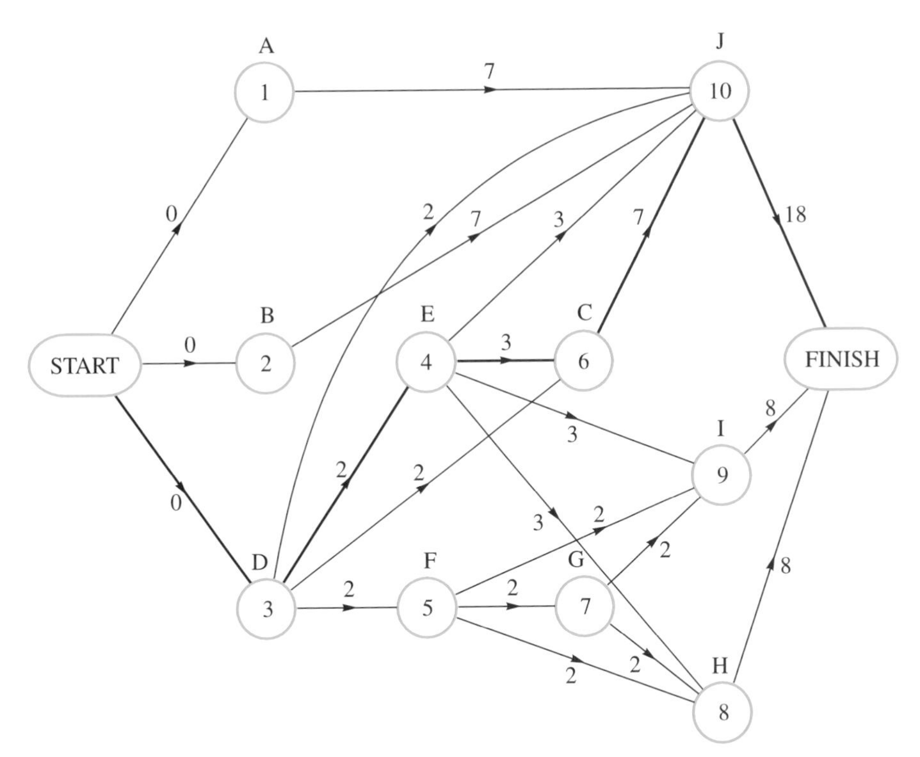

Figure 5.6
Activity network for bicycle assembly.

An ***activity network*** is a network in which each activity or operation is represented by a node. A directed link from one activity to another indicates that the first activity must be completed before the second can begin. The number on the link indicates how long that activity will take to complete. The activity network for the bicycle assembly example is shown in Figure 5.6.

This network is obtained as follows. First we mark those activities which have no preceding activity, i.e. A, B and D (shown by underlining in Table 5.3). This is the first 'layer' of activities, and it follows the start.

TABLE 5.3

Activity	Preceding activities
C	D, E
E	D
F	D
G	F
H	E, F, G
I	E, F, G
J	A, B, C, D, E

The activities just found can be numbered:

(1) A, (2) B, (3) D.

Remove A, B and D from the list and find the next layer of activities with no preceding activity: these are E and F (Table 5.4).

TABLE 5.4

Activity	Preceding activities
C	E
G	F
H	E, F, G
I	E, F, G
J	C, E

The activities found can be numbered:

(1) A, (2) B, (3) D, (4) E, (5) F.

Remove E and F from the list and find the next layer of activities with no preceding activity: these are C and G (Table 5.5).

TABLE 5.5

Activity	Preceding activities
H	G
I	G
J	C

The activities found can be numbered:

(1) A, (2) B, (3) D, (4) E, (5) F, (6) C, (7) G.

On removing C and G we are left with the last layer of activities before the finish, namely H, I and J.

The activities can now all be numbered:

(1) A, (2) B, (3) D, (4) E, (5) F, (6) C, (7) G, (8) H, (9) I, (10) J.

Then no activity is preceded by an activity with a lower number.

To obtain the activity network, each of these layers is set out in columns across the page. The activities related by precedence are then joined by an arrow, with weight the duration of the earlier activity. This method of construction ensures that all the arrows go from earlier layers to later layers.

The minimum time to complete the sequence of activities can be calculated from the activity network by finding a ***longest path*** through it. A longest path is a path between the start and finish for which the sum of the times of the activities is the largest possible. A ***critical path*** is a path for which any delay in completing an activity on that path delays the completion of the project by the same amount. In general there may be more than one critical path through an activity network. The ***minimum completion time*** is equal to the length of a critical path.

By inspection it can be seen that the critical path through the bicycle assembly project is:

$$\text{START} \xrightarrow{0} \text{D} \xrightarrow{2} \text{E} \xrightarrow{3} \text{C} \xrightarrow{7} \text{J} \xrightarrow{18} \text{FINISH}$$

and the minimum time for the assembly is 30 minutes. To complete the project in this time some of the tasks would have to be done simultaneously, for example by two or more robots.

Consider the path:

$$\text{START} \xrightarrow{0} \text{D} \xrightarrow{2} \text{F} \xrightarrow{2} \text{G} \xrightarrow{2} \text{H} \xrightarrow{8} \text{FINISH}$$

This is the longest path that can be found that passes through H, and has total time 14 minutes, which is less than the minimum time. Suppose activity H was delayed. Would this make the project overrun? In fact H could be delayed or overrun by up to 16 minutes before this path exceeded the 30 minutes minimum.

The maximum time that an activity can be delayed without delaying the project is called the ***float*** of that activity. Activities on a critical path have a float of zero. So, for example, none of activities D, E, C or J can be delayed without making the project take longer than the minimum time.

If a project is to be completed in the shortest possible time, then particular attention must be paid to activities on any critical path. For other activities there is some leeway in their starting times or durations – the float.

In general an algorithm is needed to find critical paths. This involves a ***forward scan*** in which the vertices are numbered. This is followed by a ***backward scan*** in which the critical path is found.

It is assumed that the network involves n activities (there are n vertices in the activity network, plus the start and finish vertices). The start vertex is numbered as the 0th and the finish vertex is numbered as the $(n+1)$th. In the bicycle example, $n = 10$.

The duration of the activity represented by the arrow ij is denoted c_{ij}. For example $c_{1,10} = 7$.

As will be explained, the algorithm assigns numbers p_j and e_j to each vertex j, for $j = 0, 1, 2, \ldots, n+1$.

When the algorithm is finished, e_j will be the length of the longest path to the vertex j, and p_j will be the number of the preceding vertex on this longest path.

(A) Forward scan

The forward scan effectively moves through the layers calculating the longest path length, e_j, to each vertex j.

Step 1: Label the START vertex with $p_0 = 0$ and $e_0 = 0$.

Step 2: Set $j = 1$.

Step 3: For the current vertex j: for each arow ij coming into vertex j, calculate $e_i + c_{ij}$. Choose the largest of these sums (or any of them in the event of a tie). This is to be the value of e_j. Set p_i equal to i, the value for which the sum was largest. If j is less than or equal to n, increase j by 1 and repeat Step 3.

(B) Backward scan

The backward scan effectively starts at the FINISH and picks out paths between the vertices with the largest values of e_j.

Step 4: Start with the FINISH vertex, $n+1$, and mark the link that joins this to the preceding vertex given by the number p_{n+1} which was found during the forward scan. The vertex p_{n+1} will be called the 'current vertex'.

Step 5: Suppose the current vertex is j. Mark the link joining this vertex to the preceding vertex, given by p_j, which was found during the forward scan.

Let p_j become the current vertex.

If the current vertex is not the START, repeat Step 5.

The marked arrows found in this way form a critical path. The algorithm needs some modification in order to find all the critical paths if there is more than one, but we will not consider this here.

Example of forward and backward scanning

The algorithm can be illustrated using the bicycle assembly example.

Step 1: (start) Set $e_0 = 0$ and $p_0 = 0$.

Step 2: Set $j = 1$ (A): $e_1 = 0$ and $p_1 = 0$ (p_1 is the START vertex)

Step 3: Set $j = 2$ (B): $e_2 = 0$ and $p_2 = 0$ (p_2 is the START vertex)

Set $j = 3$ (D): $e_3 = 0$ and $p_3 = 0$ (p_3 is the START vertex)
(This completes the first layer.)

Set $j = 4$ (E): $e_4 = e_3 + 2 = 2$, $p_4 = 3$ (vertex D)

Set $j = 5$ (F): $e_5 = e_3 + 2 = 2$, $p_5 = 3$ (vertex D)
(This completes the second layer.)

Set $j = 6$ (C): $e_6 = e_4 + 3 = 5$, $p_6 = 4$ (vertex E)

Set $j = 7$ (G): $e_7 = e_5 + 2 = 4$, $p_7 = 5$ (vertex F)
(This completes the third layer.)

Set $j = 8$ (H): $e_8 = e_7 + 2 = 6$, $p_8 = 7$ (vertex G)

Set $j = 9$ (I): $e_9 = e_7 + 2 = 6$, $p_9 = 7$ (vertex G)

Set $j =10$ (J): $e_{10} = e_6 + 7 = 12$, $p_{10} = 6$ (vertex C)
(This completes the last layer.)

Set $j =11$ (finish): $e_{11} = e_{10} + 18 = 30$, $p_{11} = 10$ (vertex J)
(This completes the FINISH node.)

The forward scan is now complete. We continue with the backward scan.

Step 4: Start with 11, the FINISH vertex: $p_{11} = 10$, which is vertex J.
Mark the arrow between J and the finish.

Step 5: Set the current vertex to 10 (J): $p_{10} = 6$, which is vertex C.
Mark the arrow between C and J.

Set the current vertex to 6 (C): $p_6 = 4$, which is vertex E.
Mark the arrow between E and C.

Set the current vertex to 4 (E): p_4 = 3, which is vertex D.
Mark the arrow between D and E.

Set the current vertex to 3 (D): p_3 = 0, which is the start vertex.
Mark the arrow between the START and D.

The critical path found is therefore

START $\rightarrow$ D $\rightarrow$ E $\rightarrow$ C $\rightarrow$ J $\rightarrow$ FINISH

It has length $e_{11} = 30$.

Apart from finding a critical path, the value of e_i found by this algorithm is the *earliest starting time* for activity *i*. For example, the earliest starting time for activity 8 (H) is $e_8 = 6$.

If the algorithm were 'run backwards' from FINISH to START, the values of e_i subtracted from the minimum time would be the *latest starting times*. Thus the *float* can be calculated as the latest starting time minus the earliest starting time. The float for each vertex is a measure of how sensitive it is to delay or overrunning.

Critical path analysis gives a system information which enables it to decide in which order to perform tasks. It also gives the system information which enables it to deal with uncertainty. Knowledge of the nodes along critical paths enables heuristics to be applied, such as 'monitor this node carefully and give high priority to taking action if it gets delayed or overruns'. Similarly, knowledge of the float of the nodes allows their importance to be weighted in terms of monitoring progress.

The method described here is the Critical Path Method (CPM) and was developed by the Du Pont Nemours Company in order to plan large-scale industrial projects. Their primary concern was to minimize the total cost of a project. The method is useful when activity times can be predicted with reasonable accuracy.

Another approach to planning large projects is Program Evaluation and Review Technique (PERT). This was used by the US Navy to plan the Polaris missile project, where the main objective was to complete the project in the shortest possible time. This was a complex project involving some activities whose times could not be accurately predicted. The important feature of PERT is that the technique permits probabilistic estimates of activity times and so can accommodate research and development projects in which times for the activities cannot be predicted with confidence. PERT can also be used for projects which may suffer disruption through strike action, late delivery of materials, mechanical breakdowns, and so on. Both PERT and CPM allow many subtleties which cannot be discussed further here.

5.6 Critical path activity scheduling

In the previous section we saw that the minimum completion time for a project is given by the length of a critical path in the corresponding activity network. This is the shortest time in which a project can be completed if there is no restriction on the number of workers available. However, if there is a limit on the number of machines or processors available, it may not be possible to achieve this minimum completion time.

If the product of the minimum completion time and the number of machines is less than the sum of all the durations of all activities, then it is obviously impossible to finish the project within the minimum completion time. Even if this is not the case, the precedence relations may be such that some machines must have idle periods, so that it is again impossible to finish in the minimum completion time.

In this section we investigate the problem of scheduling the activities of a project for a given number of robots in the best possible way. It is supposed that an ideal schedule satisfies the ***factory rules***:

- No machine may be idle if there is some activity which can be done.
- Once a machine starts an activity, that activity must continue until it is completed.
- The project must be completed as soon as possible with the machines available.

These rules, which may not be achieved in practice, suggest that each machine has a tightly packed schedule, and that activities on any critical path should be started as soon as possible.

There is no practical algorithm to solve the activity scheduling problem in a way that satisfies the factory rules. The following heuristic algorithm, the ***critical path scheduling algorithm***, has often been used in industry. It produces a schedule which satisfies the first two factory rules, but not necessarily the third rule.

The algorithm is illustrated for a number of processors sharing a multiple-activity computation:

START Set the project clock to zero.

Step 1: If at least one processor is free, assign to any free processor the most critical unassigned activity which can be started (this is the activity with the least latest starting time).

Repeat until no processor is free, or until no activity can be started.

Step 2: Advance the project clock until a time is reached when at least one activity is completed, so that at least one processor is free.

Step 3: If all the activities have been assigned, advance the project clock until all the activities have been completed, then FINISH. Otherwise go back to Step 1.

To illustrate this, consider the bicycle assembly project of Section 5.5. It will be assumed that two general-purpose robots are available for this project. The latest starting times are as listed in Table 5.6.

TABLE 5.6

Vertex number	Activity	Latest starting time
1	A	5
2	B	5
3	D	0
4	E	2
5	F	18
6	C	5
7	G	20
8	H	22
9	I	22
10	J	12

START Set the project clock to 0.

Step 1: The activities which can be started are A, B and D. The most critical of these is activity D, since it has the smallest latest starting time (0). Therefore we assign activity D to robot 1.

Activities A and B can both be started, and both have the same latest starting time (5), so either can be chosen. Assign activity A to robot 2.

Step 2: Advance the project clock to 2 minutes.

Activity D is now completed and robot 1 is free.

Activities E and F are now free to be started, as well as B which is waiting.

Step 1: Of activities B, E and F which can be started, E has the smallest latest starting time, so this is assigned to robot 1.

Step 2: Advance the project clock to 5 minutes (i.e. 2 + 3 minutes).

Activity E is now completed, and so robot 1 is free.

Activity C is now free to be started.

The current state of the scheduling of the activities is shown in Figure 5.7.

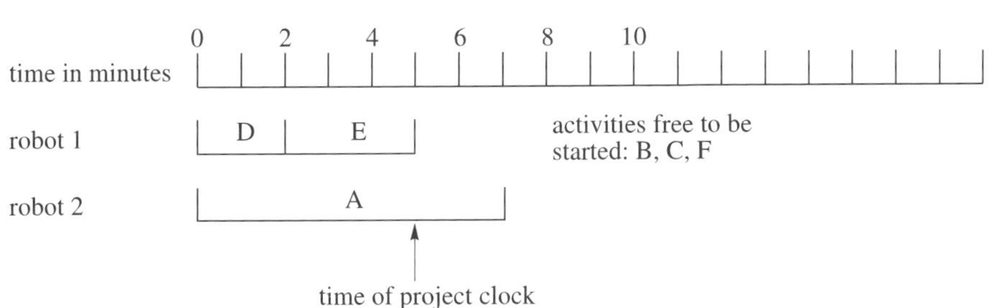

Figure 5.7
State of scheduling after 5 minutes.

Step 1: Of the three activities which are free to be started, B and C are the most critical. Both have a latest starting time of 5 minutes. Since the project clock is now at five minutes, and since we cannot assign both of these activities to a robot at this point, there will be a delay in the completion of the project. In other words the time taken with this schedule will exceed the critical path time.
Let activity C be assigned to robot 1.

Step 2: Advance the project clock to 7 minutes.

Activity A is completed, so robot 2 is free.

No further activity is made free by the completion of A, and activities B and F remain free to be started.

Step 1: Activities B and F are free to be started.

Activity B has the smallest starting latest starting time and is assigned to robot 2.

Step 2: Advance the project clock to 12 minutes.

Activity C is completed, so robot 1 is free.

Activity F remains free to be started.

Step 1: Activity F is the only activity which is free to be started, so it is assigned to robot 1.

The current state of the scheduling activities is shown in Figure 5.8.

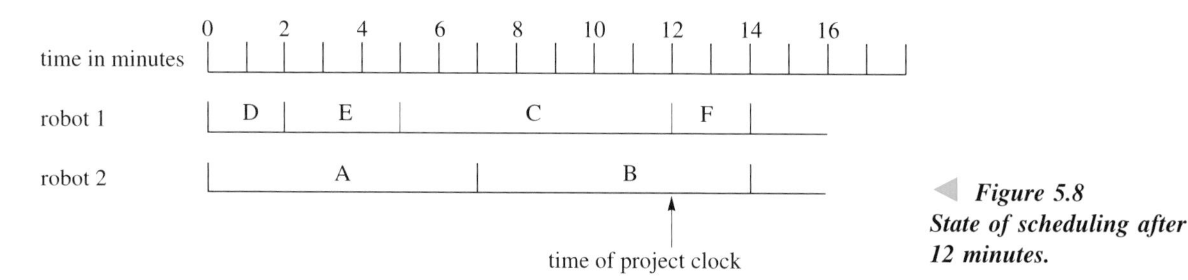

Figure 5.8
State of scheduling after 12 minutes.

Step 2: Advance the project clock to 14 minutes.

Activities B and F are completed, so both robots are free.

Activities G and J are now free to be started.

Step 1: Activity J has the smallest latest starting time, and is assigned to robot 1. The only remaining activity which can be started is G, so this is assigned to robot 2.

Step 2: Advance the project clock to 16 minutes.

Activity G is completed, so robot 2 is free.

Activities H and I are now free to be started.

Step 1: Activities H and I have the same latest starting time. Activity H is assigned to robot 2.

Step 2: Advance the project clock to 24 minutes.

Activity H is completed and robot 2 is free.

Step 1: Activity I is the only remaining activity which is free to be started, so it is assigned to robot 2.

Step 2: Advance the project clock to 32 minutes.

All activities have now been completed and both robots are free.

Step 3: FINISH.

The resulting schedule is shown in Figure 5.9.

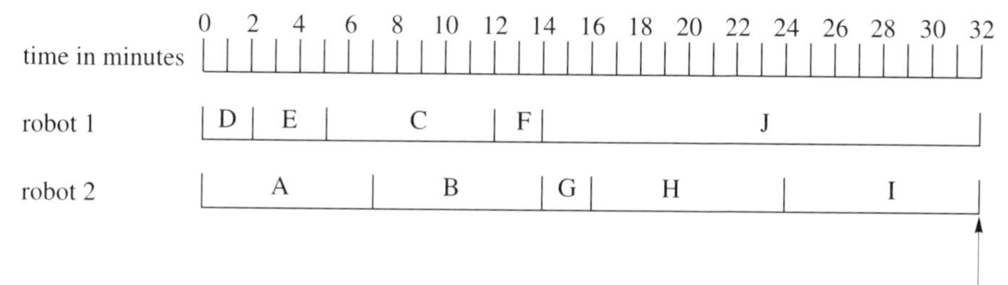

Figure 5.9
The schedule is completed at 32 minutes.

As it happens, this schedule optimizes the use of the robots, but this need not be the case. For example, if three robots were available there would be times when one of them would be idle.

In this and the previous section you have been introduced to some elementary ideas in activity scheduling. In practice one needs to take into account many more features. For example, it was assumed that the two robots were interchangeable. In general, robots will not have the same repertoire of activities and the scheduling algorithm has to be adjusted accordingly.

If a mechatronic system is to be capable of ***self-repair***, such as reconfiguring itself when parts are damaged or lose part of their functionality, it has to reschedule its activities when the damage or failure is detected. It must also revise its estimate of how long the project will take. In some safety-critical applications the system may have to be able to predict dangerous loss of synchronization long before it happens, and so signal the need for human intervention.

Activity scheduling is a highly technical subject with its own extensive literature and theory. It is yet another specialism that mechatronic engineers must draw on.

5.7 The 'travelling salesman problem'

Many of the techniques described in the previous chapters can be used to find solutions to difficult problems. A large number of problems fall into the category of being *non-polynomial indeterminate*, as described in Chapter 3, Section 3.1. The computational effort required to solve such problems grows astronomically with the 'size' of the problem. For example, in the 'travelling salesman problem' the size is determined by the number of cities to be visited. Even with modern computers, these problems cannot be solved exactly in a reasonable time, so heuristics have to be used to zoom in on inexact or sub-optimal solutions.

The travelling salesman problem is a minimization problem, which has become something of a benchmark for algorithms and artificial intelligence methods. In this section we will describe some of the more encouraging recent techniques.

The travelling salesman problem has many variants, and one which is of particular interest in mechatronics is the routeing problem. In this, a number of vehicles have to deliver items to many sites scattered around an area. The vehicles could be lorries delivering goods for shops, or autonomous vehicles in a factory delivering components to the places where they are needed. So although the travelling salesman problem is somewhat artificial, methods of finding solutions to it are applicable to many other problems.

This example is relevant to mechatronics because one of the functions of the cognitive element in a system is to plan a sequence of actions. This may take the form of planning a suitable route for a vehicle to travel in order to minimize the distance travelled.

Suppose a salesman starts in a city **a** and has to travel around N cities and return to **a** without visiting any other city more than once. The problem is to find the shortest route. Figure 5.10 shows an example where N is 6.

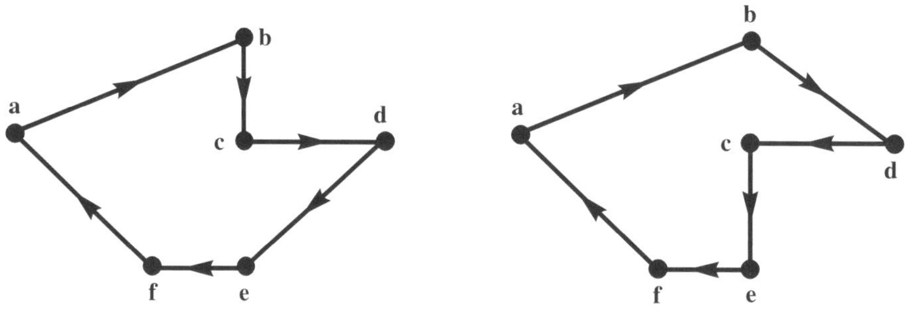

Figure 5.10 ***Travelling salesman probl*** ***with six cities, showing tw*** ***possible routes.***

Figure 5.10 is a relatively simple diagram which just shows the cities (**a** to **f**) and two closed paths between them. From Section 5.3 it should be clear that the problem could be represented as a graph, in which each of the cities is a node and the paths between the cities are links. Since a path has to be travelled in one direction, the paths would be directed links, and since each path has a distance associated with it the graph becomes a network. So the problem that we are addressing is that of finding the cycle in a network that has the shortest path.

The value that has to be minimized in the travelling salesman example is the total distance travelled. It is an interesting problem because there is no known analytical solution. It is assumed in the basic travelling salesman problem that a path exists from every city to every other city. The only way of finding the shortest distance is to consider every possible combination and measure the distance, which is clearly an unattractive proposition and a very time-consuming one. To illustrate how difficult this is, consider how many different routes there are for N cities.

When $N = 3$ there is only one route, if direction is ignored.
When $N = 4$ there are 3 different routes.
When $N = 5$ there are 12 different routes.
When $N = 6$ there are 60 different routes.
When $N = 7$ there are 360 different routes.

In general, if there are N cities there are $\frac{1}{2} \times (N-1)!$ different routes, where $N!$ (N factorial) is given by

$$N! = 1 \times 2 \times 3 \times 4 \times \ldots \times (N-2) \times (N-1) \times N$$

For example, when $N = 5$

$$N! = 1 \times 2 \times 3 \times 4 \times 5 = 120$$

$$(N-1)! = 1 \times 2 \times 3 \times 4 = 24$$

So $\frac{1}{2} \times (N-1)! = 12$ when $N = 5$

If all routes are to be examined, the computation involved becomes astronomical very quickly for increasing values of N. For example, for $N=10$ the number of routes to be examined is 181 440. For $N=20$ this increases to 6×10^{16}. Although computers are getting more powerful all the time, problems like this can still only be exhaustively solved in a reasonable time for relatively small values of N.

It has often been said that computing power has increased by a factor of 10 every five years for the same price since the 1950s. Even assuming that this continues, some problems will still remain unsolved for a long time to come. To give some idea, when $N=20$ again, if we assume that each route can be examined in a microsecond, it would still take about 2000 years to examine all the possible routes! If the trend in computing continues, in 10 years' time computers will be a hundred times faster than they are now. Still, it would take 20 years to calculate the solution for $N=20$. So it looks as though we would have to wait a considerable time before computers can handle some of these NP-hard problems if they are going to use the brute force method of looking at all the solutions first before the best one can be selected.

In order to deal with problems like this, decisions have to be made about a strategy for finding a shortest path. One strategy is to find any path, irrespective of the length, and then to try to shorten it. For example, a path can be found by the following method:

1 Start at the first city.
2 Select any other city, draw a path to that city and move to it.
3 Select any other city that hasn't been visited before, draw a path to it and move to it.
4 Repeat 3 until there are no more unvisited cities.
5 Return to the first city.

This will eventually get you around all the cities, but it almost certainly won't be the shortest path. This can be improved on using the so-called ***greedy algorithm***. This is essentially the same, except that the closest city is always selected. Thus:

1 Start at the first city.
2 Select the nearest city, draw a link to that city and move to it.

3 Select the nearest city that hasn't been visited before, draw a link to it and move to it.
4 Repeat 3 until there are no more unvisited cities.
5 Return to the first city.

This may produce a shorter path than the first method, but there are no guarantees. Another variation on the greedy algorithm that can sometimes produce an even shorter path is to 'grow' the path from both ends. For convenience, let's call the two ends of the path the *head* and the *tail*. The method then looks like this:

1 Start at the first city.
2 Select the nearest city, draw a path to that city and move to it. This is the head of the path.
3 Find the nearest cities to the head and the tail of the path that haven't been visited before. Whichever is nearest of the two, draw a path to it and move to it.
4 Repeat 3 until there are no more unvisited cities.
5 Join the head and the tail of the path.

These methods can be illustrated by a simple example. Figure 5.11 shows five cities, **a** to **e**.

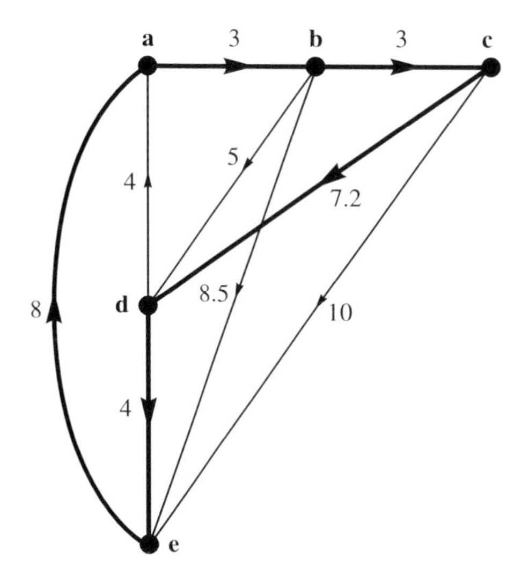

Figure 5.11
An example of five cities.

Using the first version of the greedy algorithm, and starting from **a**, the sequence of cities visited is:

$$\mathbf{a} \xrightarrow{3} \mathbf{b} \xrightarrow{3} \mathbf{c} \xrightarrow{7.2} \mathbf{d} \xrightarrow{4} \mathbf{e} \xrightarrow{8} \mathbf{a}$$

The total path length is therefore 25.2 units.

Now if the modified greedy algorithm is used, again starting from **a** we get:

$$\mathbf{a} \xrightarrow{3} \mathbf{b} \xrightarrow{3} \mathbf{c}$$

but at **c** the nearest unvisited city is **d** with a distance of 7.2, whereas the nearest unvisited city to **a** is **d** with a distance of 4, so the tail of the path moves to **d**. The next nearest unvisited city is **e**, which is closest to **d**, and then finally **e** and **c** are joined, so the final path looks like:

$$\mathbf{a} \xrightarrow{3} \mathbf{b} \xrightarrow{3} \mathbf{c} \xrightarrow{10} \mathbf{e} \xrightarrow{4} \mathbf{d} \xrightarrow{4} \mathbf{a}$$

The final path length is 24 units, which is shorter than the first method.

These examples produce solutions, but it would be very unusual if the path turned out to be the shortest in more complex examples. The following sections look at some methods which can improve on this.

5.7.1 Hill climbing

Hill climbing was described earlier in Chapter 3 on Search, where it was said that it is a form of gradient descent (or ascent) used when it is very difficult to define a gradient. The method simply ensures that the value selected at each iteration is less than (or greater than) the previous value.

For the travelling salesman problem, we start by writing down a list of the cities in any order, and call this list L_1. We may as well use the greedy algorithm to find this initial path. Measure the total distance of this path as if this was the route to be travelled and call that distance D_1. For example, where **a**, **b**, **c**, **d**, **e** and **f** represent 6 cities,

L_1 = [**a**, **b**, **c**, **d**, **e** , **f**, **a**]

Now take any pair of cities, and swap them around in the list. This process is called ***permutation***. For example, when **b** and **c** have been swapped,

L_2 = [**a**, **c**, **b**, **d**, **e**, **f**, **a**]

Measure the new total distance, D_2, and compare it to the unswapped distance. If it's shorter keep the new list, else go back to the previous list.

Let the list of the shortest path so far be L and the shortest distance so far be D. Then if the new list L_k has a distance D_k that is less than D, L becomes L_k.

If $D_k < D$ L becomes L_k, and D becomes D_k

else $D_k \geqslant D$ L is unchanged, and D is unchanged

Continue swapping in this way until a list is finally produced which has a distance that cannot be shortened by any further swapping. This list corresponds to a local minimum.

Figure 5.12 shows a set of four cities with the distances marked between them.

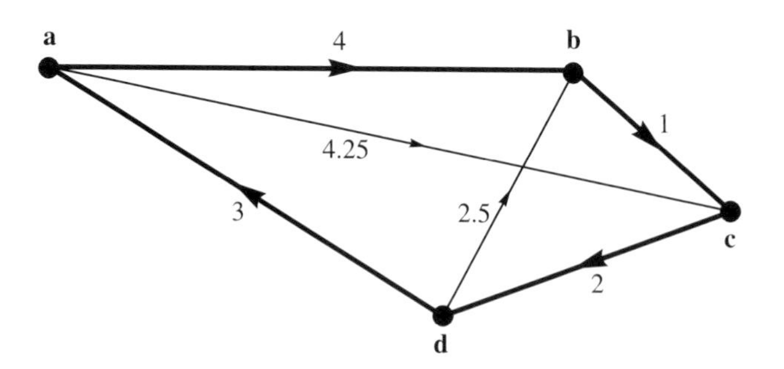

Figure 5.12
An example with four cities

The three possible permutations of the list of cities with their corresponding total distances are shown below.

abcda	10
abdca	12.75
acbda	10.75

You may be wondering why there are so few permutations. For example, where is the permutation **adcba**? The answer is that a path could be travelled in either of two directions without affecting the length of the path, so the permutation **adcba** is the same as the path **abcda** but in reverse.

Let's start the search for the shortest path with the list **acbda**, which has a distance of 10.75. By swapping any two of the middle three cities around, the possible swaps that could take place are:

abcda	10	
adbca	10.75	(same as **acbda**)
acdba	12.75	(same as **abdca**)

If **acdba** is tried, it would be rejected because its distance is greater than 10.75. If **abcda** is tried it would be accepted because its length is less than 10.75. No further swapping would produce a shorter distance, and a minimum has been reached. In this example this happens to be the global minimum. With N greater than this there are likely to be many local minima and hill climbing will most likely result in the search getting stuck in one of these. Simulated annealing is one way of improving on this, and we will look at this in the next section. First, there are some simpler heuristics that can help.

In this very simple example we have effectively done an exhaustive search, but of course in larger examples this would not be feasible.

5.7.2 Crossed paths

Very often a path will be found using a hill-climbing approach in which the path crosses itself at some point, as shown in Figure 5.13(a) for example. It is nearly always the case that a path which crosses itself is not the shortest path, and that if it could be uncrossed in some way a new path could be found which would be shorter. Sometimes, swapping two cities can uncross the path but then may cross it again at some other point, and the resulting path may be longer.

In Figure 5.13(a) the path is crossed and the total length is 27.5 units. If cities **g** and **d** are swapped, the resulting path is shown in Figure 5.13(b). This new path is also crossed and has a length of 28.5 units. Now this path can be easily uncrossed by swapping cities **e** and **f** which produces the path shown in Figure 5.13(c) which has a length of 27.5 units again. Finally, swapping cities **c** and **g** produces the path shown in Figure 5.13(d) which has a length of 23 units.

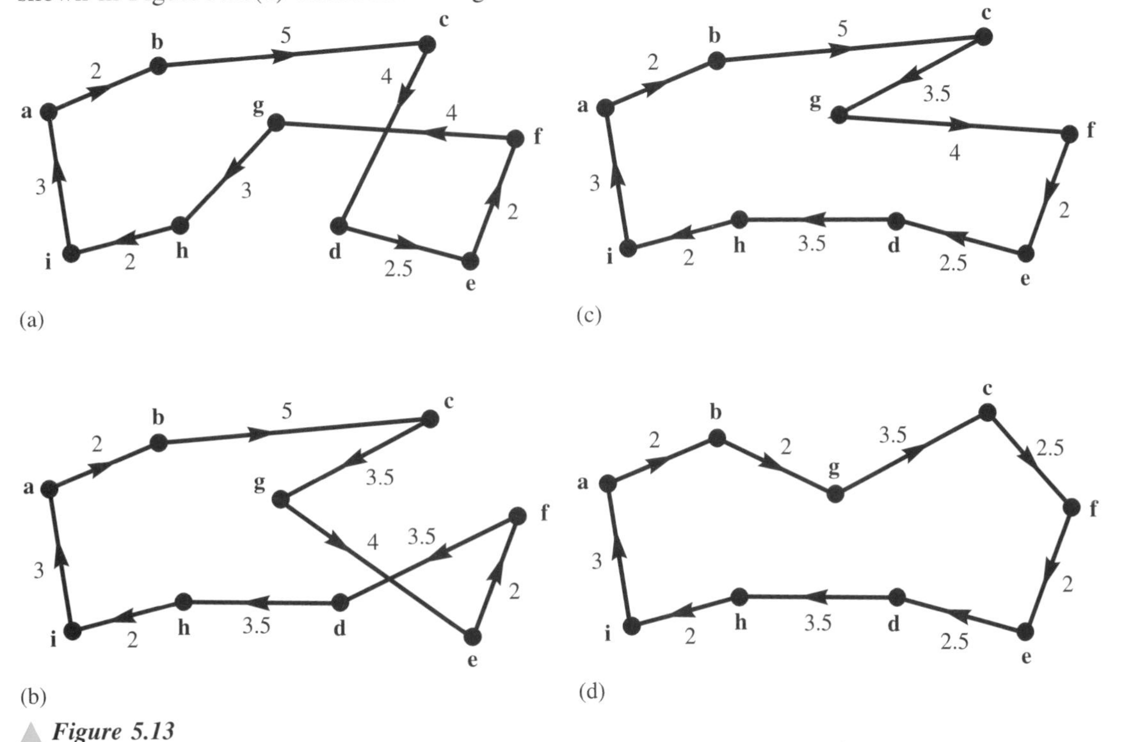

Figure 5.13
(a) Crossed path, (b) path still crossed, (c) further shortening, (d) shortest path

This example shows how uncrossing a path ultimately shortens it. However, the act of uncrossing the path sometimes increases the path length if we use the two-city swapping method. Usually, if a point is found where the path crosses itself, it is a good rule-of-thumb or heuristic to uncross the path straight away. This will nearly always produce a shorter path with the help of hill climbing.

To do this in our example, note that the path crosses between pairs of cities **f**, **g** and **c**, **d** (Figure 5.14a). Simply altering the path so that it goes from **c** to **f**, and from **d** to **g**, creates a new uncrossed path as shown in Figure 5.14(b). The new path length is 23 units, so the length is the same as in Figure 5.13(d) even though the path is slightly different. This process is not the same as swapping cities but is a new heuristic which could be called an 'uncrossing' heuristic. It produces a shorter path but doesn't have to go via a longer path to get to the solution.

5.7.3 Simulated annealing

Simulated annealing, as described in Chapter 3 on Search, mimics the process of cooling a metal. The 'energy' of a system has to be defined, and this becomes the search space for the problem, which is to find the global minimum energy. In the travelling salesman problem the energy can be equated to the distance around the path.

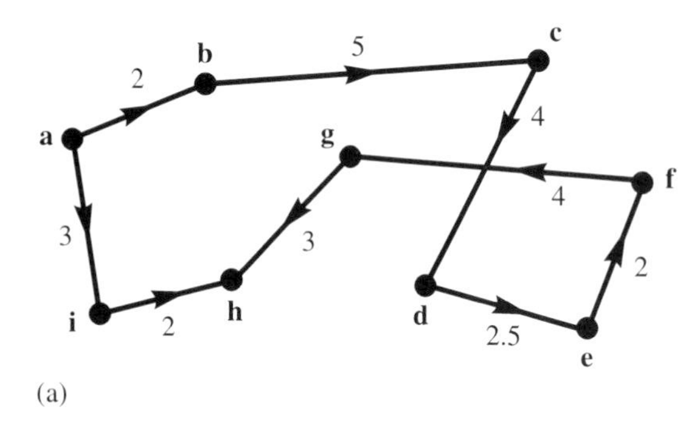

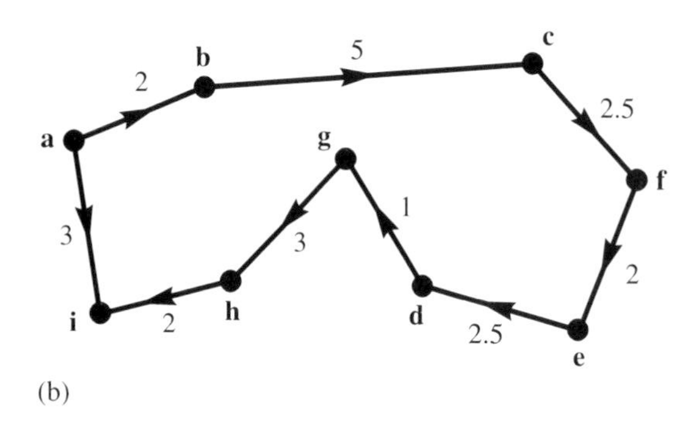

Figure 5.14 (a) Crossed path, (b) uncrossed path

The method is similar to hill climbing except that the decision about whether to keep the new list after swapping or throw it away is probabilistic. This means that sometimes the new list will be kept even though the total distance associated with it is longer than the previous list, in order to allow the search to 'jump out' of a local minimum.

The probabilities are such that if the new list, L_k, has a distance, D_k, that is less than the current best distance, D, then there is a probability of greater than 0.5 that the new list becomes the best list. Similarly, if D_k is greater than D, there is a probability of less than 0.5 that the new list becomes the current list. For example, if the probability turns out to be 0.8, then in eight cases out of ten the new list becomes the current list, but in two cases out of ten it doesn't. The values of probability, of course, always lie between 0 and 1.

The probabilities, P_k, are used as follows:

If	$D_k < D$	L becomes L_k, with a probability of P_k where $1 > P_k > 0.5$
else	$D_k \geqslant D$	L is unchanged, with a probability of P_k where $0.5 > P_k > 0$

Rearranging:

If	$(D - D_k) > 0$	L becomes L_k	$1 > P_k > 0.5$
else	$(D - D_k) \leqslant 0$	L is unchanged	$0.5 > P_k > 0$

We therefore need a function that produces a value for the probability which is between 0.5 and 1 when $(D - D_k) > 0$ and a value between 0 and 0.5 when $(D - D_k) \leqslant 0$. A function which has this property is the *sigmoid function*, described by the equation

$$y = \frac{1}{1 + e^{-x}}$$

When $x = 0$, $y = 0.5$.

When $x > 0$, $1 > y > 0.5$.

When $x < 0$, $0.5 > y > 0$.

As this method is simulating annealing, a factor equivalent to temperature, T, has to be included in the model. This is done by dividing $(D - D_k)$ by a parameter T, and then substituting for x in the equation of the sigmoid. The probability is therefore:

$$P_k = \frac{1}{1 + e^{-(D - D_k)/T}}$$

When T is very large, P_k approaches 0.5, which means that the decision about keeping the new list or throwing it away is purely random. When $T = 0$, $P_k = 1$ and the decision is not probabilistic, but is equivalent to the hill-climbing method described earlier. So if the temperature starts out high, the decisions seem arbitrary. As the temperature drops, the decision to make the new list the current list or not becomes more deterministic. The effect is that the search can jump out of local minima, and is more likely to end up at the global minimum when $T=0$.

The *cooling schedule* should be set so that thermal equilibrium is reached. To ensure this, the search has to be able to sample the entire space adequately before the temperature is dropped. In our case with N cities, the worst case is that two paths are $N-2$ permutations apart, so at least $N-2$ iterations are necessary at each temperature. For example, when $N=6$ two paths might be

abcdefa and **aedbfca**

To get from one to the other by swapping two cities around could go like this:

abcdefa	
aecdbfa	first permutation, **b** and **e** swapped
aedcbfa	second permutation, **c** and **d** swapped
aedbcfa	third permutation, **b** and **c** swapped
aedbfca	fourth permutation, **c** and **f** swapped

So to get across the search space from the first list to the second requires at least four iterations. In practice, many more iterations would be used since the decision to accept the permutation is probabilistic. Even when the temperature is high the probability is only 0.5, so that on average at least $2 \times (N-2)$ iterations are required to cross the search space.

5.7.4 Genetic algorithms

As we saw in Chapter 3 on Search, the essential features of a genetic algorithm are the chromosomes that contain the genetic information. These are strings of data that define a particular solution. In the travelling salesman problem, one way of setting up the chromosomes is to use the list of cities in the order that they are to be visited.

A population of these chromosomes, corresponding to a number of individual solutions to the problem, are created. In the travelling salesman problem there will be many possible routes that can be taken. The population of chromosomes at any one time will represent only a small number of those solutions. The population is usually created randomly although it is possible to 'seed' the initial population with individuals that are known to be good solutions.

Next we need some way of measuring the 'fitness' of the chromosomes so that the good solutions are selected to be parents more often than the not-so-good solutions. This is analogous to natural selection, in which 'survival of the fittest' is said to occur. In the travelling salesman problem the fitness measure is a function of the total distance travelled, the 'fittest' being the ones with the shortest distance. An example of a function that does this is

$$\frac{1}{\text{total distance}}$$

The first difficulty that we encounter in the travelling salesman problem is that we cannot encode the cities directly as a string or list because genetic *crossover** and *mutation* would produce strings which are not valid solutions. To overcome this we construct chromosomes consisting of binary 0s and 1s corresponding to all the possible connections between pairs of cities. For a problem with six cities and a path which can be listed as **abcdefa**, this would look like:

	ab	ac	ad	ae	af	bc	bd	be	bf	cd	ce	cf	de	df	ef
	1	0	0	0	1	1	0	0	0	1	0	0	1	0	1

In this representation, the presence of a binary 1 indicates the existence of a link between the two cities, and a 0 indicates the absence of a link.

The type of crossover that is used is called *recombination*, where individual bits of data are taken from parent chromosomes at randomly selected points. Recombination is essentially a variation of the critical path method described earlier in this chapter. For example, take the following two paths, **abcdefa** and **aefbdca**, set out as chromosomes:

	ab	ac	ad	ae	af	bc	bd	be	bf	cd	ce	cf	de	df	ef
1	1	0	0	0	1	1	0	0	0	1	0	0	1	0	1
2	0	1	0	1	0	0	1	0	1	1	0	0	0	0	1

* Note that the term crossover used here in connection with genetic algorithms is not the same as the concept of crossover on a path.

The method of recombination is as follows:

Step 1: Produce a table with all the cities in one column, and in the second column list all the cities connected to them in either of the parent chromosomes. Using the two chromosomes above, this table is:

City	Connected to:
a	**b, f, c, e**
b	**a, c, d, f**
c	**b, d, a**
d	**c, e, b**
e	**d, f, a**
f	**a, e, b**

Step 2: Randomly select a city as the current city. Let's choose **c** as the current city.

Step 3: Delete the current city from the right-hand side of the table. The table becomes:

CURRENT CITY: c

City	Connected to:
a	**b, f, e**
b	**a, d, f**
c	**b, d, a**
d	**e, b**
e	**d, f, a**
f	**a, e, b**

Step 4: Look at all the cities connected to the current city, and select the one with the fewest connections. If there are two or more cities with the same smallest number of connections, choose one randomly. In this example, **d** has two connections, while **b** and **a** have three. Therefore select **d** as the current city.

Step 5: Repeat Steps 3 and 4 until there are no cities left.

This goes as follows for this example.

Step 3: Delete the current city, **d**, from the right-hand side of the table. The table becomes:

CURRENT CITY: d

City	Connected to:
a	**b, f, e**
b	**a, f**
c	**b, a**
d	**e, b**
e	**f, a**
f	**a, e, b**

Step 4: Look at all the cities connected to the current city, and select the one with the fewest connections. In this example, both **e** and **b** have two connections, so randomly select **b** as the current city.

Step 3: Delete the current city from the right-hand side of the table. The table becomes:

CURRENT CITY: b

City	Connected to:
a	**f, e**
b	**a, f**
c	**a**
d	**e**
e	**f, a**
f	**a, e**

Step 4: Look at all the cities connected to the current city, and select the one with the fewest connections. In this example, both **a** and **f** have two connections, so randomly select **a** as the current city.

Step 3: Delete the current city from the right-hand side of the table. The table becomes:

CURRENT CITY: a

City	Connected to:
a	**f, e**
b	**f**
c	
d	**e**
e	**f**
f	**e**

Step 4: Look at all the cities connected to the current city, and select the one with the fewest connections. In this example, both **e** and **f** have one connection, so randomly select **e** as the current city.

Only **f** is left. The resulting path is the list of cities in the order that they were selected by this process of recombination:

cdbaefc

If we look at this as a chromosome and compare it with the parent chromosomes, you can see that recombination has taken data from each of the parent chromosomes. In the table below, the parent chromosomes are labelled 1 and 2, and the offspring is labelled 3. Bold digits in the parent chromosome indicate the source of the genetic material in the offspring. When the source could be either parent, both are indicated in bold.

	ab	**ac**	**ad**	**ae**	**af**	**bc**	**bd**	**be**	**bf**	**cd**	**ce**	**cf**	**de**	**df**	**ef**
1	**1**	**0**	**0**	0	1	1	0	**0**	**0**	**1**	**0**	0	1	**0**	**1**
2	0	1	**0**	**1**	**0**	**0**	**1**	**0**	1	**1**	**0**	0	**0**	**0**	**1**
3	1	0	0	1	0	0	1	0	0	1	0	**1**	0	0	1

Recombination can also cause mutation. In this example, the offspring has a 1 in the chromosome at **cf** (shown in bold) which was in neither parent. Thus recombination is a powerful method which performs crossover and mutation at the same time.

The essence of the genetic algorithm recombination technique is to construct the set of links belonging to either of the parent routes. These are then recombined to form the child route which, by construction, is made up of the links of its parents. The exception to this is when neither parent has a link between the end node and start node found by recombination: in this case a new link is created by what we have called mutation.

Genetic algorithms work in this case because relatively good routes will be made up of relatively good links. Breeding from relatively good parents will tend to recombine their relatively good links with a relatively good chance of finding a better route.

This exemplifies the genetic algorithm approach: they allow locally good parts of solutions to be combined to form globally better solutions. Mutation is essential, of course, to try to ensure that the improvements are not restricted to local optima in the search space.

5.7.5 Routeing

The travelling salesman problem is a good benchmark for testing heuristic methods of finding the shortest path. However, it is a rather simplistic problem, and although many similar problems exist to which the methods described in this chapter can be applied, it is often the case that in real applications there are some practical constraints which make these methods inapplicable. A typical example is the vehicle routeing problem.

On the surface this looks very similar to the travelling salesman problem. A depot (or sometimes more than one depot) contains several vehicles of different sizes and capacities, and the requirement is that a variety of goods are delivered to retail outlets in an area. The problem is not just one of shortening the route, which is helpful but not necessarily the best solution, but involves reducing the number of vehicles or separate journeys for each vehicle. So, for example, if a company currently requires 10 vehicles, some of which may have to make two or more journeys a day to deliver all the goods, then it would be beneficial to that company if the number of vehicles could be reduced to 8 and only one journey a day for each.

The problem is more complex than the original travelling salesman problem because there are several paths that need to be found, one for each vehicle. It is beyond the scope of this book to show how this can be achieved, but you should appreciate that essentially the same methods can be used. The route taken between retail stores has to be found which minimizes the distance, and finally, we need to decide which vehicle is going to visit which stores so as to reduce the number of journeys. All of this can be done using some variation on hill climbing with some heuristics such as paths not being allowed to cross, as described earlier.

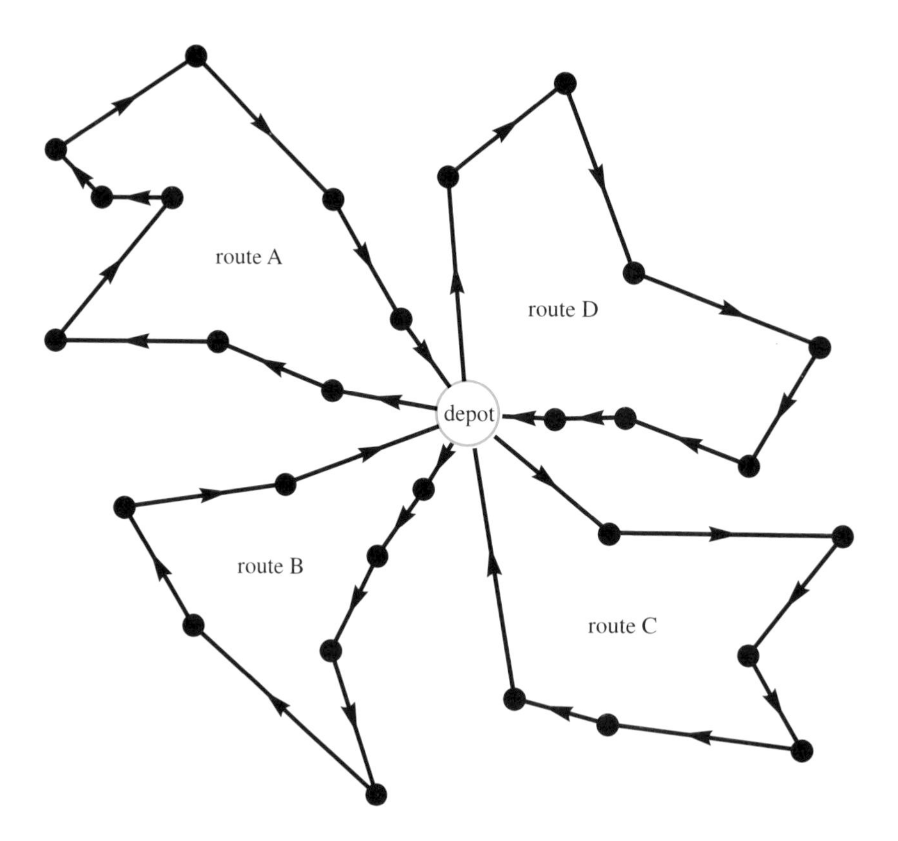

Figure 5.15 Typical solution to a vehicle routeing problem.

5.8 Intelligent scheduling

All the methods described so far for the travelling salesman problem assume that everything is known in advance, such as the cities which have to be visited, the distances between them, and the nature of the paths. What happens if, after a path has been planned, an obstruction occurs on one of the paths? With the methods described so far they would simply have to abandon the path planned and find a new one, which is both inconvenient and costly.

A situation might arise in which a path is blocked, so a detour is necessary. Suppose that one of the cities that isn't scheduled until later is nearby. If the original schedule is rigidly stuck to, that city would be ignored, since the principal goal at this point is to get back onto the original path. But this misses an opportunity to save time.

There are therefore three major drawbacks to the methods described:

1. A lot of computing has to be done in advance, which could take a lot of time.
2. All the effort could be wasted if the path changes, due to a blockage for example.
3. The methods aren't opportunistic, which means that they don't take advantage of opportunities when they arise.

One exception to this is the greedy algorithm that was described at the start of this chapter. It requires very little computing, and only needs local information to plan its next move. This doesn't produce the optimal schedule, but is still relatively good. For example, a worst case scenario for a path planner would be a situation in which several points have to be visited, and all of those points are moving in unpredictable ways. All the methods except for the greedy algorithm would not have a chance at planning a path. The greedy algorithm would at least succeed in the task, even if it does not find the optimal solution.

It would appear, then, that a method is needed which can do some planning in advance, but which makes use of local information once the plan is put into action to take account of any changes that occur. This is known as *intelligent scheduling*. The greedy algorithm seems to work for the travelling salesman problem, so it would be useful if something similar could be applied to other problems such as path planning in a complex changing environment. One such method is to use the distance transformation described earlier in Chapter 3 on Search.

Recall that forward and backward scanning are used to calculate a value for every square in a grid which represents the environment. The example given in Chapter 3 is shown in Figure 5.16.

5	4	3	2	1	2	6 = empty grid
6	[15]	[6]	1	**0**	1	[6] = object
7	[18]	[9]	2	1	2	**0** = goal
6	5	4	3	2	3	

Figure 5.16
A grid with values found by the distance transform.

The numbers in the grid represent the number of steps to the goal. Objects, which have to be avoided, end up with a relatively high value. Starting from any point on the grid, all one has to do to get to the goal along the shortest path is to move to the neighbouring square with the lowest value. We therefore have a method which calculates all the values in advance, and could, if desired, plan the whole path in advance. Alternatively, a path could be followed using only the local information that is available at the time.

If an obstacle appears that blocks the path to the neighbour with the lowest value, an alternative can be taken by looking at the neighbouring squares for the next lowest value. Having avoided the obstacle, a new path is taken to the goal based on the local information from the area that is currently occupied. In this way, opportunities to reach the goal are not missed by insisting on getting back to the original path.

Ultimately, even the distance transform method fails to be intelligent, because it also relies on there being some form of map of the environment which is known in advance. When humans, whom we might regard as the most intelligent schedulers so far, give directions, for example, it is usually in the form of instructions such as:

> 'Keep walking along this street until you come to a set of traffic lights. Turn left at the lights and then go up the hill until you come to the second turning on the right.'

These are very imprecise instructions, but somehow we all manage to follow them. We follow the road even when there are many bends; we avoid getting stuck when there is an obstruction in the way by simply going around it, and we take advantage of the fact that we might be able to see in advance the set of traffic lights and so cross the road before we get to them. Whether or not machines will ever be able to follow a similar set of instructions is debatable, but we would guess that soon they will be able to. The problem with the set of instructions is the imprecision, which is also a problem in many other areas of AI such as speech recognition, handwriting recognition, and generally in any environments that are complex or contain humans.

5.9 Conclusion

This chapter has dealt with various approaches to scheduling. One of the major tasks is formulating a problem so that the various techniques described in this chapter can be applied. Graph theory is one way that allows us to set up a scheduling problem as a graph or a network. Once the problem is represented in this way, there are methods that can be used to identify the shortest route and the critical paths in the network.

With some conventional methods the computational effort involved grows very quickly as the size and complexity of the problem increase. The result is that in many cases it is impossible to compute a shortest path in anything like a reasonable time. However, in scheduling activities, we nearly always require some value, such as the distance travelled or the time taken, to be minimized, and the results are needed quickly. This is why many of the techniques described in

Chapter 3 on Search have been applied, such as hill climbing, simulated annealing and genetic algorithms. The travelling salesman problem was used to illustrate this, and it was shown that attempting to find the shortest path by examining all the paths becomes a computational nightmare. The search techniques provide a way of finding a sub-optimal solution, and sometimes *the* optimal solution, using much less computing resource.

Although the travelling salesman problem is somewhat artificial, the methods employed to solve it can usually be applied to more complex 'realistic' problems by incorporating features such as weighting the values on the links on the graph. Yet more complex problems so far elude optimization techniques, so we have to fall back on sub-optimal solutions such as those produced by the greedy algorithm and the distance transform. Although these generally tend to be called 'intelligent scheduling', we will have to wait to see if they really are intelligent as new techniques come along that can deal with situations more in the way that humans do, with all of its imprecision.

Acknowledgement

The discussion of critical-path scheduling is based largely on that given in Dolan and Aldous (1993).

References

Dolan, A.K. and Aldous, J. (1993) *Networks and Algorithms: an introductory approach*, John Wiley & Sons, Chichester.

Lawler, E.L., Lenstra, J.K., Rinnooy Kan, A.H.G. and Shmoys, D.B. (1985) (eds.) *The Travelling Salesman Problem*, Wiley-Interscience.

CHAPTER 6
REASONING

6.1 Introduction

One fundamental item of information required by every intelligent machine is 'what to do next'. For example, a machine that can move about might contain knowledge such as 'there is an obstacle in front', and 'crashing into obstacles at a high speed may be dangerous'. From this it might deduce that its next action should be to 'slow down'. Reasoning is part of the machine's cognition subsystem, and is necessary when the sensors of the perception subsystem cannot deliver information in the required form.

Reasoning is the process of going from what is known to what is not known.

Humans are very good at reasoning, as it is something that we have to do all the time. We are constantly picking up clues from the information around us and drawing conclusions based on these clues. One of the major areas of artificial intelligence is concerned with finding ways of emulating this process. By far the most frequently used tool is logic. The logical methods that are used in artificial intelligence are drawn from many sources, including philosophy and mathematics. Of the many mechanisms for doing this, we will consider the following in this chapter:

- deterministic reasoning:

  propositional logic
  predicate logic

- dynamic reasoning:

  non-monotonic logic

- non-deterministic reasoning:

  multi-valued logic
  probability theory and Bayesian deduction
  fuzzy logic.

In the rest of this introduction we will give an overview of these logics: their origins and their applicability to machine reasoning. Each one will then be dealt with in more detail in the subsequent sections in this chapter.

6.1.1 Deterministic reasoning

Deterministic reasoning goes back to the Greeks, and Plato's student Aristotle (384–322 BC). Aristotle abstracted a set of rules called syllogisms. We can illustrate the general idea using one of the best known syllogisms which takes the form of an *If–Then* rule. First, let's state the rule in everyday language.

A moving vehicle uses computer vision to guide it through a complex environment. *If* it 'sees' an unknown object in its path, *Then* it should take action to avoid that object. This statement can be written in a more concise way as a rule:

The rule

If the camera image contains an unknown object

Then take evasive action

The first part of the rule contains a statement which can be either TRUE or FALSE. If it is TRUE, then the second part of the rule is activated. The second part of the rule is also a statement that can be either TRUE or FALSE; if evasive action is being taken then the statement is TRUE, otherwise it is FALSE. When the first part of the rule is TRUE, the system containing the rule can initiate evasive action to make the second part of the rule TRUE.

This process is deduction, where the truth value of a fact can be deduced from another. Consider this example:

The known fact

'the camera image contains an unknown object' is TRUE

The deduced fact

'take evasive action' is TRUE

Such was the respect accorded to Aristotle over the centuries that almost no further developments were made in logic until the work of George Boole (1815–1864). Boole worked out a system by which new statements can be deduced from others using the connectives AND and OR. In the previous example, the object that is 'seen' by the camera is described as being 'unknown'. We could split up the original statement into two parts: the first concerning the fact that an object has been detected, the second concerning whether or not the object can be matched to any known object in the system's database. The known facts become:

The known facts

'the camera image contains an object' is TRUE

AND

'the object cannot be matched in the database' is TRUE

This says that an object has been detected, but it doesn't match any known objects. There are two known facts, and they have to be combined into a single fact before the system can decide whether to take action or not. The connective AND is used to combine the facts.

The deduced fact

'the camera image contains an object'

AND

'the object cannot be matched in the database' is TRUE

Here the deduced fact is clearly more complicated than either of its constituents. For practical purposes we might equate the deduced fact with the proposition that the camera image contains an unknown object. Using Boole's logic (Boolean logic) we can work out the truth of this proposition given the truth values of the two sub-propositions which are connected by the AND.

Although Boolean logic is very powerful, it does not contain one of the most powerful ideas in logic, that of *quantifiers*, due to Friedrich Frege (1848–1925) in its modern form. For example, there is clearly a difference between saying 'there exists an object in the database and that object matches the object detected by the camera' and 'for all objects in the database, those objects match the object detected by the camera'. The first of these is called the ***existential quantifier*** while the second is called the ***universal quantifier***. These quantifiers are powerful devices because they save writing out the proposition many times. For example, a machine might be required to switch off its motors for all those occasions on which it reads the bar codes beginning with I II. Then instead of laboriously writing out

'if the bar code is I II I then switch off the motors'

'if the bar code is I II II then switch off the motors'

'if the bar code is I II I I then switch off the motors'

'if the bar code is I IIIIII then switch off the motors'

'if the bar code is I II I I then switch off the motors'

'if the bar code is I IIII I then switch off the motors'

and so on, we can simply write

'for all bar codes, if the bar code begins with I II then switch off the motors'

Propositional logic allows us to deduce the truth value of compound propositions made up from simpler propositions and the Boolean connectives AND, OR and NOT. ***Predicate logic***, developed mainly by Friedrich Frege and Bertrand Russell (1872–1970), goes one step further by allowing us to evaluate the truth values of compound propositions which also involve the quantifiers 'for all' and 'there exists'.

6.1.2 Dynamic reasoning

Although logical reasoning allows us to reason about things that will certainly happen, it still does not capture the richness of human thought that enables us to function in the face of incomplete and inconsistent information about our rapidly changing environment and goals. Modern research in artificial intelligence has introduced ***dynamic reasoning***, which addresses the cognitive ability of human beings to reason in the face of changing circumstances. ***Non-monotonic logic*** attempts to allow reasoning in which the truth value of a proposition is allowed to change.

The commonly cited example is that, given 'Tweety is a bird' is TRUE, and since we know that 'birds fly' is TRUE we may deduce that 'Tweety can fly' is TRUE. However, we may subsequently learn that 'Tweety is a penguin' is TRUE, and since 'penguins cannot fly' is TRUE we now deduce that 'Tweety can fly' is FALSE.

What has happened is that the original deduction was based on two propositions, plus some default knowledge. In other words, since we said that Tweety is a bird, in the absence of any other information we would assume that it could fly since typically birds can fly. In deterministic logic it is assumed that there is no default knowledge, so that when a deduction is made, any new evidence should support that deduction. However, in this example the new knowledge does not support the deduction. If we tried to resolve this problem using deterministic logic, statements like 'birds fly' would have to be qualified by a list of exceptions which could get quite cumbersome. Non-monotonic logic allows the deduction to change as new evidence arrives, so getting rid of the need for additional qualifiers.

Non-monotonic logic tries to capture the dynamic aspect of human logic in which we cope with massive uncertainty by constantly formulating working hypotheses which mostly turn out to be correct, but may sometimes have to be revised on the acquisition of new information.

6.1.3 Non-deterministic reasoning

Propositional and predicate logic have been developed in modern form in the nineteenth and twentieth centuries. However, they do not capture all the techniques of reasoning which seem to be so effective in humans. ***Non-deterministic logic***, such as ***multi-valued logic***, allows us to use predicate logic but with truth values such as 'unknown'. Furthermore, we often encounter situations in which black-and-white judgements are inappropriate. Instead of predicting future events on a TRUE–FALSE basis we often assess them in terms of likelihood.

Probability theory is an extension of the empirical notion of relative frequency. For example, if we observe that 89 identical components in a batch of 1000 fail within 1000 hours of use, the relative frequency of failure is 89/1000, or 0.089.

We can use this empirical data as a measure of the abstract probability of the component failing, p(failure) = 0.089. Although we cannot predict that the component will certainly fail at any given instant, we can predict that in 1 000 hours we expect about 89 failures in a thousand. Probability theory allows us to deduce other failure rates:

The known facts

89 components from 1000 failed within 1000 hours use.

The deduced facts

The probability of a component *not* failing within 1000 hours is

$$1 - p(\text{failing}) = 1 - \frac{89}{1000} = 0.911$$

Suppose a machine that depends on two such components, A and B, will fail if component A fails *or* component B fails. Put another way, the machine does *not* fail if component A does *not* fail *and* component B does *not* fail. Therefore the probability of the machine *not* failing within 1000 hours is

$$p = \frac{(1000 - 89)}{1000} \times \frac{(1000 - 89)}{1000} = 0.829921$$

That is, the probability of the machine failing within 1000 hours is approximately 1 – 0.83, which is 0.17. We expect 17% of machines with components A and B to fail within 1000 hours.

From this kind of calculation we can deduce that 17% of machines with two of the components will fail within 1000 hours. About one-quarter of machines depending on three of these components will fail within 1000 hours, while about one-third of machines depending on four of these components will fail within 1000 hours. Similarly, one can use probability to reason about the likelihood of complex outcomes given the probabilities of the constituent parts using the logical connectives AND, OR and NOT, just as in the case of Boolean logic.

In general, machines are constantly collecting data that enable them to update their probability estimates of various atomic and compound events. Ideally, one wants to use the new and old information in a way which optimizes the value of both. For example, the experience that a robot found the correct part in a given bin may not guarantee that the part will always be available in that place, but added to previous experience it may increase the expectation of finding the part in this place in future. A result in probability theory called Bayes' Theorem allows prior estimates of probability to be continually updated in the light of new observations.

Deterministic logic and probability theory can be combined to give rules of the form 'if the probability of collision is >0.1, initiate evasive action'. Thus, however measured, once the system has decided that the likelihood of collision exceeds the threshold value of 0.1, the rule definitely requires it to take evasive action. As we will see in this chapter, there are more subtle ways of combining intuitions about likelihood with those of predicate logic.

In 1965 Lotfi Zadeh proposed a form of reasoning using what has become known as ***fuzzy logic***. One of the main ideas is that propositions need not be classified as true or false, but their truth and falsehood can be weighted. This differs from probability theory by its dependence on the idea of a fuzzy set in which set membership is weighted. Thus a machine could be 'stationary' with a value of 0.9 and 'moving' with a value of 0.1. This is particularly useful when representing the universe in which a machine is operating. There is always some uncertainty about the position of the machine or its parts, and engineers are used to designing tolerances into their machines. Traditional engineering has tended to improve the behaviour of machines by using highly specified materials and precise, highly skilled assembly. In mechatronics, we are interested in replacing traditional solutions by incorporating machine intelligence that can compensate for low tolerances. Thus expensive manufacturing processes may be replaced by designs with less exacting specifications. Fuzzy set theory allows us to represent, say, the position of an autonomous vehicle in a 'fuzzy' way. It may be more useful to reason on the basis that the machine is 'in the corner' than to know precisely its x,y coordinates. This fuzziness in the representation allows a fuzziness in reasoning which can be very useful.

In the remaining sections of this chapter we will explain the mainstream topics from reasoning in more detail, and we will show how they can be used in designing intelligent machines.

6.2 Reasoning with certainty

6.2.1 Propositional logic

The rules that were described in the previous section under the heading of deterministic reasoning are known more precisely as ***rules of inference***. The conditional parts of the statement are known as propositions, and the theory of manipulating these statements is known as propositional logic. In a rule of the form

If (X) *Then* (Y)

the terms X and Y are called propositional symbols, and they have a truth value. In this section we shall assume that this means that they can be either TRUE or FALSE statements, but not both. (Later we look at ways of using more than two truth values.) For example, if X is the statement

(temperature of room is less than T_r)

then the proposition is TRUE when the temperature of the room is less than T_r and FALSE when the temperature is greater than or equal to T_r. Similarly, an action to be taken, *Y*, could be:

(radiator valve is ON)

which is TRUE if the radiator valve is turned on, or FALSE if the radiator valve is NOT turned on. At first sight this second statement does not look like an action statement. However, a system that uses *If–Then* rules can *force* the statement to be TRUE by taking action. Thus, if the rule states that when the temperature is below T_r then (radiator valve is ON) should be TRUE, the system will turn on the radiator valve to make the statement TRUE.

In order to make a distinction between the proposition and its truth value, the following notation is used:

$T(X)$ = TRUE if X is TRUE

$T(X)$ = FALSE if X is FALSE

This is needed because we want to make statements in which the truth values of two propositions can be compared. If, for example, the two propositions X and Y are both TRUE, then it is possible to say

$$T(X) = T(Y)$$

That is, the truth values of the two propositions are the same. This is quite different from saying that the two propositions themselves are the same, $X = Y$.

The propositional logic that we will be using in this chapter contains all the rules of Boolean logic, just as in the treatment of logic gates in Volume 1, Section 3.3.4. One of the first laws of this propositional logic is that of the 'excluded middle'. This says that a proposition has to be either TRUE or FALSE; it can't be anything else. A second law is that of 'contradiction', which says that a proposition cannot be both TRUE and FALSE at the same time.

The following are consequences of the Law of the Excluded Middle:

1. *If* X is TRUE, *Then* (NOT X) is FALSE. In other words, since a proposition has to be either TRUE or FALSE, if we know that it is NOT TRUE, then it must by definition be FALSE, and similarly if we know that it is NOT FALSE it must be TRUE.
2. The proposition that X is TRUE AND X is FALSE must be FALSE since X cannot be both TRUE and FALSE at the same time.
3. The proposition X is TRUE OR X is FALSE must be TRUE since X has to be either TRUE or FALSE, it cannot be anything else.

Clearly then, we can combine propositional symbols using Boolean operators such as AND, OR and NOT since the truth values of the propositions are logical. In propositional logic, AND, OR and NOT are known as ***connectives***. The symbols that are used for these connectives are

$X \wedge Y$ means X AND Y

$X \vee Y$ means X OR Y

$\neg X$ means NOT X

There are alternative notations for the logical operators. Other notations are in use for the negation operator, NOT, such as $\sim X$ and $-X$; students of electronics may also be familiar with the bar notation for negation, $\overline{X}$.

These connectives can generate new propositions from the combinations of individual propositions. Rules can contain combinations of propositions such as:

If ((temperature of room less than T_r) $\wedge$ (timer is ON))

Then (radiator valve is ON)

Just as the individual statements, X and Y, can be TRUE or FALSE, so a combination such as $(X \wedge Y)$ can be TRUE or FALSE. For example, if $T(X)$ is TRUE and $T(Y)$ is FALSE then $T(X \wedge Y)$ is FALSE.

The connectives AND, OR and NOT are summarized in Table 6.1, using both notations.

TABLE 6.1 SUMMARY OF THE CONNECTIVES $X \wedge Y$, $X \vee Y$ AND $\neg X$

X	Y	$X \wedge Y$	$X \vee Y$	$\neg X$	$X \rightarrow Y$
		X AND Y	X OR Y	NOT X	X implies Y
FALSE	FALSE	FALSE	FALSE	TRUE	TRUE
FALSE	TRUE	FALSE	TRUE	TRUE	TRUE
TRUE	FALSE	FALSE	TRUE	FALSE	FALSE
TRUE	TRUE	TRUE	TRUE	FALSE	TRUE

In Volume 1, Chapter 3 of this book, Boolean logic was introduced in terms of logic gates, and a statement was made that all logical functions can be implemented using only NAND gates. One of the reasons that this is true is that the OR connective can be replaced by a combination of the AND connective and the NOT connective. The relationship between the two is summarized by DeMorgan's Laws:

DeMorgan's Laws

If there are two propositions, X and Y, then

$$X \vee Y = \neg(\neg X \wedge \neg Y) \qquad X \text{ OR } Y = \text{NOT(NOT } X \text{ AND NOT } Y)$$

and similarly

$$X \wedge Y = \neg(\neg X \vee \neg Y) \qquad X \text{ AND } Y = \text{NOT(NOT } X \text{ OR NOT } Y)$$

This rule is sometimes helpful in manipulating logical expressions to get them into a form that is easy to interpret.

There is one other operation shown in Table 6.1 that plays an important role in propositional logic, namely implication. This has the symbol:

$$X \rightarrow Y \qquad X \text{ implies } Y \quad (\text{equivalent to } \neg X \vee Y)$$

This operation needs some explanation. You can see that the 'X implies Y' entry in Table 6.1 is TRUE for all cases except one, where X is TRUE and Y is FALSE. From this we can deduce that implication is equivalent to the expression $\neg X \vee Y$, but this doesn't shed any light on its usefulness. The term 'implication' often causes confusion. A lot of people expect 'X implies Y' to mean that the value of Y will be dictated by the value of X, but it does not. If it helps, imagine a small box, BOX_x, inside a larger box, BOX_y, as shown in Figure 6.1. Then let the propositional symbols, X and Y, mean:

X (object is inside BOX_x)

Y (object is inside BOX_y)

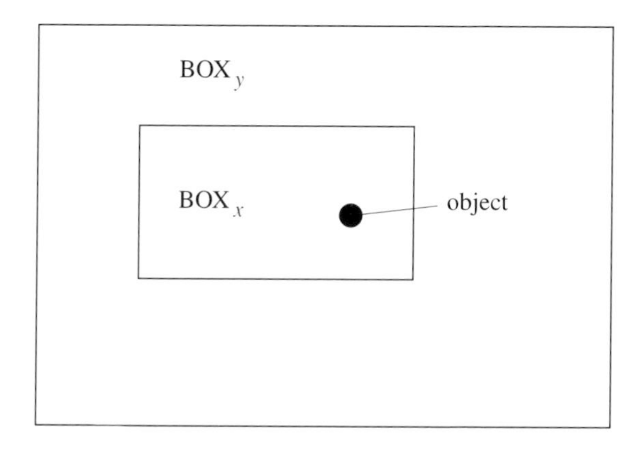

Figure 6.1
An example of implication.

We can say that $X \rightarrow Y$, because if an object is inside BOX_x then it must also be inside BOX_y. If, on the other hand, the object is not inside BOX_x, then it is valid for it to be inside BOX_y or outside; these are both situations which could be TRUE. The only situation that cannot happen is for the object to be inside BOX_x and outside BOX_y which corresponds to $T(X)$ = TRUE and $T(Y)$ = FALSE, so the combination $T(X \rightarrow Y)$ = FALSE.

For example, if X represents (lightbulb is illuminated) and Y represents (power supply connected to switch on light), then $X \rightarrow Y$. That is, if the light is illuminated then the power supply must be connected. If the light is off, the power supply may be connected or not, you can't say, as the switch controlling the light may be on or off. However, if the light is on and the power supply is *not* connected, then there is something strange going on.

Implication can appear in a rule. For example, a system which monitors the temperature of a bath might have three propositions, X, Y and Z, where

X	(hot tap turned on)
Y	(water hot)
Z	(sound alarm)

A rule which uses these propositions could be

If	$T(\neg(X \rightarrow Y))$
Then	$T(Z)$

In words, what this rule says is that if NOT(X implies Y) is TRUE – that is (X implies Y) is FALSE – sound an alarm. This means that if the hot tap is on and the water is not hot then there is something wrong: either the tap hasn't turned on or the heat sensor is faulty. In all other cases there is no need for an alarm. These cases are:

tap on, water hot	OK
tap off, water hot	OK
tap off, water not hot	OK

Propositional logic can take us quite a long way in reasoning. However, there are some situations where putting together all of the propositions required to test a situation is a long and tedious job. For example, if a system has to examine a hundred sensors and sound an alarm when any one of them is indicating a potentially dangerous situation, then the rule would have to be quite long:

If ((sensor 1 is ON) ∨ (sensor 2 is ON) ∨ … ∨ (sensor 100 is ON))

Then (sound alarm)

As you can see, this is a fairly simple construction, but it would be more convenient if there was a shorthand notation for this. Fortunately there is, but not in propositional logic. To get some new operators we need to move to predicate logic.

6.2.2 Predicate logic

The main differences between propositional logic and predicate logic are that in predicate logic we can use *variables*, and there are two new symbols called *quantifiers*. Firstly, let's look at variables.

In the previous section there were propositions like (water is hot) which had a truth value. This proposition just tests one object, the water, and so can't be used to test anything else. With predicate logic, the proposition is split into a subject (called the ***argument***) and a ***predicate*** which is a single verb phrase. In this example the argument is 'water' and the predicate is 'is hot'.

Now the predicate is independent of the argument, so it can be applied to any other argument such as 'the weather is hot' or 'ice is hot'. A shorthand way of writing this is to state the predicate and put the arguments in brackets after it. The predicate 'is hot' can therefore be written as

hot(x)

where the argument, x, could be any object, and the quality of 'hotness' is attached to it. This predicate can be used as a proposition if the object is replaced by a specific item such as water:

hot(water) TRUE if water is hot, FALSE if water is not hot

hot(ice) FALSE by most definitions

The predicate, hot(x), does not have a truth value, since x could be anything. It is only when a specific item is substituted for x that a truth value can be assigned. Even then, a logical value can be assigned only if what is substituted can be meaningfully described as hot. Formally what we mean is that the predicate 'hot' is defined for a specific *domain*, which in this case contains all objects that have a measurable temperature.

Predicates can have more than one argument. For example, the predicate 'bigger' could have two arguments, bigger(x,y), which is TRUE if $x>y$ and FALSE otherwise. For example, bigger(3,2) is TRUE whereas bigger(2,3) is FALSE. Similarly, equal(x,y) is TRUE if $x=y$ and FALSE otherwise.

The second aspect of predicate logic that makes it different from propositional logic is that there are two additional symbols. These are:

$\forall\, x\, P(x)$	means	'for all x, $P(x)$ is TRUE'
$\exists\, x\, P(x)$	means	'there exists x such that $P(x)$ is TRUE'

where x is an object and $P(x)$ is a predicate of x.

The upside-down A (for All) symbol, $\forall$, is called the *universal quantifier*, and the back-to-front E (for Exists) symbol, $\exists$, is called the *existential quantifier*. The universal quantifier has to be used with care since it is necessary to define the 'universe', or set, or domain, to which x belongs. For example, the quantified predicate

$\forall\, x,\ 1/x < x$

is false, for example, for values of x between 0 and 1. However,

$\forall\, x$, x is a number greater than 1, $1/x < x$

is true. In this case the universe is the set of numbers greater than one, and for this universe the quantified predicate is true. To see why it is necessary to specify the universe, consider the following

$\forall\, x$, x is an egg, $1/x < x$

is meaningless because the string of symbols 1/egg has no meaning. Even when the string is meaningful the universe must be carefully stated. For example,

$\forall\, x$, x is a number, $1/x < x$

is a meaningful predicate, albeit a predicate that is false. Just one exception for which a universally quantified predicate is false makes the whole thing false. Here $x = 0.1$ will do.

It is necessary to specify the universe for the existential quantifier in order to ensure that the predicate is meaningful. For example, it is certainly true that

$\exists\, x,\ 1/x < x$

since, for example, x can simply be set equal to 3. However it is not meaningful to assert that

$\exists\, x$, x is an egg, $1/x < x$

In mathematics quantifiers are essential for manipulating propositions about the members of infinite sets. In areas such as mechatronics they can be useful in presenting information about finite sets in summary forms. For example, recall the rule that sounds an alarm if any of the sensors turn on. This can now be expressed more succinctly as follows:

If ($\exists\, x$ on(sensor x))
Then on(alarm)

The predicate on(sensor x) is TRUE if sensor x is ON, and on(alarm) is TRUE if the alarm is sounding. The rule therefore says that if there exists a value of x such that the predicate on(sensor x) is TRUE then sound the alarm.

In computer implementations, the domains of quantified predicates usually have to be declared or constructed. For example, the rule which involved the predicate on (sensor x) would have been declared in such a way that the computer knew that x had to be a positive integer.

6.2.3 Rules of inference

Having established that some propositions are TRUE, it is often desirable to be able to derive further propositions and establish their truth values. This process is called ***inference*** and is based around the implication operator described earlier. There are many rules of inference – 16 in all – but two of the most frequently used rules of inference are known as ***modus ponens*** and ***modus tollens***.

Modus ponens

Assume:	$X \rightarrow Y$
and:	X
Then:	Y

This says that if we assume that $X \rightarrow Y$ is TRUE, and X is TRUE, then it follows that Y must be TRUE. Care has to be taken because nothing is said about Y if $X \rightarrow Y$ is FALSE. For example, suppose we know that if a cooling system on a car fails, X, the engine will get hot, Y. We therefore know that $X \rightarrow Y$ is TRUE. Then suppose that the cooling system fails (X is TRUE), it follows that the engine is hot (Y is TRUE). Alternatively, if we said that when the ashtray in a car is full, X, the engine gets hot, Y. In this case $X \rightarrow Y$ is FALSE. So if the ashtray is full (X is TRUE) we cannot say whether the engine is hot or not.

Modus tollens

Assume:	$X \rightarrow Y$
and:	$\neg Y$
Then:	$\neg X$

First we assume that $X \rightarrow Y$ is TRUE. Then if $\neg Y$ is TRUE (or Y is FALSE) it follows that $\neg X$ must be TRUE (or X is FALSE). Using the same examples as above, a failed cooling system (X is TRUE) implies a hot engine (Y is TRUE). If the engine isn't hot (Y is FALSE), it follows that the cooling system hasn't failed (X is FALSE). Similarly, for the example where the implication is FALSE, if the engine isn't hot (Y is FALSE) we can't say whether the ashtray is full or not.

From these simple examples you might believe that it is blindingly obvious when something implies something else. However, it is not difficult to think of historical examples where inference has been applied with the result that new discoveries are made. For example, it would be true to say that if the world is flat (X) you could fall off the edge of the world (Y), so the implication of the statement is TRUE, ($X \rightarrow Y$). People once believed that the Earth was flat (X is TRUE), so by applying *modus ponens* they believed that if you went to the edge of the Earth you would fall off (Y is TRUE). When people found that they didn't fall off the edge of the Earth (Y is FALSE) it followed that the Earth couldn't be flat (X is FALSE), even though the implication ($X \rightarrow Y$) is still TRUE.

We use rules of inference in mechatronic systems to deduce new information from existing propositions. For example, imagine that a light sensor, X, is used by a mechatronic system to determine when it is night time, Y, and if it is night time, a light is switched on, Z. The three propositions contained in the system are:

X, sensor is ON – sensor OFF when it gets dark

Y, it is night time

Z, light is ON

The system assumes that ($\neg X \rightarrow Y$) and applies a rule of the form:

If $T(Y)$

Then $T(Z)$

Some of the knowledge contained in the system is found by direct measurement or sensing, namely the state of the sensor and the state of the light, whereas the knowledge about whether it is night time or not can only be inferred from X. We will see more of this in Chapter 7 on Rule-based systems.

6.2.4 Theorem proving

We shall look at the ideas involved in ***theorem proving***, as this is also one of the main areas of logical reasoning. Theorem proving is used extensively in mathematics, but can also find a use in machine intelligence. A formal system is required which consists of *axioms*, *rules of inference* and, of course, *theorems*. The axioms are propositions which are always TRUE for a particular system, so that these axioms, together with a set of rules of inference, define a particular formal system. Essentially, theorem proving is the process of deducing whether a theorem is TRUE or not. The method is to show that a theorem can be derived from the axioms using only the rules of inference. This may all sound very abstract, so we will illustrate it with an example.

Let's take the example of the previous section, of the sensor that detects the onset of night time again. The 'theorem' that we want to prove is that it is night. First of all, let's set out the axioms. The propositions that are relevant are:

X, sensor is ON

Y, it is night

Axiom 1: Sensor is OFF ($\neg X$)

Axiom 2: Sensor OFF ($\neg X$) implies night (Y), ($\neg X \rightarrow Y$)

For the system that we are defining, these axioms are always TRUE. Next we state the rules of inference that will be used in this system. In this example, *modus ponens* will be used:

Rule of inference: *If* ($A \rightarrow B$) is TRUE, AND A is TRUE *Then* B is TRUE

Notice that we've used some 'dummy' variables here, A and B, to make the rule general.

Finally, the theorem. The theorem that we wish to prove is that it is night.

Theorem: It is night (Y)

We prove the theorem by showing that it can be deduced from the axioms using the rule of inference. In this example we only have to do this once and the theorem is proven. By substituting $\neg X$ for A and Y for B we get:

If ($\neg X \rightarrow Y$) is TRUE, and $\neg X$ is TRUE

Then Y is TRUE

We have therefore shown that Y is TRUE, so the theorem is proven. Now this is a very simple example, and in general the logical system would be much more complex. However, the process of theorem proving would be the same. You can perhaps see how this process is applied in mathematics where new theorems get proposed, and mathematicians gather to test if these new theorems are correct. But we hope that you can also appreciate that in a mechatronic system a hypothesis such as 'If I go left I will reach my goal' can only be answered by posing the question as a theorem, and testing to see if that theorem can be proven. The structure of the formal system containing axioms and rules of inference is essential for the system to prove the theorem.

Theorem proving is an important but difficult area in artificial intelligence. Some deductions are quite complicated because the interaction of quantifiers and connectives makes testing truth values difficult. In particular, different symbols may be used for what turns out to be the same variables, and sorting this out may require the application of heuristic search techniques and a great deal of computation.

6.2.5 Non-monotonic reasoning

Before we leave reasoning with certainty, the notion of monotonicity needs to be mentioned. In classical logic, if we start with a set of assumptions and then use these assumptions to deduce some new conclusions, then strictly speaking the new conclusions are expected to hold universally. No new assumptions should be discovered that alter these conclusions.

For example, a robot operates provided that there is power to the robot and that all safety devices are in place.

X	power to robot
Y	safety devices in place
P	robot operates

$$T(P) = T(X \wedge Y)$$

If $T(X)$ is TRUE and $T(Y)$ is TRUE then $T(P)$ is TRUE and the robot is operating.

At some time later it is found that the robot is not operating despite there being power available and safety devices in place. The problem is that the robot has seized up due to lack of lubricant. This changes the situation because we now have a third condition that needs to be examined before the robot operates:

Z	adequate lubricant

$$T(P) = T(X \wedge Y \wedge Z)$$

So it is now possible for $T(P)$ to be FALSE even though $T(X)$ is TRUE and $T(Y)$ is TRUE.

In classical logic it should be impossible to find a new proposition Z that would alter the original conclusions. This is because it is assumed that the system is completely understood, so any new evidence should support the original deduction. The term ***monotonic*** arises because the number of conclusions that can be drawn from a set of propositions should never decrease when new propositions are discovered. In other words, if the number of propositions is A, and the number of TRUE conclusions is C, then if new propositions are found, B, such that the number of propositions is now $A + B$, the number of TRUE conclusions should still be at least C. Mathematical functions that only ever increase (or only ever decrease) are called monotonic.

In the robot example, a new proposition causes the number of TRUE conclusions to decrease because a conclusion that was originally TRUE is now FALSE. So if we want to be able to handle conclusions that change we have to resort to non-monotonic logic.

This is all a very long-winded way of saying that we allow conclusions to change. In a system that uses logic, this would mean constantly checking to see if the conclusions that are currently held to be TRUE are still TRUE given that the propositions might have changed. So the propositions themselves are found under conditions of certainty, but they can still change as new evidence is found. This introduces an element of uncertainty. In the next sections, we discuss ways of dealing with uncertainty in the propositions.

6.3 Reasoning with uncertainty

The previous section showed how a machine could be made to follow a set of rules using *If–Then* decisions. These rely on a condition being TRUE or FALSE, which in turn rely on the data (arriving from a sensor say) being available and accurate. What happens if the data are unreliable, either because the signal is very noisy or because there are gaps in the data? We would still like to be able to (and sometimes have to) make decisions. In this section we look at a number of ways of doing this.

6.3.1 3-valued logic

Up until now we have assumed that a proposition must have a truth value which is either TRUE or FALSE. In some cases data may be missing and we find ourselves in a position where a decision still has to be made, even though there are gaps in the data. In ***3-valued logic*** a third value is allowed which is 'UNKNOWN'. The purpose of this is that it may sometimes be possible to infer some value of a compound proposition even though some of its elements are unknown. The alternative is to just give up and say that nothing can be done.

As an example, suppose we have a proposition, P, that consists of $(X \vee Y)$.

$$T(P) = T(X \vee Y)$$

Now suppose that the value of X, $T(X)$ = TRUE, but we don't know the value of Y; that is, $T(Y)$ = UNKNOWN. What is $T(P)$?

Table 6.2 is the truth table for $X \vee Y$.

TABLE 6.2 SUMMARY OF THE TRUTH VALUES OF THE CONNECTIVE $X \vee Y$

X	Y	$X \vee Y$
FALSE	FALSE	FALSE
FALSE	TRUE	TRUE
TRUE	FALSE	TRUE
TRUE	TRUE	TRUE

From this we can see that if $T(X)$ is TRUE, irrespective of the value of $T(Y)$, the value of $T(X \vee Y)$ is TRUE. So in this case we can find the truth value of P even if one of its elements were UNKNOWN. We can't always do this, but it's an improvement on never being able to determine the truth value of a compound proposition just because one of its elements is UNKNOWN.

As was the case with Boolean logic, a 3-valued logic with the truth value UNKNOWN can be defined by showing how it deals with the connectives AND, OR and NOT. This is shown in Table 6.3, where TRUE, FALSE and UNKNOWN have been abbreviated to T, F and U respectively.

TABLE 6.3 SUMMARY OF THE CONNECTIVES IN 3-VALUED LOGIC

X	Y	$X \wedge Y$	$X \vee Y$	$\neg X$	$X \rightarrow Y$
F	F	F	F	T	T
F	T	F	T	T	T
F	U	F	U	T	T
T	F	F	T	F	F
T	T	T	T	F	T
T	U	U	T	F	U
U	F	F	U	U	U
U	T	U	T	U	T
U	U	U	U	U	U

As before, the table includes implication, where we can see that implication can be true even when one of the variables is UNKNOWN in some cases.

We could go further than this and have any-valued logic – for example, 4-valued logic. All we have to do is define what happens with each of the connectives.

6.3.2 Probability theory

Most people have some intuitive idea of probability. If you were asked what are the chances of some event happening you could make a guess based on some notion of how often you think that event happens. Will it snow tomorrow? If it is summer, then your answer is 'not very likely', but if it is winter your answer could be 'yes, I think it probably will'. There is no way that you could say with complete certainty whether it will snow tomorrow, so you have to guess, but the guess is not completely wild – it will be an 'intelligent guess'. You would use some historical data, like 'it's never snowed at this time of year before', and some general knowledge like 'it doesn't snow in summer because it's too hot'. Then you would use some local or current knowledge, such as 'it has been snowing for the last two days and it doesn't look like easing up', or 'there's a low pressure front coming in from the north'. So, you would gather information from a variety of sources and combine them to make your educated guess.

All of this reasoning is done in words. Probability theory allows you to do the same but using numbers derived from statistical theory. Numerical representations are often easier to manipulate in a machine than natural language.

It has to be emphasized that probabilities are abstract quantities. The probability of an event occurring is expressed as a number between 0 and 1. A probability of 0 means that the event will never happen, and a probability of 1 means that the event will certainly happen. A value in between, 0.7 say, indicates that the number of times that the event will take place is expected to be 70% in a large number of trials. For example, when tossing a coin the probability of the coin coming down heads is 0.5, which means that we expect the number of times that the coin lands with heads up to be close to 50% in a large number of trials.

Given a set of data, the ***relative frequency*** of an event can be measured. The frequency is the number of times that an event occurs, whereas the relative frequency is the number of times an event occurs, divided by the total number of trials. If b is the number of trials (times that x *could* occur) and a is the number of times that the event actually *did* occur, the relative frequency is then:

$$\text{relative frequency, } f(x) = \frac{a}{b}$$

The relative frequency is an empirical measurement. The probability is an abstract notion of how often an event will occur in a large number of trials. In the absence of any other knowledge, the relative frequency can be used as the current estimate of the probability. It is usually assumed that as the number of events increases, the more accurate the relative frequency becomes as a measure of the probability.

Suppose that you toss a coin 10 times and note that the number of times that it was heads is 7. The relative frequency is

$$f(\text{heads}) = \frac{7}{10} = 0.7$$

We can use this value to hypothesize that the probability of heads, $p(\text{heads}) = 0.7$.

Now suppose you toss the coin another 90 times, making a hundred in total, and note that the number of additional times that it was heads was 48, making a total of 55. The relative frequency is

$$f(\text{heads}) = \frac{55}{100} = 0.55$$

We would now adjust the estimate of the probability to $p(\text{heads}) = 0.55$.

In theory, as the number of tosses of an unbiased coin increases, the probability approaches 0.5. How do we know this? The answer is that there are many situations in which the probabilities can be calculated theoretically. In the case of tossing a coin, there are two possible outcomes which, as far as we know, are equally likely. The theoretical probability is calculated by the number of outcomes that we are predicting, divided by the total number of possible outcomes. If we are trying to find the probability of an unbiased coin landing with heads up, then the number of outcomes that we are testing is one, and the total number of possible outcomes is two, namely heads or tails. The theoretical probability is therefore

$$p(\text{heads}) = \frac{1}{2} = 0.5$$

If we were throwing a die there would be six possible outcomes. The probability of the die showing a particular number, such as 6, is 1/6 = 0.167. The probability of the outcome being an even number is 3/6 = 0.5, because there are 3 possible numbers that could be counted, divided by the total number of outcomes which is 6.

Of course, this all assumes that the die is not 'loaded' and is therefore 'unbiased'. Although the theoretical probability of throwing, say, a 5 is 1/6, the only way to test this is to throw the die a number of times. If the long-term relative frequency is about 1/6 we can conclude that the die is unbiased. Otherwise we conclude that the die is biased and use the empirical probability (relative frequency) rather than the *a priori* theoretical probability.

So we have two ways of arriving at probabilities. The first is to actually measure the number of times an event takes place and use the relative frequency as an estimate of the probability. If the number of trials is large, the value that you end up with is relatively accurate, and applies to the specific event that is in question. The second is more general and more idealized. By making an assumption that all outcomes are equally likely, for example, it is possible to derive the theoretical probability of an event. Often it is a wise precaution to test the 'equally likely' assumption by experiment, and it is sometimes necessary to calibrate in order to allow for bias.

A consequence of our definitions of probability is that the sum of the probability of an event happening and the probability of an event NOT happening must be 1. A probability of 1 represents a certainty – that is, it is bound to happen. So what we are saying is that an event is either going to happen or it is not. Mathematically this is represented by

$$p(x) + p(\neg x) = 1$$

It therefore follows that if we know the probability of an event happening, then we also know the probability of the event not happening:

$$p(\neg x) = 1 - p(x)$$

6.3.3 Bayes' rule

Given that the probability of an event occurring is known, the next step is to know how to use this probability to determine how likely an event is of being TRUE, given some evidence which itself has a probability of being TRUE. For example, if a sensor detects an unusually high temperature in a system, what are the chances that this is due to a leak in the cooling system? One method which is often used is called Bayes' rule, named after the eighteenth century British cleric, the Rev. Thomas Bayes (1702–1761). Essentially, he managed to solve this problem by turning the question around to a simpler one which is usually easier to answer. In this example, the question is turned around so that it becomes a question of what are the chances of a leak in the cooling system causing an unusually high temperature? Since a leak in the cooling system will certainly cause a high temperature, this is an easier question to answer.

Mathematically, Bayes' rule can be expressed as

$$p(A|B) = \frac{p(B|A) \times p(A)}{p(B)}$$

where

$p(A|B)$	is the probability of A happening, given that B has happened
$p(B|A)$	is the probability of B happening, given that A has happened
$p(A)$	is the probability of A happening
$p(B)$	is the probability of B happening.

The first of these, $p(A|B)$, is what we want to find out. Before we do this, let's look at what this expression $p(A|B)$ means in general. To do this, let's go back to throwing dice again. Given two dice, A and B, what is the probability of throwing a double six?

We have already said that throwing a six is one outcome out of a possible six, so the probability of throwing a six is $1/6 = 0.167$:

$p(A) = 0.167$, where A means a six will be thrown with die A.

We want to throw two sixes, so we have two events, each with a probability of 0.167:

$p(A) = 0.167$, where A means a six will be thrown with die A

$p(B) = 0.167$, where B means a six will be thrown with die B.

With two dice there are 36 possible outcomes, shown in Table 6.4, and a double six is only one of those outcomes.

TABLE 6.4 THE 36 POSSIBLE OUTCOMES OF THROWING TWO DICE

		Die A					
		1	2	3	4	5	6
	1	1,1	1,2	1,3	1,4	1,5	1,6
	2	2,1	2,2	2,3	2,4	2,5	2,6
Die B	3	3,1	3,2	3,3	3,4	3,5	3,6
	4	4,1	4,2	4,3	4,4	4,5	4,6
	5	5,1	5,2	5,3	5,4	5,5	5,6
	6	6,1	6,2	6,3	6,4	6,5	**6,6**

So the probability of throwing a double six with two unbiased dice is 1/36 = 0.0278:

$$p(C) = 0.0278$$

where C means a six will be thrown on both dice A and B.

It turns out that

$$p(C) = p(A) \times p(B|A)$$

The second probability on the right-hand side, $p(B|A)$, is the probability of B given that A has already happened. In this example, since the probability of B is independent of A, $p(B|A)$ reduces to $p(B)$, and the probability of throwing a double six becomes

$$p(C) = p(A) \times p(B)$$

This is true of any two *independent* events. The probability of two independent events, A and B, happening is the product of the probabilities of each individual event happening.

Suppose, in the dice example, that die A is thrown and is a six. What is the probability of a double six now? Well, the probability is just the probability of throwing a six with a single die, namely 0.167. This probability is

$$p(C|A) = 0.167$$

This expression is the probability of throwing two sixes, given that one six has already been thrown.

Returning to Bayes' rule and the example of an overheating machine,

$$p(A|B) = \frac{p(B|A) \times p(A)}{p(B)}$$

To calculate $p(A|B)$ we have to know all the other probabilities, three in this case. So although we have a formula, quite often we can get stuck here because the information is simply not available. However, in our example of an overheating machine, let's assume that this particular machine has been monitored for most of its working life so that statistics are available about the number of times that there have been leaks in the cooling system, how often the temperature has been too high, and how often the high temperature has been caused by a coolant leak.

Total working life:	10000 hours
Number of hours temperature has been high:	42 hours
Number of hours that the cooling system has leaked:	32 hours

Now we can calculate some probabilities. First, the probability of there being a leak, $p(A)$. Over the 10000 hours, the cooling system has leaked for only 32 hours. So the probability of it leaking at any given time is

$$p(A) = \frac{32}{10000} = 0.0032$$

Over the 10000 hours the temperature has been high for only 42 hours, so the probability of there being a high temperature, $p(B)$, is

$$p(B) = \frac{42}{10000} = 0.0042$$

Finally, the probability of the system getting hot when there is a leak in the cooling system is 1 since this will definitely happen, so

$$p(B|A) = 1$$

Now we can calculate $p(A|B)$, which is the probability of overheating being caused by a leak in the cooling system:

$$p(A|B) = \frac{1 \times 0.0032}{0.0042} = 0.762$$

This figure could be used as an aid to making a decision. We could say that we are about 76% confident that the cooling system is the cause of the high temperature. So if, for example, we decided to replace the cooling system to cure overheating, then about 8 times out of 10 that would be the correct decision.

In the next section we will look at ways in which probabilities can be combined in a similar way to propositional logic so that evidence from a number of sources can be used to make a decision.

6.3.4 Probability and logic

The logical operations that have been used so far are AND, OR and NOT. If we had a Boolean expression containing propositions combined with some of these logical operations, and each proposition had a probability associated with it, we would like to know the probability of the entire expression.

The NOT operation has already been described. If an event, X, has a probability of $p(X)$, then the probability of the event not happening is $1-p(X)$. So,

$$p(\neg X) = 1 - p(X)$$

We have also seen the AND operation when we were looking at the probability of two events, X and Y, occurring. The probability is

$$p(X \wedge Y) = p(X) \times p(Y|X)$$

When these two events are independent, this reduces to

$$p(X \wedge Y) = p(X) \times p(Y)$$

This just leaves the OR operation. This turns out to be

$$p(X \vee Y) = p(X) + p(Y) - p(X \wedge Y)$$

The probability of event X OR event Y taking place is the sum of the two probabilities, together with a compensating factor which takes into account the possibility that the two events may occur together. This is needed because $p(X)$ is the probability that event X will occur which can be split into two parts: the probability that X will occur when Y is not happening, plus the probability that event X will occur when Y is happening. Similarly, $p(Y)$ can be split into the probability that Y will occur when X is not happening plus the probability that Y will occur when X is happening. When these two probabilities are added, $p(X)+p(Y)$, the probability of X and Y happening together has been double counted, so the amount $p(X \wedge Y)$ has to be subtracted to redress the balance.

To summarize, the three logical operators can be replaced by arithmetical operations when probabilities are used as follows:

Negation

$$p(\neg X) = 1 - p(X)$$

Conjunction

$$p(X \wedge Y) = p(X) \times p(Y|X)$$

When these two events are independent, this reduces to

$$p(X \wedge Y) = p(X) \times p(Y)$$

Disjunction

$$p(X \vee Y) = p(X) + p(Y) - p(X \wedge Y)$$

Again, when X and Y are independent, this reduces to

$$p(X \vee Y) = p(X) + p(Y) - p(X) \times p(Y)$$

As an example, let $p(X) = 0.8$ and $p(Y) = 0.9$, where X and Y are independent events. The combinations of these events are

$$p(\neg X) = 1 - p(X) = 1 - 0.8 = 0.2$$

$$p(X \wedge Y) = p(X) \times p(Y) = 0.8 \times 0.9 = 0.72$$

$$p(X \vee Y) = p(X) + p(Y) - p(X) \times p(Y) = 0.8 + 0.9 - 0.8 \times 0.9 = 0.98$$

Earlier it was shown that the AND and OR operations are related by DeMorgan's Laws. Does this still apply to probabilities? DeMorgan's Law for the OR operation was stated as:

$$X \vee Y = \neg(\neg X \wedge \neg Y)$$

Rewriting this probabilistically for independent events X and Y gives

$$\begin{aligned} p(\neg(\neg X \wedge \neg Y)) &= 1 - p(\neg X \wedge \neg Y) \\ &= 1 - p(\neg X) \times p(\neg Y) \\ &= 1 - (1 - p(X)) \times (1 - p(Y)) \\ &= 1 - (1 - p(X) - p(Y) + p(X) \times p(Y)) \\ &= p(X) + p(Y) - p(X) \times p(Y) \\ &= p(X \vee Y) \end{aligned}$$

A similar proof can be shown for the AND connective, so DeMorgan's Laws still apply to probabilities when events are independent.

This gives us a method of finding the probabilities of Boolean expressions given the individual probabilities of the propositions contained in the expressions. However, great care must be taken when applying these equations, as it is very easy to be misled. To finish this section, and before we go on to look at fuzzy reasoning, some examples will be given from the dice example again, just to show that care must be taken when applying probabilities.

Example 1

What is the probability of throwing a 5 and a 6 with two dice?

The probability of throwing a 5 with one die is 1/6, and similarly the probability of throwing a 6 with one die is 1/6. The probability of throwing a 5 AND a 6 is therefore:

$$p(5 \text{ AND } 6) = p(5) \times p(6) = 1/6 \times 1/6 = 1/36$$

However, if you look back to Table 6.4 you will see that there are two possibilities of throwing a 5 and a 6, namely 5,6 and 6,5, so the probability should have been 2/36. What's gone wrong? What has happened is that we haven't expressed the problem well enough. The question should have been interpreted as:

What is the probability of throwing either a 5 followed by a 6 OR a 6 followed by a 5?

The answer becomes:

$$p((5 \text{ AND } 6) \text{ OR } (6 \text{ AND } 5))$$

$$= p(5 \text{ AND } 6) + p(6 \text{ AND } 5) - p((5 \text{ AND } 6) \text{ AND } p(6 \text{ AND } 5))$$

$$= p(5) \times p(6) + p(6) \times p(5) - 0$$

$$= 1/6 \times 1/6 + 1/6 \times 1/6$$

$$= 2/36$$

The term

$$p((5 \text{ AND } 6) \text{ AND } (6 \text{ AND } 5))$$

is zero because you cannot throw both a 5 followed by a 6 AND a 6 followed by a 5 with the same two dice.

Example 2

What is the probability of throwing a 6 with two dice?

This time we'll go straight for the right answer. The question has to be interpreted as:

What is the probability of throwing a 6 with the first die OR throwing a 6 with the second?

$$p(6 \text{ OR } 6) = p(6) + p(6) - p(6) \times p(6) = 1/6 + 1/6 - 1/36 = 11/36$$

Check with Table 6.4 to see if there are actually 11 possibilities of throwing a 6.

Example 3

What is the probability of not throwing a 6 with two dice?

It is tempting to apply the OR operation to this problem as in Example 2, but that would give the wrong answer (35/36). This question should be interpreted as:

What is the probability of NOT throwing a 6 with the first die AND NOT throwing a 6 with the second?

$$\begin{aligned} p(\text{NOT } 6 \text{ AND NOT } 6) &= p(\text{NOT } 6) \times p(\text{NOT } 6) \\ &= 5/6 \times 5/6 \\ &= 25/36 \end{aligned}$$

Check with Table 6.4 to see if there are actually 25 possible ways of throwing two dice and not getting a 6.

6.3.5 Possibility and fuzzy reasoning

Possibilistic logic was first proposed by Zadeh in 1965, and has become more popular under the name of ***fuzzy logic***. The ideas involved in fuzzy logic allow us to combine in a 'logical' way some weighting factors associated with propositions from different sources. In fuzzy logic the truth value can vary between zero and one, i.e. $0 \leqslant T(X) \leqslant 1$. Earlier we saw that propositional logic uses the logical operators AND, OR and NOT to combine logical inputs. In fuzzy logic we have equivalent operations, namely MIN, MAX and $(1 - T(X))$ as follows:

TABLE 6.5 FUZZY LOGIC EQUIVALENTS TO THE BOOLEAN CONNECTIVES

Boolean	Fuzzy
$T(X \wedge Y)$	MIN$(T(X), T(Y))$
$T(X \vee Y)$	MAX$(T(X), T(Y))$
$T(\neg X)$	$(1 - T(X))$
$T(X \rightarrow Y)$	MAX$((1 - T(X)), T(Y))$

Fuzzy logic is consistent with Boolean logic; for example, the fuzzy values $T(X) = 0$ and $T(X) = 1$ work in exactly the same way as Boolean variables in this table.

Remember that X and Y are the propositions, and $T(X)$ and $T(Y)$ are their truth values respectively. Any Boolean logic expression can now be converted to a fuzzy logic expression by substituting MIN, MAX and $(1 - T(X))$ for AND, OR and NOT, respectively. For example, the last entry, X implies Y, shows how a complicated logical expression can be made using fuzzy logic. Earlier you were shown that $X \rightarrow Y$ is equivalent to $\neg X \vee Y$. In fuzzy logic, the truth value of $\neg X$ is replaced by $1 - T(X)$, and similarly $T(X \vee Y)$ is replaced by $\text{MAX}(T(X), T(Y))$. So $\neg X \vee Y$ is replaced by $\text{MAX}((1 - T(X)), T(Y))$. There are other definitions of fuzzy implication, but this one uses the substitutions that we have defined.

If we assign 0 to FALSE and 1 to TRUE as is usually done, then we can show that the operations are equivalent. This is done in Tables 6.6 and 6.7.

TABLE 6.6 BOOLEAN CONNECTIVES

X	Y	$X \wedge Y$	$X \vee Y$	$\neg X$	$X \rightarrow Y$
0	0	0	0	1	1
0	1	0	1	1	1
1	0	0	1	0	0
1	1	1	1	0	1

TABLE 6.7 FUZZY CONNECTIVES

X	Y	$\text{MIN}(X, Y)$	$\text{MAX}(X, Y)$	$(1 - X)$	$\text{MAX}((1 - X), Y)$
0	0	0	0	1	1
0	1	0	1	1	1
1	0	0	1	0	0
1	1	1	1	0	1

So when the truth values are 0 and 1 there is no difference between Boolean logic and fuzzy logic. Fuzzy logic, however, can generalize so that the truth values can be any number between 0 and 1, as we shall see in the next section.

To conclude this section we shall check to see if one of the rules of Boolean logic still applies to fuzzy logic. Earlier you were introduced to DeMorgan's Laws, which related the OR and AND operators. The first of these was stated as

$$X \vee Y = \neg(\neg X \wedge \neg Y) \qquad X \text{ OR } Y = \text{NOT(NOT } X \text{ AND NOT } Y)$$

Now let's try this with the fuzzy operators MAX, MIN and $1 - T(X)$. Suppose $T(X) \geqslant T(Y)$, so that $1 - T(X) \leqslant 1 - T(Y)$. Then

$$\begin{aligned}\text{MAX}(T(X), T(Y)) &= T(X) \\ &= 1 - (1 - T(X)) \\ &= 1 - \text{MIN}((1 - T(X)), (1 - T(Y)))\end{aligned}$$

which is analogous to $X \vee Y = \neg(\neg X \wedge \neg Y)$. This derivation works equally well if we start with $T(Y) \geqslant T(X)$ and $1-T(Y) \leqslant 1-T(X)$; you can try this for yourself.

The second of DeMorgan's Laws was stated as

$X \wedge Y = \neg(\neg X \vee \neg Y)$ $\qquad$ X AND Y = NOT(NOT X OR NOT Y)

Suppose $T(X) \geqslant T(Y)$, so that $1-T(X) \leqslant 1-T(Y)$. Then

$$\begin{aligned} \text{MIN}(T(X), T(Y)) &= T(Y) \\ &= 1 - (1-T(Y)) \\ &= 1 - \text{MAX}((1-T(X)), (1-T(Y))) \end{aligned}$$

which is analogous to $X \wedge Y = \neg(\neg X \vee \neg Y)$. This derivation also works if we start with $T(Y) \geqslant T(X)$.

So DeMorgan's Laws still apply in fuzzy logic, even when the values of $T(X)$ and $T(Y)$ are between 0 and 1.

6.3.6 Fuzzy sets and membership functions

Logic and set theory are closely related by the concept of the *characteristic function*, otherwise called the ***membership function***. In conventional set theory the membership function of the set A is denoted by χ_A (using the Greek letter 'chi') and has the property

$\chi_A(x) = 1$ if x is a member of A

$\chi_A(x) = 0$ if x is not a member of A

This can be rewritten as

$\chi_A(x) = 1$ if (x is a member of A) is TRUE

$\chi_A(x) = 0$ if (x is a member of A) is FALSE

This concept can be extended to let the value of χ_A lie *between* zero and one. In this way one can represent uncertainty about set membership. For example, will tomorrow belong to the set of 'sunny days'? If the available information suggests that it is more likely to be sunny tomorrow than not, we might estimate that $\chi_{\text{sunny days}}(\text{tomorrow}) = 0.7$.

A ***fuzzy set*** is a set whose membership function takes values between zero and one. There is a close relationship between fuzzy sets and fuzzy logic, as you will see. Fuzzy set membership is probably easier to explain by looking at an example.

The terms 'cold', 'warm' and 'hot' are all descriptions related to temperature. Figure 6.2 shows how you might set some arbitrary thresholds T_1 and T_2 so that as you start from a low temperature and move up the scale you cross the thresholds. Starting from COLD, you cross the first threshold, T_1, and now you are WARM.

Continuing up the scale you cross the second threshold and become HOT.

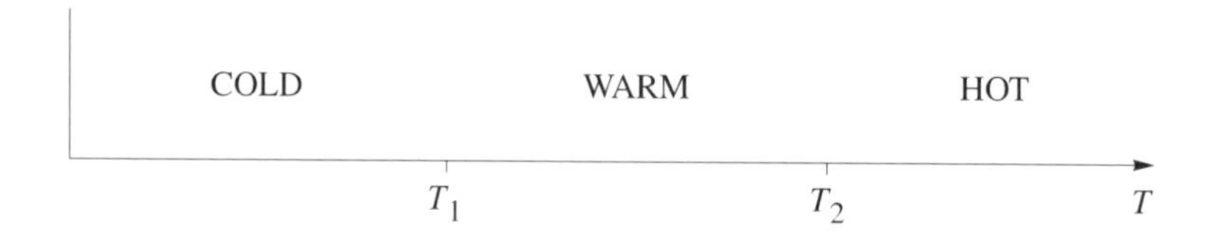

Figure 6.2 Temperature scale with cri[tical] thresholds.

At any temperature, T, you could say that you belong to a particular region, either COLD, WARM or HOT.

$T \leqslant T_1$	COLD
$T_1 < T \leqslant T_2$	WARM
$T_2 < T$	HOT

We can think of these regions as sets of temperatures, and ascribe a membership to these sets. Any given temperature T would be a member of only one set. Membership of a set is given the value 1 and non-membership of a set is given the value 0, as shown in Table 6.8.

TABLE 6.8 BINARY SET MEMBERSHIPS FOR THE SETS COLD, WARM AND HOT

Temperature	COLD	WARM	HOT
$T \leqslant T_1$	1	0	0
$T_1 < T \leqslant T_2$	0	1	0
$T_2 < T$	0	0	1

In other words, the set called COLD contains all temperatures which are less than or equal to T_1. These membership functions are shown in Figure 6.3. The membership can be used as the truth value so that, for example, if the membership of the set COLD is 1, then this is equivalent to the proposition (temperature is COLD) being TRUE. At the same time, membership of the other two sets is 0 is equivalent to the propositions (temperature is WARM) and (temperature is HOT) being FALSE.

The problem with these sets is that on the boundary between two sets there is a very 'crisp' change from membership of one set to membership of another. For example, if T_1 is 10°, say, then at a temperature of 5° it is clearly COLD. However, it is not so obvious whether a temperature of 9.999° should be interpreted as COLD or WARM: it is somehow both.

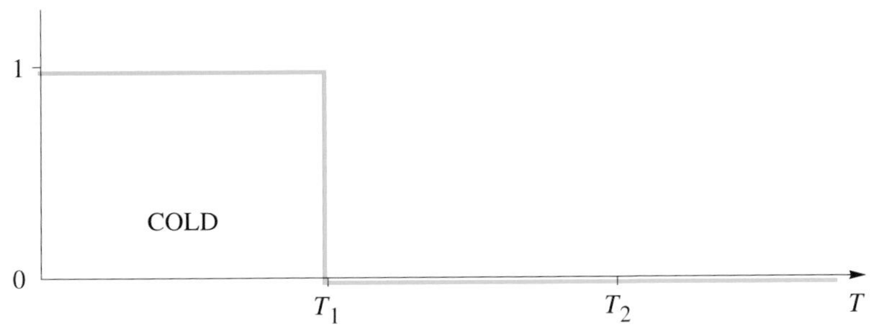

(a) Binary set membership for COLD

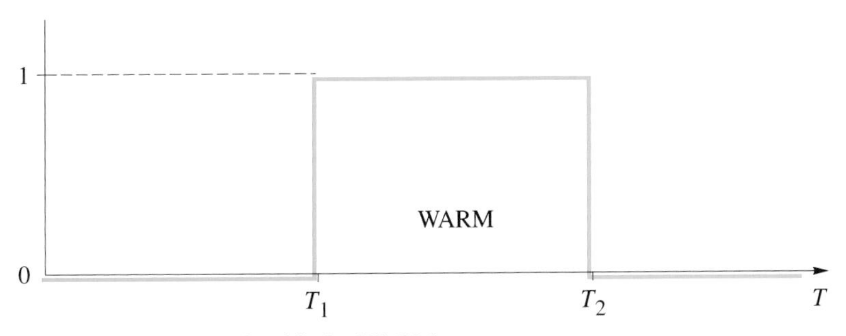

(b) Binary set membership for WARM

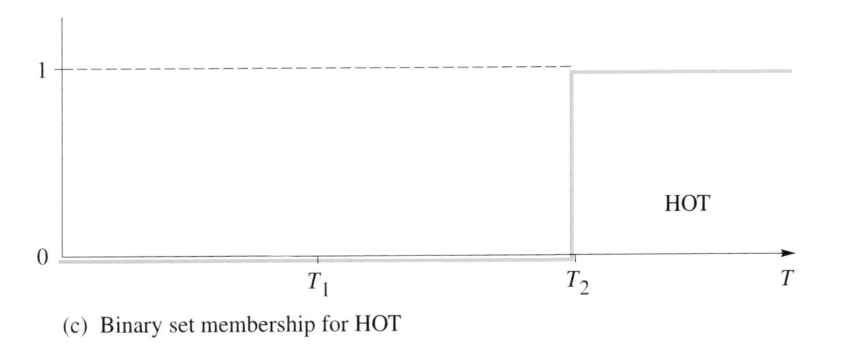

(c) Binary set membership for HOT

Figure 6.3
Membership functions for **COLD, WARM** ***and*** **HOT.**

In fuzzy logic we can change the shape of the membership functions so that the boundary is not so crisp. Figure 6.4 shows an example, using the commonly found triangular functions.

The most obvious difference is that the sets overlap, so that at some temperatures it is possible to be a member of two different sets. At the temperature of 8° shown in Figure 6.4(d), the memberships are:

COLD	0.7
WARM	0.3
HOT	0.0

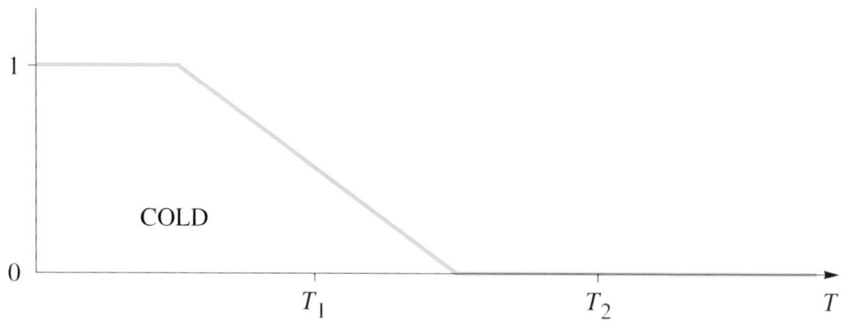

(a) Fuzzy set membership for COLD

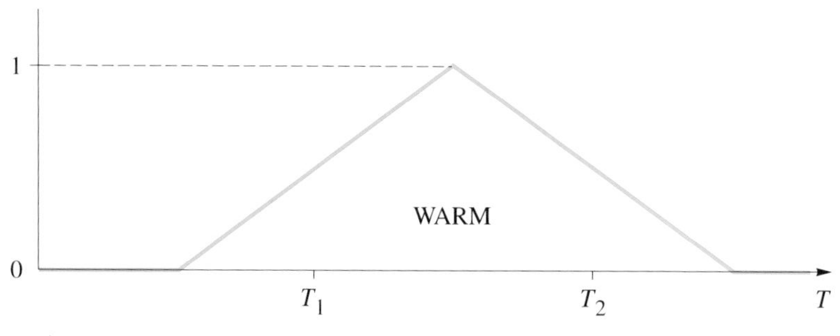

(b) Fuzzy set membership for WARM

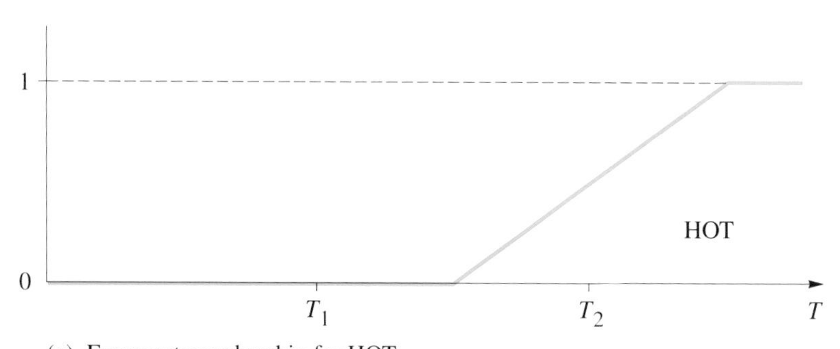

(c) Fuzzy set membership for HOT

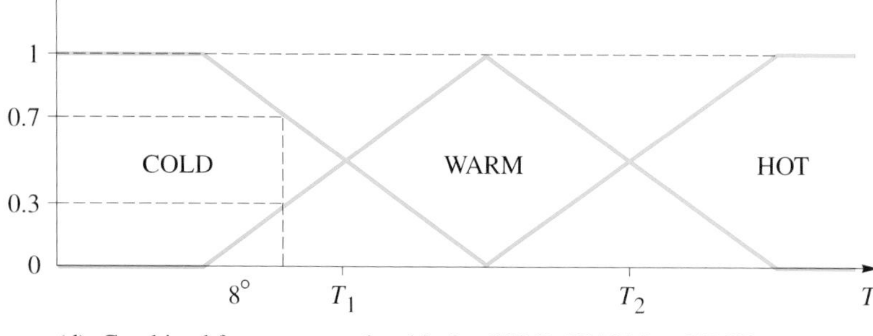

(d) Combined fuzzy set membership for COLD, WARM and HOT

Figure 6.4 Fuzzy set membership functions.

So at 8° the temperature is a member of the COLD set with a membership value of 0.7, and a member of the WARM set with a membership value of 0.3. In some ways these memberships are similar to probabilities, and have been described as the 'probability of an event being possible'. However, it is important not to read too much into fuzzy set membership values, as they may be rather arbitrary. Whereas there is a clear relationship between theoretical probability and relative frequency, fuzzy set membership values may be derived in a less rigorous and more empirical way.

These memberships can be combined or processed using the MIN, MAX and $(1 - T(X))$ operations so that logical *If–Then* rules can be applied. For example:

If (temperature is COLD)

Then (turn heating on HIGH)

If, at a particular temperature, the membership of COLD is 0.7, then the heating is turned on HIGH with a membership of 0.7. In other words, the membership value is passed on to the action part of the rule. The set HIGH could be only one of a number of options, such as OFF, LOW, MEDIUM or HIGH. This rule asks for the heating to be turned on HIGH. Other rules might ask for the heating to be turned on LOW, for example:

If (temperature is WARM)

Then (turn heating on LOW)

Membership of the set WARM could be 0.3, so now this rule wants the heating to be turned on to LOW with a membership of 0.3. Figure 6.5 shows the membership functions for the heater. Here too, the membership functions are triangular, and are spread over the operating range of the heater from 0 to 15.

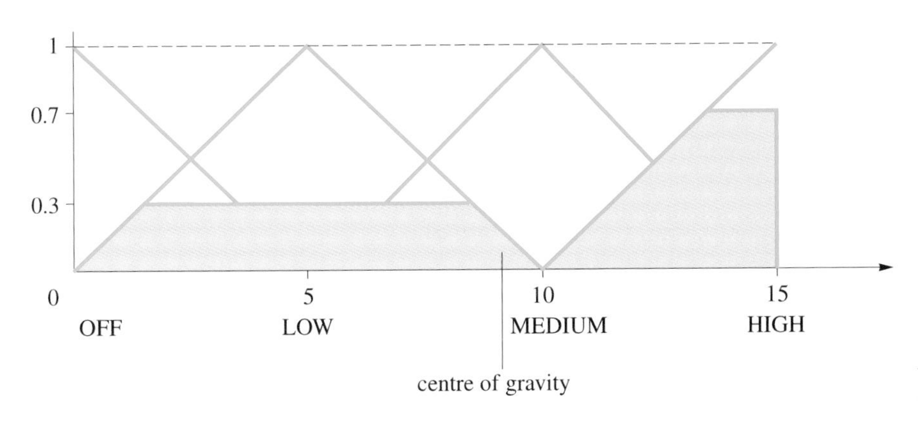

Figure 6.5 Membership functions of a heater.

The membership values that are passed to the output appear as the shaded areas on the diagram. To calculate the final setting for the heater a process called *defuzzification* is used which finds the 'centre of gravity' of the shaded area. This is shown on the diagram, and is the point at which the shaded area to the left of the point equals the shaded area to the right. This point turns out to be approximately 9, so the heater would be turned up to a setting of 9 or 60% of its full power.

6.3.7 Defuzzification

The usual method for defuzzification involves taking the centre of gravity for all the areas shaded under the fuzzy set membership curves. Suppose it is known that curve 1 has its centre of gravity at $x = x_1$, and that curve 2 has its centre of gravity at $x = x_2$. Let the areas under these curves be A_1 and A_2 respectively. Then the centre of gravity of both areas is the point c for which $(c - x_1)A_1 = (x_2 - c)A_2$, i.e. a 'weight' of A_1 at x_1 would balance a 'weight' of A_2 at x_2 at the 'fulcrum' point c. From this it follows that

$$cA_1 - x_1A_1 = x_2A_2 - cA_2$$

so that

$$c(A_1 + A_2) = x_1A_1 + x_2A_2$$

and

$$c = (x_1A_1 + x_2A_2) / (A_1 + A_2)$$

In general,

$$c = (\Sigma\, x_i\, A_i) / \Sigma\, A_i$$

Finding the centre of gravity of a general curve can be quite complicated, but it can be simplified considerably if one uses symmetric set membership functions since the centre of gravity lies on the axis of symmetry of the curve.

The centre of gravity of the areas under the curves χ_i for $i = 1, \ldots, n$ is given by the formula

$$\text{centre of gravity} = \frac{\sum_{i=1}^{n} \text{centre of gravity}_i \times \text{area under curve}_i}{\sum_{i=1}^{n} \text{area under curve}_i}$$

The area under a symmetric triangular curve has its centre of gravity under the apex, and is a trapezium which is easy to calculate, as shown in Figure 6.6.

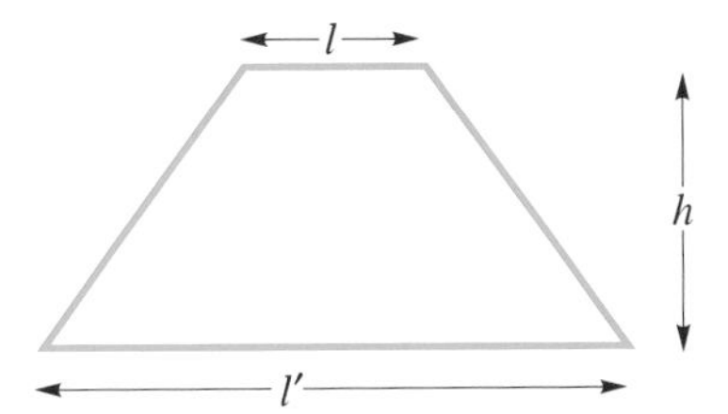

(a) A trapezium of altitude h and sides l and l'

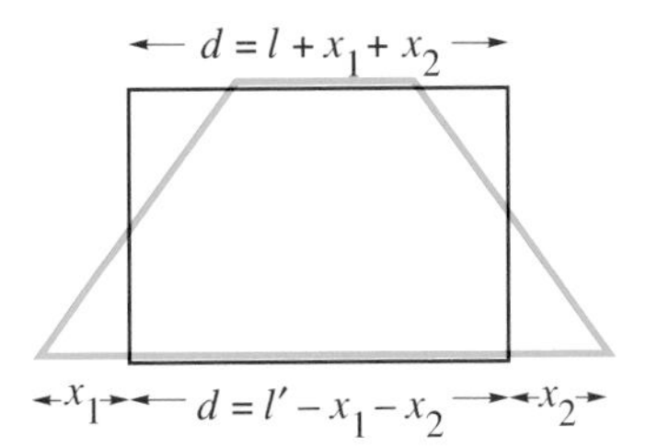

(b) $d + d = l + x_1 + x_2 + l' - x_1 - x_2$, therefore $d = \frac{1}{2}(l + l')$, therefore area $= \frac{1}{2}h(l + l')$

Figure 6.6 The area under a trapezium of altitude h and sides of length l and l′ is $\frac{1}{2}h(l + l')$

So, using triangular fuzzy set membership functions means that defuzzification can be calculated according to the formula

$$\text{area}_i = \frac{(l_i + l_i')h_i}{2}$$

If x_i is the centre of gravity of the area under curve χ_i, then

$$\text{centre of gravity} = \frac{\sum_{i=1}^{n} x_i \frac{(l_i + l_i')h_i}{2}}{\sum_{i=1}^{n} \frac{(l_i + l_i')h_i}{2}}$$

In the case of the curve in Figure 6.7 the calculation is

$$\text{area}_\text{A} = 0.75\left(\frac{0.5 + 2.0}{2}\right) = 0.9375\ , \quad x_\text{A} = 0.00$$

$$\text{area}_\text{B} = 0.25\left(\frac{1.5 + 2.0}{2}\right) = 0.4375\ , \quad x_\text{B} = 1.00$$

$$\text{centre of gravity} = \frac{(0.0 \times 0.9375) + (1.00 \times 0.4375)}{0.9375 + 0.4375} = 0.318$$

So in this case the defuzzification of $\chi_\text{A} = 0.75$ and $\chi_\text{B} = 0.25$ gives a value of $x = 0.318$.

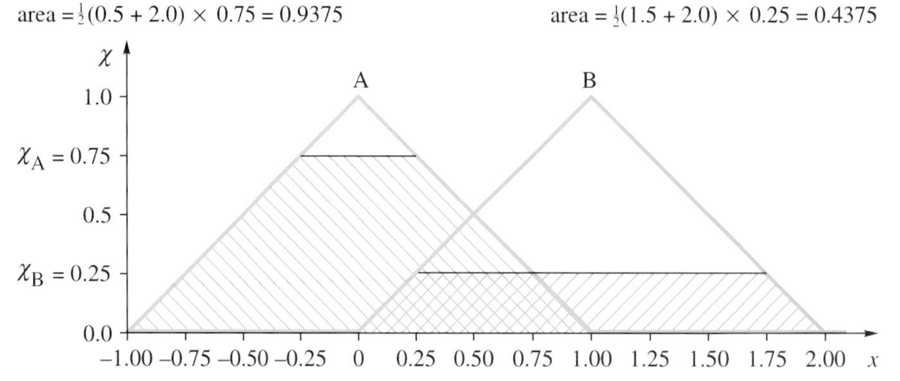

Figure 6.7 Defuzzifying the set membership values $\chi_A = 0.75$ and $\chi_B = 0.25$.

In the following example, the use of symmetric functions suggests negative salaries, which may seem odd, but it does no harm and greatly facilitates the defuzzification calculation.

To illustrate defuzzification, consider the rules

If	a person has high skills ______
and	that person has high responsibility ______
and	that person gets new business ______
Then	that person gets a high salary ______

If	a person has high skills ______
and	that person does their job well ______
Then	that person gets a medium salary ______

If	a person has low skills ______
and	that person is not experienced ______
Then	that person gets a low salary ______

Suppose the fuzzy constructs of being highly paid or lowly paid have the set membership functions shown in Figure 6.8.

Now suppose the rules are applied to Mr A who has low skill (0.9) and high skills (0.1), little responsibility (0.1), gets no new business (0.0), does his job OK (0.5) but is not a very experienced worker (0.6). Then his fuzzy values can be calculated as

If	a person has high skills ________________	(0.1)
and	that person has high responsibility ________________	(0.1)
and	that person gets new business ________________	(0.0)
Then	that person gets a high salary ________________	(0.0)

since 0.1 *and* 0.1 *and* 0.0 means $\min\{0.1, 0.1, 0.0\} = 0.0$ in this rule.

If	a person has high skills ________________	(0.1)
and	that person does their job well ________________	(0.5)
Then	that person gets a medium salary ________________	(0.1)

since 0.1 *and* 0.5 means $\min\{0.1, 0.5\} = 0.1$ in this rule.

If	a person has low skills ________________	(0.9)
and	that person is not experienced ________________	(0.6)
Then	that person gets a low salary ________________	(0.6)

since 0.9 *and* 0.6 means $\min\{0.9, 0.6\} = 0.6$ in this rule.

Mr A has an area of

$$0.6\left(\frac{40.0 + 16.0}{2}\right) = 16.8$$

under the LOW SALARY curve and

$$0.1\left(\frac{60.0 + 54.0}{2}\right) = 5.7$$

under the MEDIUM SALARY curve. His area under the HIGH SALARY curve is zero. His defuzzified salary is therefore

$$\text{salary for Mr A} = \frac{16.8 \times 0 + 5.7 \times 30}{16.8 + 5.7} = \frac{171.0}{22.5} = £7600 \text{ p.a.}$$

So Mr A earns a crisp salary of £7600 per annum.

Now consider Ms B who has good skills (0.8), has some responsibility (0.6), gets some new business (0.6), does her job well (0.8), and is quite experienced (0.8).

If	a person has high skills ______	(0.8)
and	that person has high responsibility ______	(0.6)
and	that person gets new business ______	(0.6)
Then	that person gets a high salary ______	(0.6)

since 0.8 *and* 0.6 *and* 0.6 means $\min\{0.8, 0.6, 0.6\} = 0.6$ in this rule.

If	a person has high skills ______	(0.8)
and	that person does their job well ______	(0.8)
Then	that person gets a medium salary ______	(0.8)

since 0.8 *and* 0.8 means $\min\{0.8, 0.8\} = 0.8$ in this rule.

If	a person has low skills ______	(0.2)
and	that person is not experienced ______	(0.2)
Then	that person gets a low salary ______	(0.2)

since 0.2 *and* 0.2 means $\min\{0.2, 0.2\} = 0.2$ in this rule.

Ms B has an area of

$$0.2\left(\frac{32.0 + 40.0}{2}\right) = 7.2$$

under the LOW SALARY curve, an area of

$$0.8\left(\frac{12.0 + 60.0}{2}\right) = 28.8$$

under the MEDIUM SALARY curve, and an area of

$$0.6\left(\frac{48.0 + 120.0}{2}\right) = 50.4$$

under the HIGH SALARY curve. Therefore:

$$\text{salary for Ms B} = \frac{7.2 \times 0 + 28.8 \times 30 + 50.4 \times 100}{7.2 + 28.8 + 50.4} = \frac{5904}{86.4} = \text{£}68{,}333 \text{ p.a.}$$

So Ms B gets a salary of £68,333 per annum. You may feel that these fuzzy rules do not represent this situation very well. If so, try changing them to see if you can get a fairer set of rules for financial rewards, including fuzzy characteristics that you might think important if you were an employer.

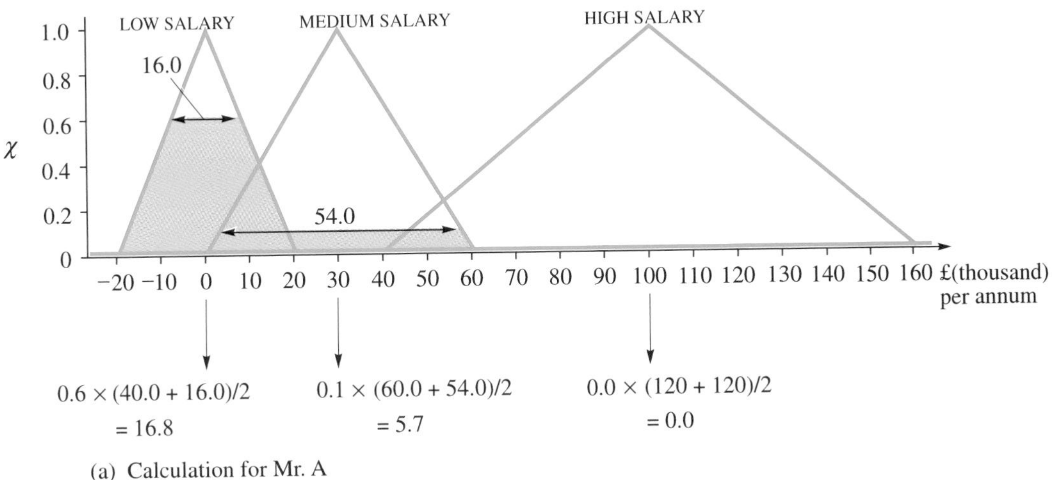

(a) Calculation for Mr. A

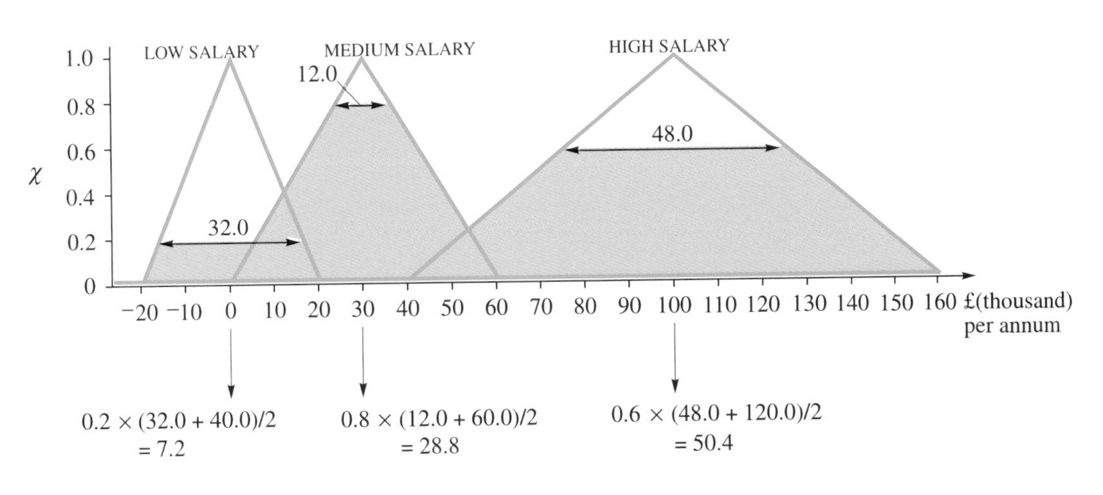

(b) Calculation for Ms. B

***Figure 6.8** Defuzzifying peoples' salaries: calculating areas under fuzzy curves.*

6.3.8 Paradoxes in applying fuzzy sets

One of the useful features of fuzzy sets is that they can be built up on the basis of minimal information, and then adjusted until they are more consistent with observation. However, we should not be careless, as the following examples show.

Consider the fuzzy set membership function shown in Figure 6.9. When $x = 0.0$, $\chi = 1$ and the area under the curve is the whole triangle, the centre of gravity for this triangle is approximately at the point $x = 0.29$. When x is just below 1.0, χ is just above 0.0 and the area is approximated by a thin bar between $x = 0.0$ and $x = 1.0$ (ignoring the small triangular piece which should be removed from the right of the bar). Therefore the centre of gravity is approximately at the centre of the bar, and in the limit the centre of gravity corresponding to $x = 1.0$ is at the point $x = 0.5$. Therefore, whatever the original value of χ is (between 0.0 and 1.0), the defuzzified value of χ always lies between 0.29 and 0.50.

Of course, if the fuzzy values are defined by a symmetric curve about the point x_c, then for all values of x, χ will defuzzify to the constant value x_c.

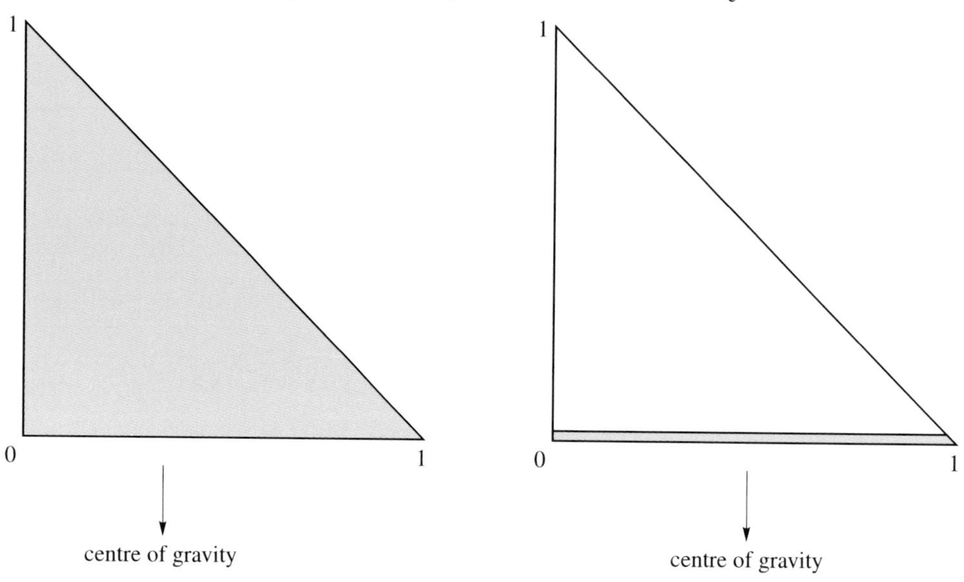

Figure 6.9

A more perplexing property of fuzzy sets is exhibited by what we call *Hopgood's Paradox*. This is illustrated in Figure 6.10. Suppose there is little information available about this system, and due to calculations elsewhere in the system the fuzzy values are:

TINY	0.1
MEDIUM	0.0
HUGE	0.1

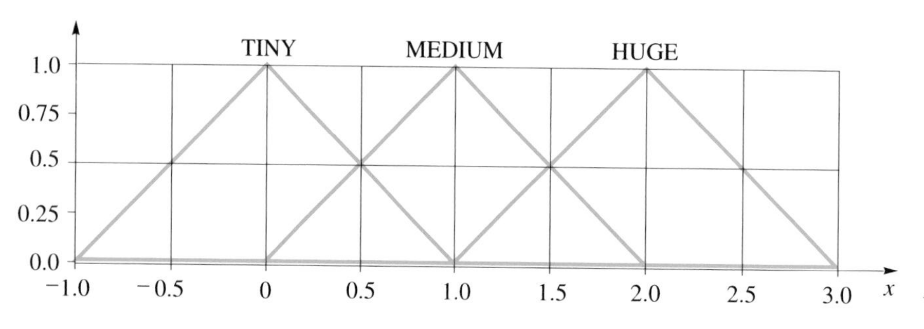

Figure 6.10 Fuzzy sets which lead to Hopgood's defuzzification paradox.

If these values are defuzzified, the result is $x = 1.00$, and so the new fuzzy values become

TINY	0.0
MEDIUM	1.0
HUGE	0.0

In this case both TINY and HUGE started with small values of 0.1 and finished up with values of 0.0, but MEDIUM started out with a small fuzzy value of 0.0 and finished with 1.0.

The paradox here is that apparently 'weak' information can result in apparently 'strong' information on defuzzification.

6.3.9 Defuzzification is not the inverse of fuzzification

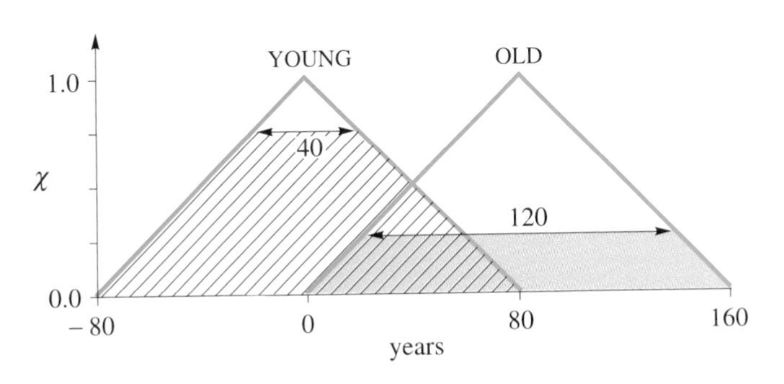

Figure 6.11 *Membership set for young and old.*

Figure 6.11 shows the fuzzy set membership functions for YOUNG and OLD in years. For any value of age there are fuzzy values of YOUNG and OLD; for example, if age = 20 years those values are 0.75 and 0.25, respectively. These two values can be used to recover a defuzzified value of age. In this case the fuzzified age 20 defuzzifies to age 25.5. This is one of the worst distortions caused by defuzzifying the fuzzified value, as shown in Figure 6.12. In a perfect system the defuzzification of the fuzzified value would equal the original value, but, as Figure 6.12 shows, this is not always the case.

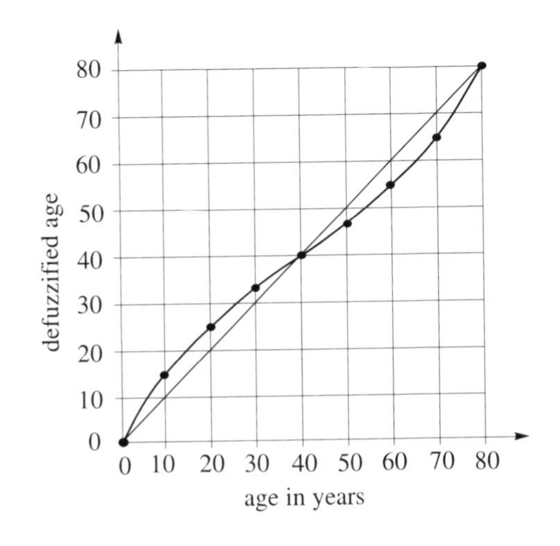

Figure 6.12 *Graph of defuzzified values of young and old.*

6.4 Conclusion

This chapter has described ways of reasoning which can be incorporated into the cognitive subsystem of a mechatronic system. The main tool that is available is logic, which takes many forms, and in this chapter we have introduced some of the more commonly found examples. Propositional and predicate logic allow statements to be evaluated as being either TRUE or FALSE. These statements can be combined using the Boolean connectives, AND, OR and NOT, to form more complex propositions, and the truth or falsity of these complex statements can also be evaluated. The *If–Then* rule shows how a decision can be made based on some conditions being TRUE or FALSE. These conditions would generally be determined by sensor inputs, although some of the higher level propositions would be inferred from lower level propositions rather than directly from sensor inputs. Theorem proving can also be used by applying rules of inference to propositions, to determine whether they are TRUE or FALSE, and thus hypotheses can be tested.

Variations on these ideas have been shown in the form of non-monotonic logic, where deductions are allowed to change in the light of new evidence, and multi-valued logic where a new truth value, UNKNOWN, can be processed logically, giving a system the ability to reason even when information is missing.

In situations where the data are less certain, probabilities can help to make decisions. Bayes' rule showed how the probability of a proposition can be calculated indirectly from data which are easiest to collect. Finally, it is often desirable to 'fuzzify' some of the rigid thresholds that are used to determine propositions. The result is a membership function whose value lies in the range 0 to 1. Fuzzy logic can also handle this information logically.

In the next chapter we shall see how these ideas of reasoning can be implemented in a rule-based system, so that some of the more practical aspects of reasoning in machines will become clearer.

References and further reading

For further reading, we recommend:

Hopgood, A. (1993) *Knowledge Based Systems for Scientists and Engineers*, CRC Press, London.

Chuen Chien Lee (1990) 'Fuzzy logic in control systems: Fuzzy Logic Controller – Part I', *IEE Transactions on Systems, Man and Cybernetics*, Vol 20, No. 2, March/April 1990, pp. 405–418.

Chuen Chien Lee (1990) 'Fuzzy logic in control systems: Fuzzy Logic Controller – Part II', *IEE Transactions on Systems, Man and Cybernetics* Vol 20, No.2, March/April 1990, pp. 419–435.

CHAPTER 7
RULE-BASED SYSTEMS

7.1 Knowledge-based, rule-based and expert systems

Rule-based systems are sometimes called by other names, some authors using the terms *knowledge-based systems* and *expert systems* interchangeably. In this chapter we use the term rule-based system because, at the heart of all these systems, is a set of *rules*.

Figure 7.1 shows a typical structure of a rule-based system. The major components are the ***knowledge base***, which contains facts as well as rules, and the ***inference engine***. The inference engine is the part where the reasoning takes place: where input information is combined with the rules and facts in the knowledge base to make decisions and construct new information.

Figure 7.1(a) shows what is involved in building a rule-based system. The rules usually come from humans via a user interface, most commonly a graphic user

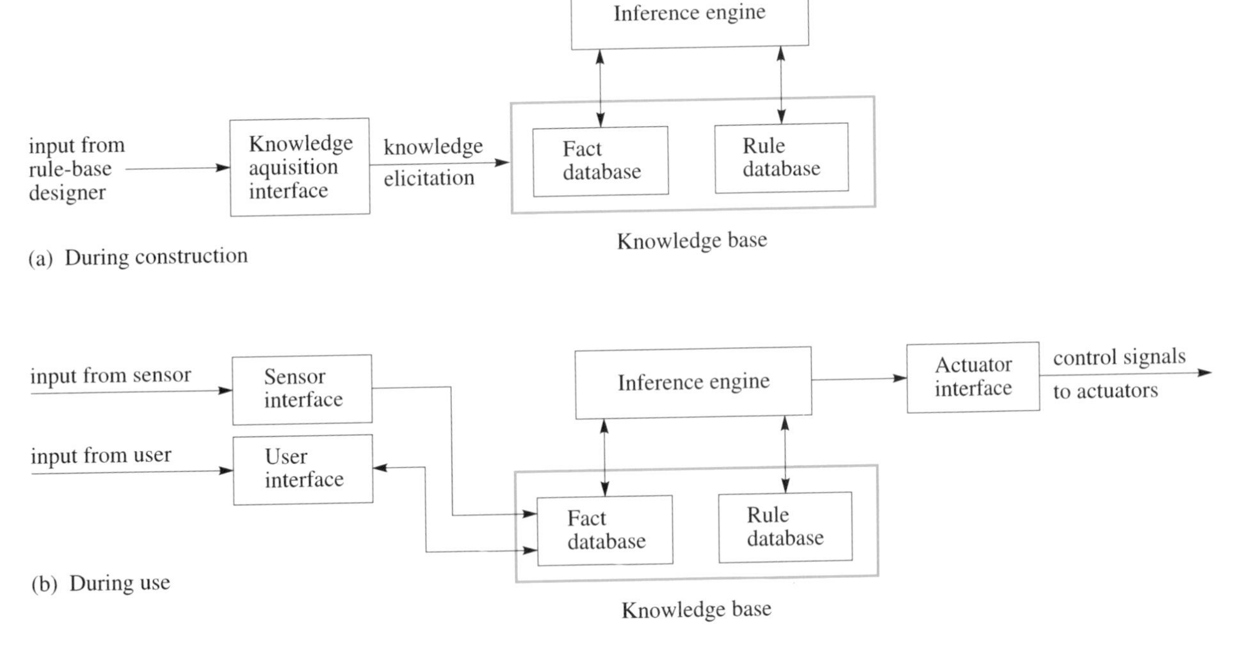

Figure 7.1
Architecture of a rule-based system.

interface (GUI). The *a priori* facts which are built into the system also enter the system in this way. The process of getting knowledge from humans to put into machines is called ***knowledge elicitation***. In expert systems, which attempt to capture and emulate some area of human expertise, knowledge elicitation is very important. Usually an expert's knowledge is not explicit: he or she 'just knows'. It can be very difficult to get this implicit information from an expert and convert it into the explicit form required to store and manipulate it in a computer. For this reason, specialists called *knowledge engineers* may be used to guide and facilitate the knowledge elicitation process. In mechatronics it is likely that the 'expert' on designing intelligent machines is also an expert on building rule-based systems, and therefore acts as his or her own knowledge engineer.

Figure 7.1(b) shows a rule-based system in use. In this case information is supplied to the system by sensors or human users. In mechatronic systems the output will include control signals to actuators, such as electrical logic levels which switch motors on and off. The user may or may not supply information to such a system when it runs. When they do it will be through some interface, which may be a full GUI in the case of a power station controller, or it may be a few buttons in the case of a washing machine. Usually the users of rule-based systems are not the people who designed them, and an interface which makes them intuitive and easy to use is an important feature.

The term ***expert system*** would usually be applied to systems which are specifically designed to be operated by humans. The human operators would be less expert in a particular field than the person who designed the system, but the combination of the human operator and the expert system acts as an expert in that particular field. One example is in medical diagnosis where information is fed into the system and analysed by applying the rules. These rules would have been extracted from one or more human experts on medical matters by analysing their decisions given similar sets of data. So an expert system is just a rule-based system, but the rules tend to have been extracted from human experts, and the system is expected to give the same diagnoses or answers as human experts, or to improve the performance of a lesser expert. Paradoxically, expert systems based on many human experts sometimes have worse performances than systems based on a single expert. This is because any given expert is likely to be consistent in their knowledge and beliefs, but combining expertise may introduce inconsistency.

In mechatronic systems, a rule-based system may also be required to behave like a human expert. But it is often the case that the rules that it contains are not extracted from experts because experts in that particular field do not exist. The rules are found by trial and error and may be adapted as time goes by, by the machine itself. We will therefore not use the term expert system in this chapter, but understand that in some cases the rules may have been extracted from human experts.

Rule-based systems differ in many respects from conventional computer programs. The main features are as follows:

- In a particular application, all the knowledge about that application is kept separate from the control structure of the programs. In a conventional program the two would be intermixed, typically through the design of application-specific data structures.
- Rule-based systems have the significant advantage that new knowledge can be added or unwanted knowledge taken away relatively easily. In a conventional program, if some new knowledge became available the program would probably have to be re-written.
- Rule-based systems have mechanisms to explain their conclusions and lines of reasoning.

When a computer system is operating it may take some action which seems odd. With rule-based systems a user can interrogate the system and ask for reasons to be given for that action. A conventional system could only supply some sort of explanation if a continuous record or log has been kept of all the actions it has taken. In rule-based systems the method called backward chaining (which will be described later) is particularly well suited to work out and indicate which rule has been used and why it has been used, by way of an explanation for its action. It would be very difficult to incorporate this facility into any other architecture such as conventional programs or neural networks for example.

For all their advantages, there are some drawbacks which have to be borne in mind. The first is that rule-based systems tend to be used in narrowly defined applications. If a rule-based system is constructed for one application and the same system is then applied to a similar but not identical application, it often fails catastrophically because some of the rules are too specific. In other words, rule-based systems do not *degrade gracefully* when they reach the edges of their understanding. Also, they have to be periodically updated to check that the knowledge with which they are working is still relevant. This means that they cannot be left to work autonomously for long periods of time (a year, for example). The virtue of being able to add knowledge incrementally to rule-based systems can also lead to the introduction of subtle inconsistencies. And finally, when they are used in very complex situations where a large number of rules are stored they can sometimes be relatively slow. This latter aspect is not so crucial in many applications, but in real-time control, where decisions and actions have to be taken typically in milliseconds, this is very important.

In Chapter 9 you will see how a system can be controlled using rules, and in particular fuzzy rules. In this chapter we want to concentrate on some of the particular issues relevant to rule-based systems irrespective of their application. These are mainly concerned with the *control* of the rule-based system, which means the order in which things happen and the general house-keeping that needs to go on. In this chapter we will concern ourselves with small rule-based systems which would be *embedded* in a system, rather than the larger rule-based systems which involve much more house-keeping such as truth maintenance, where the truth value of the deductions that have been made are checked to ensure that no contradictions occur.

7.1.1 The knowledge base

The knowledge base of a rule-based system consists of facts and rules. Since the rule-based system is going to be applied to a specific problem, the facts and rules will be specific to that problem. This is called the problem ***domain***, and the facts and rules can only be used in that domain.

The assertion that something is true or false is called a ***proposition***, or a ***predicate*** when variables are involved. For example 'the heating is switched to ON is TRUE' is a predicate. TRUE is the ***truth value*** of the predicate. The part of the predicate which is asserted to be true will be called a ***clause*** (this term has a more precise meaning which is beyond the scope of this book). The clause is often written as a letter, so that p could mean 'the heating is switched on'. We will use the notation $T(p)$ to mean the truth value of p. For example, $T(p)$ = FALSE means 'the heating is switched on is FALSE'.

Unfortunately the notation used in the literature is rather loose. Sometimes you will see 'p = TRUE' to mean '$T(p)$ = TRUE', and sometimes you will see things like 'if p then q' to mean 'if p is TRUE then q is TRUE', or equivalently, 'if $T(p)$ = TRUE then $T(q)$ = TRUE'. Usually it is clear what is meant, but you are warned that these different usages can be confusing.

In the following we abbreviate '$T(p)$ = TRUE' to $T(p)$, and '$T(p)$ = FALSE' to $\neg T(p)$, unless stated otherwise.

Facts and rules can be divided into two types – deep and surface. ***Deep knowledge*** concerns the basic principles such as the laws of physics, which we assume are not going to change. ***Surface knowledge*** concerns ***heuristics*** that are known to work from experience of similar problems but which may change.

The sort of rules that are kept in the knowledge base are often in the form of the *If–Then* statements that were described in Chapter 6 on Reasoning, although there can be other types. These *If–Then* rules are called ***production rules*** and typically look like this:

If (something is TRUE)
Then (something else is TRUE)

where new knowledge is deduced from old. In mechatronics some rules result in *actions*:

If (something is TRUE)
Then (do something else)

For example,

If T(cold)

Then T(heating ON)

Remember that the terms in brackets are called propositions and that they have a truth value associated with them. So terms like 'cold' would have to be more precisely defined, such as:

T(cold) = TRUE if temperature from a sensor is less than 10°C

= FALSE if temperature from a sensor is greater than or equal to 10°C

where $T(x)$ is the truth value of x, and

T(heating ON) = TRUE if the heating is switched to ON

= FALSE if heating is switched to OFF

The left-hand side of this rule contains a condition that has to be satisfied and is called the ***antecedent***. The right-hand side of this rule contains the consequence of the antecedent being TRUE and is called the ***consequent***. So if the antecedent is TRUE it follows that the consequent is TRUE. Recall from Chapter 6 that this process is called *modus ponens*, and follows from the inference rules of implication. If $(A \rightarrow B)$ is TRUE and A is TRUE, it follows that B is TRUE.

7.1.2 Forward chaining

In a rule-based system, a rule is said to be ***triggered*** if the antecedent of the rule is TRUE. If the rule goes on to be used it is said to have been ***fired***. If the rule does not fire it ***fails***, which could be due to the antecedent being FALSE or UNKNOWN or because the rule wasn't selected to fire.

Often in a rule-based system more than one rule could be triggered, so there has to be a strategy for selecting which rule to fire. The inference engine is in control of the rule firing, but it can work in two quite distinct ways called *forward chaining* and *backward chaining*. In many applications a system would only ever need to use one or the other approach, but there are instances where both are used.

In ***forward chaining*** the inference engine works in cycles. In each cycle the facts in the working memory are updated from information that has been input or deduced since the last cycle. Next the rules are examined and all the rules whose antecedents are satisfied are triggered. The collection of triggered rules is called the ***conflict set***, and this conflict has to be resolved so that only one rule fires. Only one rule is fired in a cycle because by the time it has fired the conditions that led to the other rules in the conflict set being triggered may have changed.

To illustrate forward chaining, consider a rule-based security system with the following rules and facts:

Rule Database

Rule 1

If T(image contains a face)

and T(face recognized)

Then T(open door)

and $\neg T$(image contains a face)

and $\neg T$(face recognized)

Rule 2

If T(image contains a face)

and $\neg T$(face recognized)

Then T(alert security guard)

and $\neg T$(image contains a face)

Rule 3

If $\neg T$(image contains a face)

Then $\neg T$(open door)

and $\neg T$(alert security guard)

Fact Database

$\neg T$(image contains a face)

$\neg T$(face recognized)

$\neg T$(open door)

$\neg T$(alert security guard)

When this system starts running none of the antecedent predicates *matches* the first two facts in the Fact Database. The inference engine begins at Rule 1, finds that it is not triggered, moves on to Rule 2, finds that it is not triggered, moves on to Rule 3, finds that it is triggered, fires Rule 3 setting $\neg T$(open door) and $\neg T$(alert security guard) (initially they are already set with these truth values) and goes back to Rule 1 to begin the cycle all over again.

Now suppose that the security system has a pattern-recognition system which, through its hardware and software interface, can alter the truth value of (face recognized) to make it True. Suppose also that it can automatically alter the truth value of (face recognized) to make it True or False depending on whether the face can be matched in the image database.

Rule 2: the first consequent

First suppose that a visitor comes to the door and the vision system changes the Fact Database to contain T(image contains a face), but the face recognition test fails so the predicate $\neg T$(face recognized) remains unchanged in the Fact Database.

When forward chaining, the inference engine starts with Rule 1. The first antecedent predicate of Rule 1 now matches the Fact Database, but the second does not. The inference engine then tests Rule 2. In this case both antecedent predicates are matched and the second rule is triggered. Then the inference engine tests Rule 3, which is not triggered because its antecedent predicates do not match the Fact Database. Since Rule 2 is the only rule to be triggered, it fires, and the predicate $\neg T$(alert security guard) is changed to T(alert security guard) in the Fact Database. It will be supposed that the security guard is alerted by a hardware–software interface that sounds a buzzer whenever the Fact Database contains the predicate T(alert security guard).

Rule 2: the second consequent

Rule 2 resets T(image contains a face) to $\neg T$(image contains a face) to stop itself being triggered indefinitely. The vision system will of course change this back to T(image contains a face) until the visitor is either admitted by the manual intervention of the security guard or goes away.

After Rule 2 has fired, the system will again begin at Rule 1 and test Rule 2 and Rule 3. If the visitor can still be seen by the vision subsystem, Rule 2 will again fire so the security guard's buzzer will sound until this person has been dealt with.

When the person has been dealt with, the predicate $\neg T$(image contains a face) will be in the Fact Database. This means that on subsequent cycles Rule 1 will not be triggered, Rule 2 will not be triggered, but Rule 3 will be triggered and the system will ensure that the predicates $\neg T$(alert security guard) and $\neg T$(open door) are in the Fact Database. Thus the guard's buzzer will no longer sound, and the door will be locked.

If the vision system senses that the image contains a face and it recognizes the face, the predicates T(image contains a face) and T(face recognized) will be added to the Fact Database. On the next forward chaining cycle, Rule 1 will be the only rule triggered and it will fire, changing the predicate $\neg T$(open door) to T(open door). Assuming that the Fact Database is interfaced to the door lock, the presence of the predicate T(open door) will cause the door to be unlocked so that the recognized person can enter. The consequent predicates of Rule 1, $\neg T$(image contains a face) and $\neg T$(face recognized), are used to update the Fact Database.

Thus, while forward chaining, the system goes through all the rules in sequence. In this example the conflict set has only ever contained at most one member, which is the one selected to fire. However, in general, the conflict set will contain more than one triggered rule, and the conflict as to which one should fire must be resolved by *conflict resolution*.

7.1.3 Conflict resolution

Resolving the conflict set is a skill in its own right. There are no hard and fast methods that are guaranteed to be the 'best', but some of the methods that are commonly used are as follows:

First-come, first-served

The first rule that is found which has its antecedent satisfied is fired. This has the advantage that there is no need to create a conflict set at all, so the method is fast. It has the disadvantage that the rule that fires might not be the most important rule in some respect.

Prioritizing the rules

The rules are rank-ordered so that the most important rule (as decided by the rule-base designer) is placed first in the list, and the least important is placed last. This has all the advantages of the first-come, first-served method and none of the disadvantages. This approach is also called *rule ordering*.

Prioritizing the data

The data or facts are rank-ordered by the designer, and the rule that uses the highest ranked data in its antecedent is fired. There may still be conflict here if more than one rule uses the same data.

Recency ordering

The least recently fired rule in the conflict set is fired, or the rule which uses the most recently updated data is fired. An alternative strategy is to fire the most recently fired rule.

Generality ordering

The rule that is most specific to the situation is fired. This usually means the rule that has the most conditions that have to be satisfied. For example, given two rules in the conflict set where the first rule requires condition A to be satisfied before it will fire, and a second rule which requires both condition A and condition B to be satisfied, then the second rule is chosen under this strategy. This approach is also called *size ordering* or *specificity ordering*.

Context limiting

Separate the rules into groups. At any one time, only one group of rules will be active, so the chances of conflict arising are reduced.

Buggins' Turn

Another conflict resolution strategy, which we have called ***Buggins' Turn***, works well in many circumstances. This strategy involves cycles in which every rule gets a chance of firing if it can. Each cycle start with the first rule and goes through all the other rules to form a conflict set. After this, the remaining rules in the conflict set are examined, excluding the rule that just fired, and a new sub-conflict set is formed. The first rule in this sub-conflict set is chosen to fire. After this the remaining rules in the original conflict set are examined, excluding any that have fired, and the first of these to be triggered is fired. This continues until all the rules in the original conflict set have been fired, or are no longer triggered.

Then the Buggins' Turn cycle starts again with the first rule. The term 'Buggins' turn' comes from a system in which career promotion is based on how long people have been waiting rather than their merits: everyone gets a turn in the end no matter how mediocre they are.

One problem that can arise in most of these strategies, but particularly in the 'first-come, first-served', is that the same rule can be chosen in every cycle. This may be correct, but it could be the case that the conditions are satisfied for two rules – one which does very little and one which causes a profound change in the system. In the next cycle it is selected again, and so on for every future cycle. So care has to be taken to select the correct strategy for a particular application. Recency ordering and Buggins' Turn do not suffer from this defect. It is also permissible to use more than one strategy. For example, a rule consequent may explicitly change the mode of conflict resolution. Such a facility would have to be built into the inference engine.

Example

Let's look at a hypothetical rule-based temperature controller. Valve 1 and valve 2 are responsible for circulating hot water from the boiler around the radiators and the hot-water tank respectively.

There are many different ways that rules could have been formulated; here the criteria are simplicity and brevity. The set of rules might look something like this:

Rule Database

Rule 1: *If* (*T*(room temperature < 20) AND *T*(timer ON))
Then *T*(boiler ON)

Rule 2: *If* (*T*(water temperature < 40) AND *T*(timer ON))
Then *T*(boiler ON)

Rule 3: *If T*(boiler ON)
Then T(pump ON)

Rule 4: *If* (*T*(pump ON) AND *T*(room temperature < 20))
Then T(valve 1 OPEN)

Rule 5: *If* (*T*(pump ON) AND *T*(water temperature < 40))
Then T(valve 2 OPEN)

Rule 6: *If* ¬*T*(timer ON)
Then ¬*T*(boiler ON)

Rule 7: *If* ¬*T*(room temperature < 20)
Then ¬*T*(valve 1 OPEN)

Rule 8: *If* ¬*T*(water temperature < 40)
Then ¬*T*(valve 2 OPEN)

Rule 9: *If* ¬*T*(boiler ON)
Then ¬*T*(pump ON)

Rule 10: *If* (¬*T*(room temperature < 20) AND ¬*T*(water temperature < 40))
Then ¬*T*(boiler ON)

Notice that there are a number of rules which are similar but produce the opposite effects, such as Rule 3 and Rule 9. You might think that you just need Rule 3, that if the boiler is ON then the pump is ON, but this only turns the pump ON. If the conditions change and the boiler is no longer ON, Rule 3 does not fire, which is not the same as turning the pump OFF. A separate rule has to be constructed to make sure that the pump can be turned OFF.

All the propositions, or facts, make up the rest of the knowledge base.

Fact Database

(room temperature < 20)
(water temperature < 40)
(timer ON)

(valve 1 OPEN)
(valve 2 OPEN)

(boiler ON)
(pump ON)

In forward chaining, the inputs are taken from the user or the sensors and combined with the rules in the knowledge base. Given a new set of facts, the rules are examined, a conflict set produced from all the rules whose antecedents are satisfied (are triggered), and a conflict resolution strategy applied to select a single rule for firing. In the heating controller a sequence of events could be as follows:

(room temperature < 20)	TRUE
(water temperature < 40)	FALSE
(timer ON)	TRUE
(valve 1 OPEN)	FALSE
(valve 2 OPEN)	FALSE
(boiler ON)	FALSE
(pump ON)	FALSE

The following rules are triggered:

Rule 1: *If* (*T*(room temperature < 20) AND *T*(timer ON))
Then *T*(boiler ON)

Rule 8: *If* ¬*T*(water temperature < 40)
Then ¬*T*(valve 2 OPEN)

Rule 9: *If* ¬*T*(boiler ON)
Then ¬*T*(pump ON)

If the first-come, first-served strategy is used then Rule 1 fires, and *T*(boiler ON) = TRUE. The propositions now look like this:

(room temperature < 20)	TRUE
(water temperature < 40)	FALSE
(timer ON)	TRUE
(valve 1 OPEN)	FALSE
(valve 2 OPEN)	FALSE
(boiler ON)	TRUE
(pump ON)	FALSE

In the next cycle, assuming that the external conditions have remained constant, the inference engine triggers the following rules:

Rule 1: *If* (T(room temperature < 20) AND T(timer ON))
Then T(boiler ON)

Rule 3: *If* T(boiler ON)
Then T(pump ON)

Rule 8: *If* $\neg T$(water temperature < 40)
Then $\neg T$(valve 2 OPEN)

Now if we just used first-come, first-served then Rule 1 would fire again, which gets us nowhere. Instead, let's use first-come, first-served with recency, so that the rule that was least recently fired is selected, and if there are several rules that have never been fired, choose the first one that is encountered. In this example, Rules 3 and 8 have not been used, and Rule 3 is the first one in the list, so this is the one that is fired. The result is that T(pump ON) = TRUE. Assuming that there are no external changes, the data in the knowledge base becomes:

(room temperature < 20)	TRUE
(water temperature < 40)	FALSE
(timer ON)	TRUE
(valve 1 OPEN)	FALSE
(valve 2 OPEN)	FALSE
(boiler ON)	TRUE
(pump ON)	TRUE

The inference engine produces a new conflict set. To keep track of how long ago a rule was fired, we add a number in brackets to the left of each rule, as shown below. Rule 3 was fired 1 cycle ago and Rule 1 was fired 2 cycles ago. There have only been 2 cycles so far, so rules which have not yet been fired just have to have a number which is 1 more than the number of cycles so far. The new conflict set is:

(2) **Rule 1:** *If* (T(room temperature < 20) AND T(timer ON))
Then T(boiler ON)

(1) **Rule 3:** *If* T(boiler ON)
Then T(pump ON)

(3) **Rule 4:** *If* (*T*(pump ON) AND *T*(room temperature < 20))
Then *T*(valve 1 OPEN)

(3) **Rule 8:** *If* ¬*T*(water temperature < 40)
Then ¬*T*(valve 2 OPEN)

In this case, Rule 4 and Rule 8 are triggered, but Rule 4 is first so it fires.

The knowledge base is updated and the numbers by each rule are incremented. So it now looks like this, assuming no changes to the time and temperature propositions:

(room temperature < 20) TRUE
(water temperature < 40) FALSE
(timer ON) TRUE

(valve 1 OPEN) TRUE
(valve 2 OPEN) FALSE

(boiler ON) TRUE
(pump ON) TRUE

In the next cycle the new conflict set is:

(3) **Rule 1:** *If* (*T*(room temperature < 20) AND *T*(timer ON))
Then *T*(boiler ON)

(2) **Rule 3:** *If* *T*(boiler ON)
Then *T*(pump ON)

(1) **Rule 4:** *If* (*T*(pump ON) AND *T*(room temperature < 20))
Then *T*(valve 1 OPEN)

(4) **Rule 8:** *If* ¬*T*(water temperature < 40)
Then ¬*T*(valve 2 OPEN)

This time Rule 8 fires, which doesn't actually change anything since valve 2 is already closed. If no more changes occur to the antecedent predicates, these four rules will be triggered each time the inference engine cycles round. In the next cycle, Rule 1 will be selected again. Following that Rule 3, then Rule 4, then

Rule 8 again, and so on. Although a rule will fire each cycle, there will be no more changes to the knowledge base until there is some external change. The most likely change is that the room temperature will rise, so that the predicate *T*(room temperature < 20) will become FALSE. Rule 1 would no longer be triggered, and a new sequence of actions would take place.

In forward chaining a sequence of events takes place which is determined by sensors changing the facts or data in the knowledge base, so the whole process is described as ***data-driven reasoning***. The alternative process of backward chaining is a method where you start with the goal, and work backwards to see how that goal can be achieved. Backward chaining is therefore an example of ***goal-driven reasoning***.

7.1.4 Backward chaining

Forward chaining followed from the inference rule *modus ponens*: if $(A \rightarrow B)$ is TRUE and A is TRUE, it follows that B is TRUE. The repeated application of the rules in the rule base to the facts in the fact database will drive the system forward, producing new facts. If the system had no time dependency, it would deduce every possible consequence of the facts and rules in its knowledge base. As such the process is not particularly goal-oriented – the system is just producing knowledge as it goes along. Usually the system designer focuses the activity of a forward-chaining system by judicious use of conflict resolution, knowledge of the data entering the system through time, and the introduction of predicates which help to control the process.

For example the predicate 'I am in path planning mode is TRUE' could be used to direct a forward-chaining system to the task in hand, namely path planning. The last rule to fire during path planning would then probably hand over to another task using consequent predicates such as 'I am in path planning mode is FALSE' and 'I am in sensing and motor control mode is TRUE'.

Whereas forward chaining goes from antecedent predicates, ***backward chaining*** goes the other way. It is a goal-oriented strategy which assumes that something specific must be deduced.

For example, in his well known textbook *Artificial Intelligence*, Winston (1984) presents a backward-chaining system called Identifier. This system identifies animals, and can be implemented using a predicate such as 'the animal is identified as something', with the system working out what that 'something' is. This system contains rules of the form

Rule 10

If the animal is a carnivore ______________ is TRUE

and the animal has a tawny colour ______________ is TRUE

and the animal has dark spots ______________ is TRUE

Then the animal is identified as a cheetah ______________ is TRUE

This rule has a consequent predicate following '*Then*' which matches the question and can become the system's *goal*. Now the system can work *backwards* from consequent predicates to antecedent predicates: if it can show that the animal is a carnivore, has a tawny colour, and has dark spots, then it can achieve its goal and identify the animal as a cheetah. So the system looks in its fact database to see if it contains the antecedent fact that the animal is a carnivore.

Initially it does not know this, but it may find a rule of the form

Rule 5

If the animal is a mammal ______________ is TRUE

and the animal eats meat ______________ is TRUE

Then the animal is a carnivore ______________ is TRUE

Finding out if the animal is a carnivore can become a new *intermediate goal*. Then the system works backwards from this as a consequent predicate in this rule, and tries to find out if the animal is a mammal and if it eats meat.

Identifier contains the rule

Rule 2

If the animal gives milk ______________ is TRUE

Then the animal is a mammal ______________ is TRUE

So now the system works backwards from the consequent predicate 'the animal is a mammal _ is TRUE', and has the new goal of finding out if the animal gives milk.

Eventually the system will find the necessary facts in its fact database, or it will have to ask for information. In this case the system might ask you if the animal gives milk. Assuming you answered that it does, the rule would fire and the fact that the animal is a mammal would be added to the fact database. You might then be asked if the animal eats meat, in which case on answering 'yes' the rule would fire which deduces that the animal is a carnivore.

Having successfully matched the first antecedent predicate necessary to identify the animal as a cheetah, the system might ask you if the animal has a tawny colour. Suppose you answer 'yes'. The system then has one more antecedent predicate to go, and asks you if the animal has dark spots. Suppose you answer 'no'. Then the rule identifying cheetahs cannot fire after all.

Having failed to prove that the animal is a cheetah, the system looks for the next rule with an antecedent consequence which matches the original goal. In Identifier it finds

If	the animal is a carnivore ________________	is TRUE
and	the animal has a tawny colour ____________	is TRUE
and	the animal has black stripes ______________	is TRUE
Then	the animal is identified as a tiger __________	is TRUE

Now the goal becomes that of proving that the animal is a tiger. The first antecedent predicate is already in the fact data, as is the second. It was determined that the animal is a carnivore and has a tawny colour when the system was trying (incorrectly) to prove the animal was a cheetah. So now Identifier asks if the animal has black stripes. Suppose it does, and you answer 'yes'. Then this rule fires and the system concludes that the animal is identified as a tiger.

Note that during backward chaining the system accumulates intermediate facts such as the fact that the animal is a carnivore. Although the system may seek these facts in order to try to make one rule fire, they may be useful even when that rule does not fire. For example, the knowledge that an animal has black stripes is more general than tigers: it could be used to help identify zebras.

Winston writes:

> Having one if–then rule in for each animal in the zoo is possible, albeit dull. The consequent side of each rule would be a simple statement of animal name, and the antecedent side would be a bulbous enumeration of characteristics large enough to reject all incorrect indentifications. In operation, the user would first gather up all facts available and then scan the antecedent-consequent rule list for an antecedent–consequent rule that has a matching antecedent part.
>
> A better idea is to generate intermediate facts, making the reasoning procedure more interesting. The advantage is that the antecedent–consequent rules involved can be small, easily understood, easily created. Using this approach, the Identifier procedure produces chains of conclusions leading to the name of the animal.
>
> *(Winston, 1984)*

The difference between forward chaining and backward chaining is illustrated in Figure 7.2.

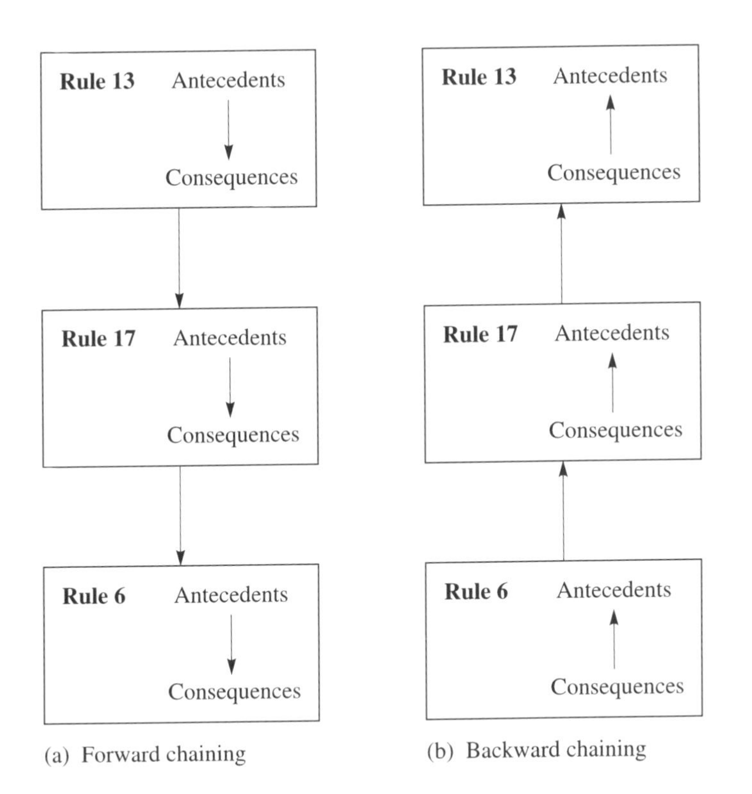

Figure 7.2
Examples of the forward chaining and backward chaining deduction mechanisms. (Note that control does not always pass to consecutive rules.)

7.1.5 Rule-based systems can explain their reasoning

One of the defining features for an *expert system* is that it is a rule-based system which can explain the way it reasons and why it makes the deductions it does. For example, if you were running Identifier and looking at what, unbeknown to you, was a tiger, then you might be surprised to be asked if the animal gives milk, thinking this is a totally inappropriate question when trying to identify such a ferocious beast.

Any well-designed expert system interface would allow you to answer this question by asking why it has been asked. In this case the system can invoke a ***rule inventory*** or ***rule trail*** mechanism giving an answer such as:

I am trying to find out if:	the animal gives milk
to deduce from rule 2 if:	the animal is a mammal
I am trying to find out if:	the animal is a mammal
to deduce from rule 5 if:	the animal is a carnivore
I am trying to find out if:	the animal is a carnivore
to deduce from rule 10 if:	the animal is identified as a cheetah

From this rule trail the user can see which rules are being invoked to make deductions leading towards the eventual goal. In this case the user may be satisfied that the 'explanation' is satisfactory, and that asking about milk was relevant after all. However, it is possible that the rule inventory will expose a highly suspect line of reasoning that the user does not accept as valid. In this case the user can ignore the expert system and fall back on human intelligence and common sense, or can suggest that the rule(s) and/or facts resulting in this odd behaviour be modified or removed from the knowledge base. In this way rule-based systems offer diagnostic information which allows human operators to intervene when it is clear that the system is not working properly or making flawed deductions and giving bad advice.

Usually, conventional computer programs do not have this explanation feature, and when they behave oddly or malfunction it is difficult to find out why.

The explanation facility of expert systems can also be used for teaching and training purposes, since novices interact with the system and quiz it when they do not understand the basis of a deduction. From this they may have rules brought to their attention of which they were previously ignorant.

7.1.6 Diagnosis in rule-based systems

Backward chaining can also be used in diagnosis. In the central heating example in the previous section, an example of backward chaining would be when you notice that a room heater is OFF but the room is cold and you want to know why. Three possibilities are:

1. The boiler is not ON
2. The pump is not ON
3. Valve 1 is not OPEN

Let's look at these in turn. The only fact that you know is that T(room temperature < 20) is TRUE. First, let's check the possibility that the boiler is not ON by making the first goal the predicate, $\neg T$(boiler ON). The rules are examined to see if this goal is the consequent of any rule, and we find that Rule 6 and Rule 10 apply.

Rule 6: *If* $\neg T$(timer ON)
Then $\neg T$(boiler ON)

Rule 10: *If* ($\neg T$(room temperature < 20) AND $\neg T$(water temperature < 40))
Then $\neg T$(boiler ON)

The antecedent for Rule 10 is that ($\neg T$(room temperature < 20) AND $\neg T$(water temperature < 40)) is TRUE, but we know that it is FALSE, so Rule 10 could not have fired. The antecedent for Rule 6 is $\neg T$(timer ON), but we do not know the state of the timer. The predicate $\neg T$(timer ON) now becomes the goal, and we examine the rules to find if $\neg T$(timer ON) is a consequent of any rule, which it is not. So we cannot go any further with this search, and leave the hypothesis that $\neg T$(timer ON) is TRUE could be one explanation of why the heating isn't on.

Let's do that again for the pump. The goal is $\neg T$(pump ON), which is the consequent for Rule 9.

Rule 9: *If* $\neg T$(boiler ON)
Then $\neg T$(pump ON)

For this rule to have fired, the antecedent $\neg T$(boiler ON) must be TRUE, so this becomes the next goal. Check the rules to see if $\neg T$(boiler ON) is a consequent and the search follows the same pattern as the previous goal, so the hypothesis is again that $\neg T$(timer ON) is TRUE.

Finally, let the goal be $\neg T$(valve 1 OPEN), which is the consequent of Rule 7.

Rule 7: *If* $\neg T$(room temperature < 20)
Then $\neg T$(valve 1 OPEN)

The antecedent of this rule is $\neg T$(room temperature < 20), which we know is FALSE, so this rule couldn't have fired. The conclusion, then, is that the reason the heating isn't ON is because the timer is OFF. However, the timer is ON.

This gives a demonstration of the diagnostic ability of backward chaining. There are variations on this technique, which is essentially a *tree-search* mechanism. As described in Chapter 3 on Search, it is possible to search spaces of this sort by *breadth-first* or *depth-first*. In a depth-first search, of the kind just described, each goal is checked until the knowledge base is reached and the goal can be shown to be satisfied or not. In the example just given, at one point there were two rules that had the goal as a consequent. One of those goals is pursued until it either succeeds or fails, and in our case the first attempt failed. The search then backtracks to where there was a division and searches down the next branch until a conclusion can be made. This is the most commonly used method of goal-driven reasoning, and is illustrated in Figure 7.3.

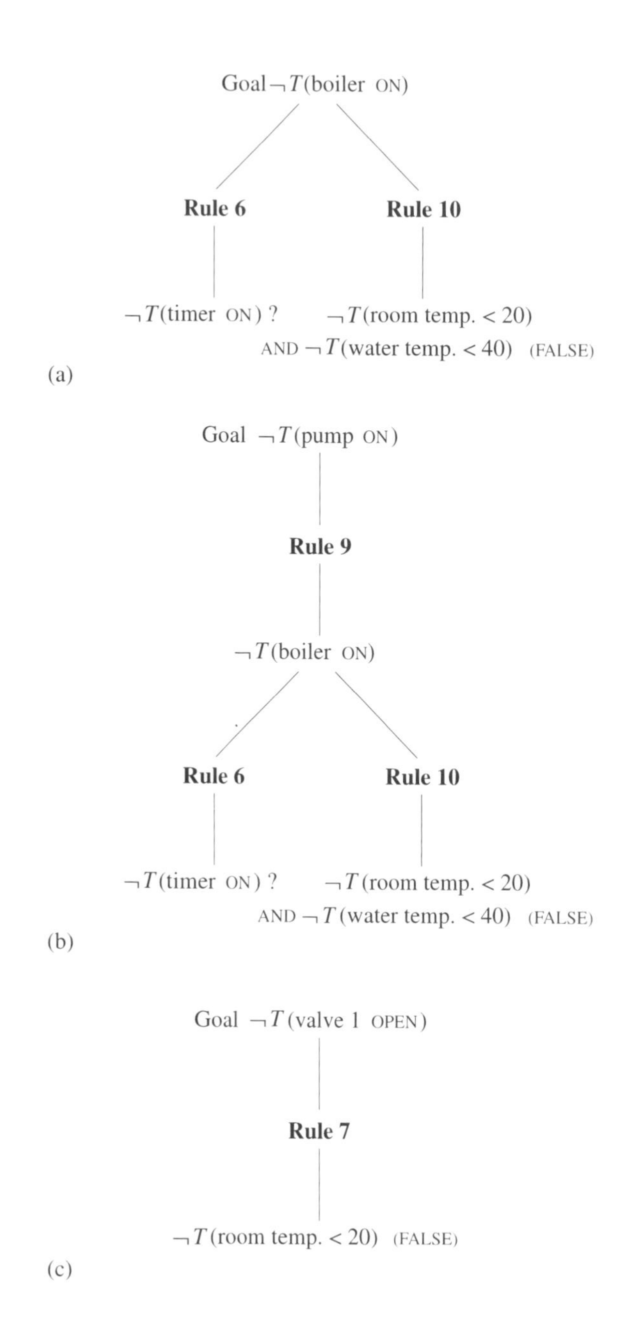

***Figure 7.3* Backward chaining as a tree search.**

7.1.7 Variables and instantiation

Predicates allow the use of variables and make use of the quantifiers introduced in Chapter 6: ∃ (there exists) and ∀ (for all). An example of this might be where there are a number of sensors which test for dangerous conditions and can be either ON or OFF. Each condition that the sensors are monitoring is indicated by a red light being ON or OFF, so that if, for example, sensor 1 detects a dangerous situation, red light 1 will come ON if it is not already ON. A rule for doing this is:

If (on(sensor 1) $\wedge$ $\neg$on(red light 1))

Then on(red light 1)

Note the use of the predicate on(x), which is TRUE if x is ON and FALSE if x is OFF.

We could have a single rule for each sensor, but if there were a hundred sensors this would get very tedious and difficult to read. Alternatively we could use the existential quantifier as follows:

If ($\exists$ x (on(sensor x) $\wedge$ $\neg$on(red light x)))

Then on(red light x)

This says that if there exists a value for x such that sensor x is ON and the red light x is OFF, then turn red light x ON.

One problem with this is that there may be more than one value for x which satisfies the antecedent; in other words, more than one sensor which is ON. The designer of the rule-based system has to decide whether or not to select one value or all the values whose antecedents are TRUE. The first of these methods is called ***single instantiation***, and simply substitutes the first value for x that it finds that has a TRUE antecedent. The rule with that value of x is then placed into the conflict set. The second method is called ***multiple instantiation***, and finds all the values of x that have TRUE antecedents, and places all those instantiated rules into the conflict set.

Example

For another example of instantiation, consider the rule-based system run by Intelligent Marriage Brokers Inc. which contains the rule

If 'a first client' is male

and 'that first client' is single

and 'a second client' is female

and 'that second client' is single

Then 'that first client' and 'that second client' are eligible and has the Fact Database

Fred is male
Fred is single
Anne is female
Anne is not single
Pat is single
Pat is female
Jim is male
Jim is single

Suppose that the inference engine will match 'a first client' or 'a second client' with the first word of any predicate. So, initially the first antecedent predicate is instantiated with 'a first client' replaced by Fred, so the first part of the rule becomes instantiated as:

If Fred is male

and Fred is single

The inference engine then tries to match 'a second client' and comes up with the third antecedent predicate

Anne is female

which has a match in the Fact Database. The system continues on the assumption that 'a second client' is Anne. However, on trying to match

Anne is single

it fails because the only other data about Anne is that

Anne is not single

Since the rule cannot fire with this instantiation, the inference engine abandons Anne, and seeks to instantiate 'a second client' with the next item in the Fact Database. It matches

Pat is female

and having instantiated 'a second client' with Pat, the rule becomes

If Fred is male

and Fred is single

and Pat is female

and Pat is single

Then Fred and Pat are eligible

Since all of these instantiated antecedent predicates match the Fact Database, the rule is triggered. In single instantiation it would be the only rule in the conflict set and fire. As a result, the facts that Fred and Pat are eligible would be added to the Fact Database, possibly to be accessed by other rules which list possible partners for introductions.

With multiple instantiation the following rule would also be obtained:

If	Jim is male
and	Jim is single
and	Pat is female
and	Pat is single
Then	Jim and Pat are eligible

and the conflict set would contain the two instantiated rules. In this case a variant of random ordering might be considered appropriate in order to ensure that clients at the bottom of the list are not disadvantaged.

7.2 Implementation

Having considered the theoretical basis of rule-based systems, we will now address some important issues of implementation. These include:

- How is the knowledge represented inside the machine?
- Where does the knowledge come from?
- How does the knowledge get into the machine?
- How are the rules selected for the conflict set?
- How does the knowledge get updated?

7.2.1 Knowledge representation

In order to explain how knowledge can be represented in a rule-based system, we will describe some of the features of the rule-based system that we created for the Open University's home experiment laboratory, which we call SmartLab.

In SmartLab, a predicate is made up of a sequence of words which represent things of interest. For example, *Dogs are fun is* a clause which can have one of the truth values True, False, or Unknown. The *predicate* is made up by attaching a truth value to the clause, which we write as:

Dogs are fun ______________ is True

Inside the computer, this could be represented by a sequence of numbers, for example

3 72 18 91 1

and a list of text strings such as

17 is

72 dogs

91 fun

where 3 means there are three words, 72 is the numerical *token* for the string *dogs*, 18 is the token for the string *are* and 91 is the token for the string *fun*. The strings are stored in a character array, and they are only used for user interface display and printing purposes. The 1 at the end is a number representing the truth value True.

7.2.2 Editors, parsing, and inputting knowledge

With SmartLab, users can type in clauses from the keyboard using our Rule and Fact editors. These accept a line of text made up of keyboard characters, and *parse* it to find words and other syntactically correct constructions such as formulae. The words are tested by the parser against the existing tokenized strings, and any new words are given a new token number and stored. The user is unaware of the numbers that the editors give the words, since these are part of the system's *internal representation*.

Users can build rules using the Rule Editor which has boxes for predicates to be entered as antecedents, and boxes lower down for predicates to be entered as consequents. The editors allow the truth values of the predicates to be set as appropriate. So, for example, the user might input a rule which looks similar to

If dogs are fun

and you like dogs

and you want fun

Then get a dog

On parsing your input the database might include the following data:

String data list

5	a
7	you
17	is
18	are
47	want
56	get
71	dog
72	dogs
91	fun
96	like

Antecedent predicate data list

	1		3			
−1	3	72	18	91	1	−2
−1	3	7	96	72	1	−2
−1	3	7	47	91	1	−2

Consequent predicate data list

1		1		
3	56	5	71	1

The numbers 1 and 3 before the antecedents record the information that this is Rule 1 and it has three antecedent predicates. The numbers 1 and 1 before the consequent records that this is Rule 1 and it has 1 consequent predicate.

The numbers −1 at the start of each antecedent predicate indicate conjunction, i.e. the consequent predicates are connected by the word **and**. The numbers −2 at the end of each consequent predicate indicate that the system is allowed to ask the user to supply that information.

Suppose the Fact Database contained the single fact represented by the numbers

1				
3	72	18	91	1

The 1 at the beginning shows there is just one fact. This fact can be translated as meaning that it is true that dogs are fun.

7.2.3 Pattern matching

When this system is run, our inference engine takes the first antecedent predicate in the first rule and tries to *match* it with the facts in the Fact Database.

The 3s match, which shows that the first antecedent predicate and the first (and only) fact have the same size. They both have 1 (True) as truth value. The first number in the antecedent predicate is 72 which matches that of the first fact, the second number in the first antecedent predicate is 18 which also matches the second number of the first fact, and the third number in the first antecedent predicate is 91 which also matches the first fact. So, the first antecedent predicate matches the first fact perfectly, and the inference engine goes on to the next antecedent predicate:

−1	3	7	96	72	1	−2

The −1 tells the inference engine that **and** is being used, which means that this antecedent must also be perfectly matched for the rule to fire. The inference engine starts with the first fact in the database. Although the size (3) and truth value (1) match, the first token in the antecedent predicate is 7 while that in the fact is 72. This is a mismatch, and this antecedent predicate does not match this fact.

7.2.4 Dynamic data acquisition

There are no more facts in the Fact Database and normally this would mean that the rule is not triggered. Without a mechanism for obtaining new data, this rule-based system will get no further. The two main mechanisms involve (a) asking humans questions, or (b) getting the sensors to update the Fact Database automatically through suitable hardware–software interfaces. In this case the symbol −2 at the end of the antecedent predicate tells the inference engine to ask you a question. So you would see a message on your screen asking

Is it true that: you like dogs

and you could click the mouse on a 'yes' box, a 'no' box', a 'don't know' box or an 'explain' box. If you click on 'no' or 'don't know' the inference engine will realize that this rule cannot fire. If there were another rule it would move on to that. In this case the system would give you a message saying that it cannot deduce anything and that it has finished.

However, suppose you clicked on the 'yes' button. Then the Fact Database would be updated to become

2				
3	72	18	91	1
3	7	96	72	1

The inference engine would then move on to the next antecedent predicate.

−1	3	7	47	91	1	−2

Again it would interpret the −1 as **and**; again it would not be able to match the numbers 7, 47, 91 with the facts in the Fact Database; and again it would interpret the −2 as meaning that it could ask you a question:

Is it true that: you want fun

and again if you were to click the mouse on the 'yes' button, and the Fact Database would contain the following data:

3				
3	72	18	91	1
3	7	96	72	1
3	7	47	91	1

By now the inference engine would know that the antecedent predicates had been successfully matched, and so it would know that the rule had been triggered and would add the number of the rule to a list of numbers which represents the conflict set. In this case the conflict set would only contain this rule, and so it would be selected to fire.

7.2.5 Updating the Fact Database when rules fire

On firing a rule, the inference engine takes the consequent predicates of the rule and tries to match them against the Fact Database. If a match is found, the inference engine changes the truth value of the fact in the Fact Database to that of the rule. In this case there is one consequent predicate:

3	56	5	71	1

which does not match any of those in the Fact Database. Therefore the consequent predicate is added as a new fact to the Fact Database which ends up as

4				
3	72	18	91	1
3	7	96	72	1
3	7	47	91	1
3	56	5	71	1

The system then tries to fire another rule. Since there is only one rule, the inference engine goes back to the beginning and starts all over again. This time all the antecedent predicates match the Fact Database, and the rule would fire again indefinitely. We have a special consequent predicate call, end(), which is used to halt the system when it has done its work.

Note that the only part of the knowledge base to change during this run is the Fact Database. The Rule Database is usually not changed when the system is run. Rules are usually changed only by the rule-based system designer when the rule-based system is built. Exceptions to this may occur when a system *learns* new rules from data.

7.2.6 Arithmetic and mathematical calculations

In most applications it will be necessary to have numerical variables, and be able to manipulate them to perform numerical calculations. Consider a rule which might be part of the control system of a mobile robot:

If	temperature > 20	is True
and	x<50	is True
Then	x=x+1	Assign
	theta=arctan(x,y)	Assign

When our parser encounters symbols like > it assumes that the tokenized string temperature is a variable and allocates appropriate memory to store the value it takes. Similarly the parser will realize that the string 20 is a constant, and convert it to a number. Both these tokenized strings become antecedent predicates which can either be true or false.

The consequent predicates, however, are rather different. They are not logical statements but numerical imperatives telling the system it must *assign* new values to variables. Usually these arithmetic assignments will allow mathematical functions such as sine, cosine, arc tangent, square root, and so on. Thus in the last consequent predicate the parser will recognize the string arctan as meaning the arc tangent function, test to see that it is syntactically correct with the necessary brackets and two parameters, and store it in tokenized form along with the tokens for its parameters.

7.2.7 Interfacing a rule-based system to sensors and actuators

In designing intelligent machines, how can we ensure that sensor data enter the Fact Database and how can a rule-based system switch motors on and off?

There are many possible implementations of sensor and actuator interfaces, including the following two methods.

One approach to reading sensors is to have special variables in the database. The idea is that the sensor interface hardware and software are constantly updating these variables, independently of the rules.

Another approach (the one we adopted for the Open University's SmartLab rule-based system) involves defining special functions which can assign values to variables. For example, the consequent p = pressure() might set the variable p to the current value of a pressure sensor, using a function called pressure().

The same approaches can be used for controlling actuators. For example, variables could be defined which are directly interfaced to actuators. Suppose that 'motorA' is such a variable. Then an assignment such as motorA = 1 might send motor A forward, motorA = –1 would send it backward, and motorA = 0 would stop it.

In the SmartLab rule-based system we use a different approach based on functions such as go() and stop(). For example, a consequent such as go(forward) sends two motors in a 'forward' direction, while go(clockwise) sends the left motor 'forward' and the right motor 'backward'.

The rules

If pressure() = 0

Then go(forward)

direction = forward

and

If pressure() > 0

Then stop()

p = pressure()

illustrate how functions such as those described might be used to interface the rule-based system.

The antecedent predicate of each rule requires the pressure sensor hardware to be 'read' by the pressure() functions. The first consequent predicate results in control signals being sent to the actuators. It is supposed that 'forward' is a system variable which is preset to a meaningful value. The last consequent predicates in the rules update the values of the variables 'direction' and 'p' in the knowledge base.

7.2.8 Knowledge elicitation

The rules in the knowledge base have to be obtained somehow. In expert systems these rules would have been *elicited* from human experts by a person called a knowledge engineer. This is a highly skilled operation because the expert often finds it very difficult to verbalize the reasons for an action that has been taken or a decision that has been made. One attempt at making this process easier is to let a non-expert in the field attempt to construct a set of rules. Then the system is shown to an expert who corrects the decisions. This means that the expert doesn't have to say why a rule is wrong, or why the substituted rule is correct. In mechatronics the system designer is likely to act as his or her own knowledge engineer.

Attempts at automating the knowledge elicitation process, particularly where an expertise does not already exist, usually involve *learning*, as described in the next chapter. One example of this is Quinlan's TDIDT algorithm which constructs a set of rules to classify a set of objects.

7.3 Confidence levels and fuzzy rules

It is a relatively simple process to modify the propositions in the knowledge base so that they are no longer either TRUE (1) or FALSE (0) but a number in between such as the number 0.72 for example. This says that the proposition is TRUE with a confidence of 72% (or FALSE with a confidence of 28%). This figure comes from the confidence that the system has in the input data and could be a probability of an event happening or a fuzzy set membership, as described in Chapter 6 on Reasoning. If the figures are the probabilities of certain events occurring, they can be updated as new data arrive using Bayesian statistics. If an unusual or unexpected input arrives, it would get a low confidence rating to start with. If it persisted, then the probability associated with it would increase.

Rules can be selected from the conflict set on a priority basis – for example, the rule with the highest confidence value is selected. The confidence level of a rule is found from the confidence levels of its antecedents. If an antecedent has connected propositions, then the rules for combining probabilities of events connected by OR, AND and NOT can be calculated as described in Chapter 6.

In a fuzzy rule-based system the value of the fuzzy set membership could be stored as a real value between 0 and 1. You should recall that the fuzzy set membership is calculated from the membership function, which is quite often a triangular function. These functions are stored in the rule-based system, sometimes as look-up tables in the memory.

For example, if the temperature is read from a sensor as 9°C, this gives the predicate 'temperature = 9'. However, it may be desirable to interpret this in fuzzy terms as the predicates temperature is cold _____ 0.3

temperature is warm _____ 0.4

temperature is hot _____ 0.1

Any rule which involves 'hotness' would go into the conflict set. For example,

If temperature is warm

Then component has failed

results in the predicate 'component has failed_0.4' being added or updated to the fact database.

In the case of fuzzy rule-based systems, *defuzzification* is used to determine the output described in Chapter 6 and Chapter 9 on Intelligent control.

Fuzzy rule-based systems have proved to be very useful in control systems. Fuzzy rule-based controllers have become very popular, even to the extent that several companies now produce fuzzy-controller integrated circuits. These have been used in all sorts of consumer devices such as cameras and washing machines, as well as in large industrial process control systems. For a more detailed exposition of fuzzy control see Hopgood (1993).

7.4 Programming language and rule-based system shells

In principle, any computer language can be used to build a rule-based system. For example, the SmartLab shell is written in the C++ language. However, we had to build our own inference mechanisms and pattern matcher as discussed in Section 7.2. Of all the high-level languages available, Prolog is particularly suitable for building rule-based systems since the language itself has a pattern matcher built in. In Prolog, facts can be declared with statements such as on(pump), and rules can be built using the words *if* and *then*; connectives such as *and, or, not* and the logical quantifiers 'there exist' and 'for all'. These too are part of the language. For these reasons, many rule-based systems are programmed in Prolog.

The architecture of a rule-based system allows the knowledge base to separate from the inference engine. This means, in principle, that the same inference engine can be used for applications in many different knowledge domains. In practice this makes it worthwhile to develop 'empty' rule-based systems which have very good user-interfaces and make it easy to enter new facts and rules. The programmer therefore does not have to design new data structures or re-program the inference mechanism every time a new system is built. Indeed the system builder does not even need to know how to program the computer and can enter facts and rules using everyday language. As discussed in Section 7.2, SmartLab provides an example of what is called a rule-based system ***shell*** into which domain-specific knowledge can be entered to build rule-based systems. It is used by our students to build rule-based systems which do many things, including controlling an autonomous vehicle over a wireless communications link. Even students who do not know how to program a computer learn how to create working rule-based systems in a few hours.

7.5 Conclusion

Rule-based systems are becoming very common in mechatronic systems. The architecture allows new information to be quickly added to the knowledge base without having to make major alterations to the system, as would be the case in conventional programming. Some of the more recent developments, such as fuzzy logic, have revolutionized some parts of the engineering industry, and this is expected to continue into the future. As you will see in Chapter 11, rule-based systems are especially useful when integrated into the blackboard system architecture.

References

Hopgood, A. (1993) *Knowledge Based Systems for Scientists and Engineers*, CRC Press, London.

Winston, P.H. (1984) *Artificial Intelligence*, Addison-Wesley.

CHAPTER 8
LEARNING

8.1 Introduction

Since the earliest days of artificial intelligence it has been realized that machines with a fixed knowledge base are much more limited than those that can extend and change their knowledge base by learning. *Machine learning* involves:

- acquiring new information and knowledge
- acquiring new skills
- finding new ways of organizing existing knowledge.

When it is built, a machine will have a certain amount of information and knowledge designed into it. To learn it must also have some *meta-knowledge* built in, i.e. knowledge about knowledge. In particular, the machine must be able to absorb new data and operate on them so that they can be used in a purposeful way. This assumes that the machine is able to store this accumulating knowledge in appropriate data structures, that it has techniques for transforming raw data from its sensors into knowledge, and that it is able to manage its information base. For example, it may be necessary to overwrite old or redundant data.

Learning can be thought of as adaptation to the environment based on experience. This inevitably requires new knowledge, new skills, or the reorganization of existing knowledge. Usually the act of learning is motivated by attempts to improve a system by enabling better performance or avoiding poor performance.

The process of learning in human beings is very complex and imperfectly understood. It is clear that human beings have fabulous learning abilities, both in the control of their bodies and use of their minds. These can be observed from watching young children play and study at school. Although it is an area of intense research, machines do not have human-like learning abilities. It is important to realize that currently machines have a rather limited capacity for learning. The following categories of learning will be discussed in this chapter:

- learning by memory
- learning by updating parameters
- Bayesian learning
- learning from examples
- learning by analogy
- learning by observation and discovery.

There are other aspects to learning such as *learning by instruction*, *concept learning*, *learning by deduction*, and *learning by induction*, but they will not be considered here.

Learning cannot easily be separated from other aspects of machine intelligence. Pattern recognition (Chapter 2) frequently involves learning by example. Search (Chapter 3) will be seen to be particularly important in learning. Neural networks (Chapter 4) exemplify learning from examples. Scheduling (Chapter 5) relates to learning since successful activity schedules and paths can be learnt and used to evaluate new alternatives. Reasoning (Chapter 6) is important in deducing new knowledge from old, and rule-based systems (Chapter 7) provide an architecture which allows new facts to be deduced from existing facts. In this chapter you will see how a machine can learn new rules from data. Intelligent control (Chapter 9) also involves learning, as does computer vision (Chapter 10). Many of these connections will become clearer as we proceed.

8.2 Learning by memory

By ***learning by memory*** we mean the process of new data being stored in an unprocessed form and later used by the system. For example, a stream of data from a sensor may be stored in sequential memory to be processed when appropriate, or an image might be stored in a two-dimensional array. Although learning by memory is very simple, it plays an important role in machines.

Usually the memory will be digital, and the designer of an intelligent machine must estimate how much data will be stored in order to allocate enough memory hardware, which is usually implemented as random-access memory (RAM). Memories soon fill up, and some kind of memory management is required. In general, this means keeping records of what data are stored where, and which data are no longer required and can be overwritten.

In a machine, the most important data are usually those acquired most recently, and the memory management may involve cycling round a fixed allocation of memory overwriting the oldest data. Sometimes the old data may be converted and stored in a more compact summary form before being erased.

For example, consider the circuit board at the heart of an autonomous vehicle which can read 16 sensors. Suppose two of the sensors are special, since they are used to count the wheel rotations for dead-reckoning the vehicle's position. One wheel on each side of the vehicle has a cam mechanism which lifts a lever which makes and breaks a circuit twice per revolution. The microprocessor examines the data stream produced by each sensor through time, such as

...0100000011111110000001111110000000001110111101000000...

Inevitably, the lever mechanism has some degree of bounce, and the data stream is not a clean sequence of zeros (no contact) and ones (contact). This means that sufficient historical samples have to be logged as data for the debounce subroutine. This subroutine is able to detect the correct on–off sequence and so count the wheel revolutions backwards and forwards in time. The logged data are stored in memory and kept until the counting subroutine has done its work. The memory allocated to logging the data is fixed, and the 'start position' of the logged data cycles around, with the oldest sensor readings being overwritten by the current reading. The wheel rotation counts are stored passively until the system accesses them in order to calculate the vehicle's position.

Memory can be *distributed* over a system. For example, an autonomous vehicle which has very limited memory may be able to communicate information to a host PC which has much more memory, especially when its disk capacity is taken into account. However, the amount of data that can be stored elsewhere will depend on the bandwidth (defined in Chapter 3 of Volume 1) of the communication channels used.

Machines can use their environments to store information. To see this, consider the story of Theseus in the labyrinth under the palace of Knossos. Although it was impossible to learn a route through the myriad passages and openings, Theseus used the thread given to him by the king's daughter, Ariadne, to store the route information. Similarly, it might be useful for a machine to put down markers in its environment to store positional information.

8.3 Learning by updating parameters

Logging data is the simplest way for a machine to accumulate new knowledge. ***Parameter updating*** goes one step beyond this, by using the stream of incoming data to modify parameters within the machine. The wheelcounts discussed in the previous section provide a simple example of parameters which are updated as a result of processing an incoming data stream.

In Chapter 2 on 'Pattern recognition' you were told about data-to-data transformation. It was said that in many applications, the data are easier to manipulate in one form than in any other, so a transformation is used. One example is where data have to be sent along a communication channel which doesn't have a sufficiently large bandwidth to send the raw data in real time. The data therefore have to be compressed, and as more data are received the system has to learn the best form of compression. One way is to convert the data into its Fourier spectrum and to send only a relatively small number of the largest spectral components. The signal can be reconstructed at the receiving end and will be approximately the same as the original signal. In other words, the signal is modelled using the Fourier transform, which enables it to be represented by a small number of parameters.

The knowledge in this case is embedded in the spectral coefficients and is extracted as parameters from the data. More data may result in further changes to the spectral representation parameters and therefore to the model.

Earlier in this volume (Chapter 4) we described neural networks. Although these are relatively recent inventions, they are essentially parameterization networks. Learning consists of taking a set of known input–output data and using a form of gradient descent to search for the set of parameters (called *weights* in neural networks) that will best describe the input–output relationship of a given set of data.

In neural networks, and parameterization in general, what is described as 'learning' quite clearly involves searching. The behaviour of the networks can be described as learning, while the mechanism employed is searching. The search space is defined by some form of error function between the data received and the parameterized model of the data. The aim of the search is usually to minimize this error, and preferably to reduce it to zero. Therefore in parameterization, the machine sometimes learns by performing a search such as gradient descent.

8.4 Learning during execution using Bayesian updating

At any time a machine will be receiving new streams of data as it works. Some of these data will require that a parameter be updated in a relatively incremental way. Bayesian updating can be used for this.

In Chapter 6 on Reasoning, uncertainty was defined in terms of probabilities. A probability of 1 meant that an event was certain to happen, whereas a probability of 0 meant that an event was never going to happen. In between, the probability represents the certainty or confidence that an event will take place. Bayes' rule was shown as a way of calculating the probability of an event happening in the light of evidence from simpler probabilities that are known. This effectively gives us a way of learning 'on the hoof', and updating the confidence that we have of an event taking place.

Bayes' rule can be written as

$$p(H \mid E) = \frac{p(E \mid H) \times p(H)}{p(E)} \tag{8.1}$$

where

H is a hypothesis;

E is an example;

$p(H|E)$ is the probability of the hypothesis H being TRUE given that an example E has been found;

$p(E|H)$ is the probability of an example E being found given that the hypothesis H is TRUE;

$p(H)$ is the probability of a hypothesis H being TRUE;

$p(E)$ is the probability of an example E being found.

We will use these definitions in what follows:

$p(\neg H)$ is the probability of the hypothesis being FALSE;

$p(E|\neg H)$ is the probability of finding an example when the hypothesis is FALSE.

Let's look first at $p(E)$. This can be expanded by noting that the probability of finding an example, $p(E)$, equals the probability of the hypothesis being TRUE *and* finding an example when it is TRUE, $p(E|H) \times p(H)$, plus the probability of the hypothesis being FALSE and finding an example when it is FALSE, $p(E|\neg H) \times p(\neg H)$:

$$p(E) = p(E|H) \times p(H) + p(E|\neg H) \times p(\neg H) \tag{8.2}$$

Substituting this into the Bayesian expansion of $p(H|E)$ gives

$$p(H|E) = \frac{p(E|H) \times p(H)}{p(E)} = \frac{p(E|H) \times p(H)}{p(E|H)p(H) + p(E|\neg H)p(\neg H)} \tag{8.3}$$

But $p(\neg H) = 1 - p(H)$, so

$$p(H|E) = \frac{p(E|H) \times p(H)}{p(E|H)p(H) + p(E|\neg H)(1 - p(H))} \tag{8.4}$$

Suppose the initial, unknown, probability of H being true is p_0.

When the first example is found, E_1, the probability of the hypothesis being TRUE is updated to p_1 from the known or estimated initial value of p_0:

$$p_1 = p(H|E_1) = \frac{p(E_1|H) \times p_0}{p(E_1|H)p_0 + p(E_1|\neg H)(1-p_0)} \tag{8.5}$$

Note, it is assumed that $p(E_i|H)$ and $p(E_i|\neg H)$ are known for any new evidence E_i. The formula allows the probability of H to be updated every time new information comes in.

This is not a very elegant expression, so a new term called ***the odds*** is defined as the ratio of the probability of an event happening and the probability of the same event not happening:

$$O(H) = \frac{p(H)}{p(\neg H)} = \frac{p(H)}{1 - p(H)} \tag{8.6}$$

Thus

$$O(H|E) = \frac{p(H|E)}{1 - p(H|E)} \tag{8.7}$$

Equation (8.3) yields

$$p(H|E) = \frac{1}{1 + \dfrac{p(E|\neg H)\ p(\neg H)}{p(E|H)\ p(H)}}$$

which by equation (8.6) becomes

$$p(H|E) = \frac{1}{1 + \dfrac{p(E|\neg H)}{p(E|H)} \times \dfrac{1}{O(H)}}$$

From this

$$\frac{1}{p(H|E)} = 1 + \frac{p(E|\neg H)}{p(E|H)\ O(H)}$$

so that

$$\frac{1 - p(H|E)}{p(H|E)} = \frac{p(E|\neg H)}{p(E|H)\ O(H)}$$

By equation (8.7) this gives

$$O(H|E) = \frac{p(H|E)}{1 - p(H|E)} = \frac{p(E|H)\ O(H)}{p(E|\neg H)}$$

Thus we have

$$O(H|E) = \frac{O(H) \times p(\mathrm{E|H})}{p(E|\neg H)} \tag{8.8}$$

and updating gives us

$$O_1 = \frac{O_0 \times p(E_1|H)}{p(E_1|\neg H)} \tag{8.9}$$

At any stage, the odds can be converted back to a probability using the equation

$$p(H) = \frac{O(H)}{1 + O(H)} \tag{8.10}$$

Example 1

To illustrate these ideas, consider a bag which contains 10 coins, one of which is double-headed. Take out one coin and toss it a number of times. At each toss, what is the probability of the coin being the double-headed one?

Here the hypothesis, H, is that the double-headed coin has been pulled out of the bag.

The probability of pulling the double-headed coin out of the bag is 0.1. So initially, the probability of it being the double-headed coin is 0.1.

$$p_0 = 0.1$$

From equation (8.5)

$$p_1 = p(H|E_1) = \frac{p(E_1|H) \times p_0}{p_0\, p(E_1|H) + (1 - p_0)p(E_1|\neg H)}$$

where $p(E_1|H)$ is the probability of it being heads when it is tossed given that it is the double-headed coin, which is therefore 1. The term $p(E_1|\neg H)$ is the probability of heads, given that it is not the double-headed coin, which is 0.5. So, assuming the coin comes down heads,

$$p_1 = p(H|E_1) = \frac{1 \times 0.1}{0.1 \times 1 + (1 - 0.1) \times 0.5} = 0.18$$

Using equation (8.5) again,

$$p_2 = p(H|E_2) = \frac{p(E_2|H) \times p_1}{p_1 p(E_2|H) + (1 - p_1)\, p(E_2|\neg H)}$$

So, assuming the coin lands as heads the second time, we have

$$p_2 = p(H|E_2) = \frac{1 \times 0.18}{0.18 \times 1 + (1 - 0.18) \times 0.5} = 0.30$$

We could keep going like this until p_n approaches 1; that is, when n is large and the coin has not once come down tails we become almost certain that the coin is double-headed.

Now do the same calculation using odds. From equation (8.6) the odds are

$$O(H) = \frac{p(H)}{1 - p(H)}$$

$$O_0 = \frac{p_0}{1 - p_0} = \frac{0.1}{1 - 0.1} = 0.11$$

Equation (8.9) then becomes

$$O_1 = O_0 \times \frac{p(E_1 | H)}{p(E_1 | \neg H)} = O_0 \times \frac{1}{0.5} = 2.0 O_0$$

Using odds we simply double the odds every time the coin comes down heads when it is tossed. Odds therefore do not range between 0 and 1, but range from 0 to infinity. They are perhaps more difficult to interpret, but simpler to update.

It is easy to convert odds back to probabilities. In this case $O_1 = 2.0O_0 = 0.22$, and $O_2 = 2.0O_1 = 0.44$. By formula (8.10),

$$p(H) = \frac{O(H)}{1 + O(H)} = \frac{0.44}{1 + 0.44} = 0.3$$

so that $p_2 = 0.3$, which is the value found previously.

In the cases of either probabilities or odds, if the tossed coin ever lands tails then the probability or odds should go to zero. The term $p(E|H)$ means the probability of an event happening given that the hypothesis is TRUE. Well, if the hypothesis is TRUE that the coin is double-headed, the probability of it landing tails is 0. Since the probability or odds are multiplied by this term, if the coin ever lands tails the value drops to 0 as expected.

Example 2

This method is used in expert systems for diagnosis. Consider a machine which monitors itself and tries to diagnose potential faults before they cause major damage. Initially the probability of any particular machine having any one of these faults is derived from statistics of the whole population of machines tested to date.

Next, the symptoms of each fault are stored as a probability. For example, suppose some of the faults usually involve a high temperature in some part of the machine. Let's say we have three faults, A, B and C, and the probability of a machine having any one of these faults is derived from the test population and found to be

$$p(A) = 0.01, \quad p(B) = 0.05, \quad p(C) = 0.10$$

Converting to odds gives

$$O(A) = 0.010, \quad O(B) = 0.053, \quad O(C) = 0.111$$

A and C are usually accompanied by high values of temperature, T, so they might have probabilities of

$$p(T|A) = 0.8, \quad p(T|B) = 0.04, \quad p(T|C) = 0.7$$

Similarly, B and C are often accompanied by fluctuations of voltage, V, with probabilities

$$p(V|A) = 0.03, \quad p(V|B) = 0.75, \quad p(V|C) = 0.60$$

We also need to know the probability of high temperatures and voltage fluctuations when a machine does not have any of these faults. Again these statistics would be available for a population of test machines, and might be empirically determined to be

$$p(T|\neg A) = 0.30, \quad p(T|\neg B) = 0.28, \quad p(T|\neg C) = 0.32$$

$$p(V|\neg A) = 0.40, \quad p(V|\neg B) = 0.30, \quad p(V|\neg C) = 0.35$$

If we work with odds, then the updating factor for A when a high temperature, which we shall call K_T, is found is

$$\frac{p(T|A)}{p(T|\neg A)} = \frac{0.8}{0.3} = 2.667 = K_T(A)$$

Similarly,

$$K_T(B) = 0.143, \; K_T(C) = 2.188$$

When a machine detects voltage fluctuations

$$\frac{p(V\;|A)}{p(V\;|\neg A)} = \frac{0.03}{0.4} = 0.075 = K_V\;(A)$$

Similarly,

$$K_V(B) = 2.50, \; K_V(C) = 1.714$$

So, if a machine detects a high temperature with voltage fluctuations the probabilities of each of the faults is updated in two steps.

First, the high temperature:

$$O_1(A) = O_0(A) \times K_T(A) = 0.010 \times 2.667 = 0.027$$

$$O_1(B) = O_0(B) \times K_T(B) = 0.053 \times 0.143 = 0.008$$

$$O_1(C) = O_0(C) \times K_T(C) = 0.111 \times 2.188 = 0.243$$

Next, the voltage fluctuation:

$$O_2(A) = O_1(A) \times K_V(A) = 0.027 \times 0.075 = 0.002$$

$$O_2(B) = O_1(B) \times K_V(B) = 0.008 \times 2.50 = 0.020$$

$$O_2(C) = O_1(C) \times K_V(C) = 0.243 \times 1.714 = 0.417$$

After calculating these odds, fault C has the highest probability. We can convert these odds into the more conventional probability using formula (8.10):

$$p(C) = \frac{O(C)}{1 + O(C)} = \frac{0.417}{1 + 0.417} = 0.294$$

Therefore the probability of there being a fault C is 0.294. At this stage in the self-diagnosis the machine could instigate the repair of fault C, update parameters or rules concerning potential problems with C, or seek more data.

Thus we have a method of updating probabilities as evidence accumulates which a machine can use for self-diagnosis. When indications of faults appear it can update the probabilities of the various causes.

More generally, this method gives mechatronic systems the ability to update parameters about the environment when the environment is complex and the data are somewhat imprecise. In Chapter 6 on Reasoning we saw how machines could reason with probabilities. Bayesian learning provides a method for a machine to update the probability values during operation.

8.5 Learning from examples

8.5.1 Classification through training

A machine *learns from examples* when it infers relationships between things on the basis of examples.

In Chapter 4 of Volume 1 we described images on an 8×8 grid of pixels obtained from a wire scanner. Suppose a character **3** had been scanned. Then the data for the grid are stored in memory, along with the number **3**. In this way the system learns the image produced by a particular scan of the **3**. By matching this with subsequent scans the system is able to decide if the character was a **3**.

Neural networks also learn by example. In this case, examples of input–output pairs are shown during training. The network's weights (parameters) change as it learns, and subsequently it is able to classify the inputs in terms of the outputs. Subsequently it can generalize from the training data, and classify further examples of inputs.

8.5.2 Learning rules by searching for relationships

How can a machine abstract knowledge from data in the way that we do? For example, people say 'red sky at night, shepherd's delight; red sky in the morning, shepherd's warning'. This kind of heuristic for weather forecasting is obviously based on years of observation, and generalizing from those data.

To illustrate the general idea, suppose an autonomous vehicle uses dead-reckoning to determine its position in the environment, i.e. it calculates its position according to the number of revolutions of its wheels. Since the wheels may slip this dead-reckoning is subject to errors which may accumulate over time. The vehicle can correct these errors when it encounters a known object, but sometimes the errors will be too great for this: expected objects will not be found, and the machine will not know whether this is because they have moved or because it is lost.

Suppose the vehicle has two strategies to deal with this situation: the first strategy (A) assumes that the known object has moved, and seeks another known object; the second (B) assumes that the vehicle is lost and re-maps the environment in the hope that the new map can be matched against the old map and the vehicle can relocate itself.

When it works, strategy A consumes less time, but when it fails the machine reverts to strategy B and the time spent on A is wasted. Suppose the success of strategy A depends on the distance travelled since a known object was successfully recognized.

TABLE 8.1

Distance travelled	**Strategy A successful**
175.4	Yes
293.3	Yes
805.9	Yes
930.5	No
1001.8	No
1123.2	Yes
1305.6	No
1565.4	No

These data do not give a clear-cut value of distance below which strategy A is always successful and above which A is always unsuccessful. However, the rules

If (distance < 868.2)

Then adopt strategy A

If (distance $\geqslant 868.2$)

Then adopt strategy B

will result in these data being correctly classified for seven out of the eight sample data. The number 868.2 is obtained here by taking the mean of 805.9 and 930.5.

This is a simple example of how a machine can learn a rule by inspecting a data set. Such rules can then be used in rule-based systems, as discussed in Chapter 7. We will now show how this idea can be extended to a more complex data set which generates a multi-branch decision tree.

Here we will describe a mechanism for abstracting rules from tabulated data which is based on ***Quinlan's TDIDT method*** proposed in the 1980s. TDIDT stands for top-down induction of decision trees. This form of learning uses *best-first search* to build a decision tree.

Consider a complex mechatronic system involving many autonomous mobile machines engaged in construction work in a hostile environment. When it rains the machines cannot work on some parts of the site, so to optimize the scheduling of the machines it is necessary to know if it will rain or be fine the next day.

One way to do this is to take some measurements of today's weather and see if there are some rules that could be applied that will predict tomorrow's weather.

Recall the old saying 'red sky at night, shepherd's delight'. This is a heuristic that is based on observation, but can also be supported by scientific knowledge. In the UK the weather fronts tend to come from the Atlantic in the west and the sun sets in the west. So if the sky is red at sunset it means that light is able to get through the atmosphere after the sun has set below the horizon because there isn't much cloud. The lack of cloud means that it probably won't rain in the morning.

As with all rules, there will be exceptions. The rule itself is not 100% accurate, but it might be possible to add more qualifying rules to improve the overall accuracy. So you might add to the saying 'red sky at night' the qualifier 'and the wind is light' to give delight to the shepherd.

One method is to use a best-first search (defined in Chapter 3 on Search). Starting with a rule that is generally useful but not 100% accurate, one looks for exceptions to the rule and finds new qualifying rules to supplement the original one. The search method uses a database of statistics and gradually finds a set of rules that describe the statistics as correctly as possible.

Table 8.2 gives weather statistics for each day in March 1992: the rainfall, hours of sunshine, maximum and minimum temperatures. It also shows what the weather was like on the day after the data were collected.

TABLE 8.2 WEATHER FOR THE 31 DAYS IN MARCH 1992 FOR THE LONDON REGION

Day in March	T_{min} °C	T_{max} °C	Rainfall mm	Sunshine hours	Weather next day
1	9.4	11.0	17.5	3.2	Rain
2	4.2	12.5	4.1	6.2	Rain
3	7.6	11.2	7.7	1.1	Rain
4	5.7	10.5	1.8	4.3	Dry
5	3.0	12.0	0.0	9.5	Dry
6	4.4	9.6	0.0	3.5	Dry
7	4.8	9.4	0.0	10.1	Rain
8	1.8	9.2	5.5	7.8	Rain
9	2.4	10.2	4.8	4.1	Rain
10	5.5	12.7	4.2	3.8	Rain
11	3.7	10.9	4.4	9.2	Rain
12	5.9	10.0	4.8	7.1	Rain
13	3.0	11.9	0.2	8.3	Dry
14	5.4	12.1	0.0	1.8	Rain
15	8.8	9.1	8.8	0.0	Rain
16	2.4	8.5	3.0	3.1	Rain
17	4.3	10.8	4.2	4.3	Dry
18	3.4	11.1	0.0	6.6	Rain
19	4.4	8.4	5.4	0.7	Rain
20	5.1	7.9	3.0	0.1	Rain
21	4.4	7.3	1.0	0.0	Dry
22	5.6	14.0	0.0	6.8	Dry
23	5.7	14.0	0.0	8.8	Dry
24	2.9	13.9	0.0	9.5	Dry
25	5.8	16.4	0.0	10.3	Dry
26	3.9	17.0	0.0	9.9	Dry
27	3.8	18.3	0.0	8.3	Dry
28	5.8	15.4	0.0	7.0	Rain
29	6.7	8.8	6.4	4.2	Dry
30	4.5	9.6	0.0	8.8	Rain
31	4.6	9.6	3.2	4.2	Rain

The aim of the weather forecasting system is to find a rule that predicts if the next day will be raining (or ¬DRY). DRY in this context is defined as no rain.

The best-first search algorithm constructs a decision tree. The root node is the single rule that best describes the data. For the data in Table 8.2 there are a number of rules that could be found. Probably the simplest are rules of the form:

If (variable > constant)

Then Prediction

For example:

If (Sunshine $\geqslant$ 4.2 hours)

Then (Nextday DRY)

This is correct 20 times out of 31, or 65%, as shown in Table 8.3. The value of the constant is chosen so that the number of times that the rule is correct is maximized.

TABLE 8.3 PREDICTIONS FROM *If* (Sunshine $\geqslant$ 4.2 hours) *Then* (Nextday DRY)

Day	Sunshine hours	Actual	Predicted	Correct
1	3.2	Rain	¬DRY	Yes
2	6.2	Rain	DRY	No
3	1.1	Rain	¬DRY	Yes
4	4.3	Dry	DRY	Yes
5	9.5	Dry	DRY	Yes
6	3.5	Dry	¬DRY	No
7	10.1	Rain	DRY	No
8	7.8	Rain	DRY	No
9	4.1	Rain	¬DRY	Yes
10	3.8	Rain	¬DRY	Yes
11	9.2	Rain	DRY	No
12	7.1	Rain	DRY	No
13	8.3	Dry	DRY	Yes
14	1.8	Rain	¬DRY	Yes
15	0.0	Rain	¬DRY	Yes
16	3.1	Rain	¬DRY	Yes
17	4.3	Dry	DRY	Yes
18	6.6	Rain	DRY	No
19	0.7	Rain	¬DRY	Yes
20	0.1	Rain	¬DRY	Yes
21	0.0	Dry	¬DRY	No
22	6.8	Dry	DRY	Yes
23	8.8	Dry	DRY	Yes
24	9.5	Dry	DRY	Yes
25	10.3	Dry	DRY	Yes
26	9.9	Dry	DRY	Yes
27	8.3	Dry	DRY	Yes
28	7.0	Rain	DRY	No
29	4.2	Dry	DRY	Yes
30	8.8	Rain	DRY	No
31	4.2	Rain	DRY	No

Other possible rules include

If (Rainfall $\geqslant$ 2 mm)

Then (Nextday $\neg$DRY)

which is correct 24 times out of 31, or 77%;

If ($T_{min} \geqslant 6°C$)

Then (Nextday DRY)

which is correct 16 times out of 31, or 52%; and

If ($T_{max} \geqslant 13°C$)

Then (Nextday DRY)

which is correct 23 times out of 31, or 74%.

Of all these rules, the rainfall measure is the best predictor, correctly classifying 77% of the data. This divides the tree into two branches as shown in Figure 8.1.

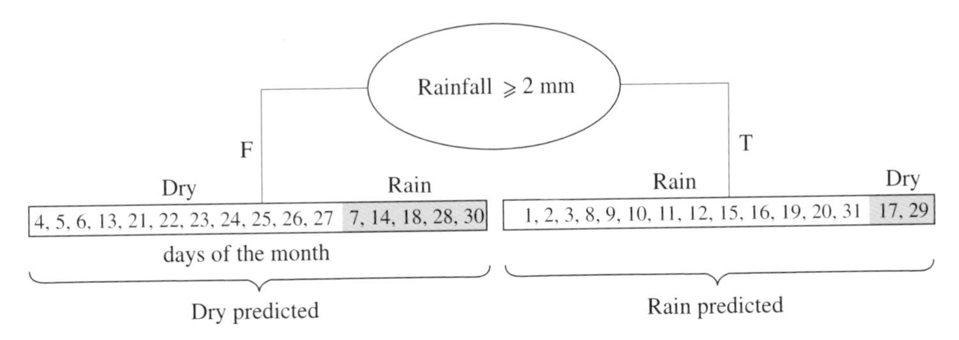

Figure 8.1
Dividing the data using the rule Rainfall $\geqslant$ 2 mm. (Incorrect predictions are shaded.)

The next step is to move down one of the branches and find the rule that works best at classifying the data again. If we follow the right-hand branch, a set of possible rules with the constants chosen to maximize the discriminatory power of the rules are:

Rule		Correct out of 15	%
If (Sunshine ⩾ 4.3 hours)	*Then* (Nextday DRY)	10	67%
If (Rainfall ⩾ 6.5 mm)	*Then* (Nextday ¬DRY)	5	33%
If (T_{min} ⩾ 6.7°C)	*Then* (Nextday DRY)	11	73%
If (T_{max} ⩾ 10.8°C)	*Then* (Nextday DRY)	9	60%

Notice that all the variables are examined. In this instance the third rule is correct most often, so is used as the second test and creates two new branches in the tree as shown in Figure 8.2.

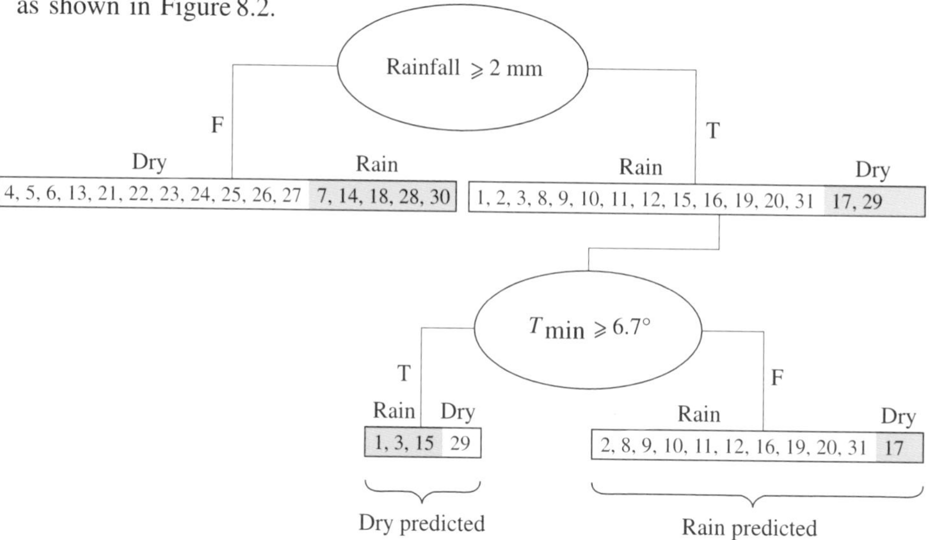

▲ ***Figure 8.2***
A further division using the rule T_{min} ⩾ 6.7°C. (Incorrect predictions are shaded.)

Next we will follow the new left-hand branch. There are only four entries, three which are ¬DRY and one which is DRY. The rule:

If (Sunshine ⩾ 4.2 hours)

Then (Nextday DRY)

is correct 4 out of 4 times, or 100%, as shown in Figure 8.3.

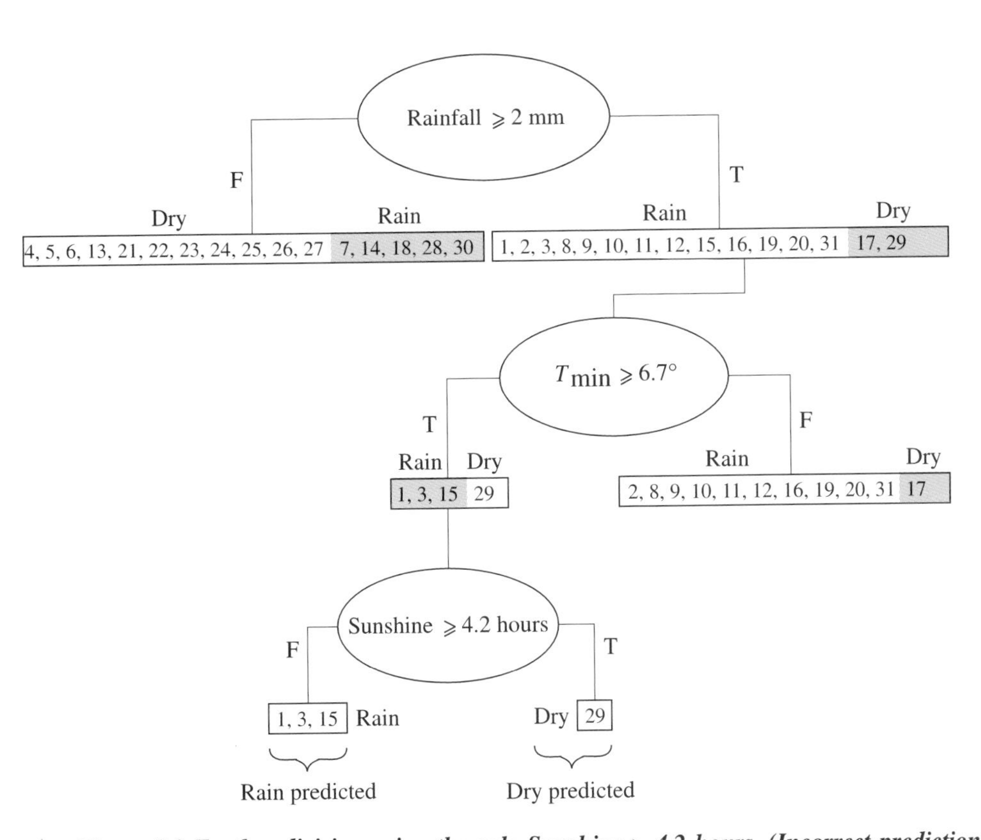

▲ ***Figure 8.3 Further division using the rule Sunshine ⩾ 4.2 hours. (Incorrect predictions are shaded.)***

Figure 8.4 shows the tree half complete. It is left for you as an exercise to complete the other half. The tree of Figure 8.4 is interpreted as:

If		(Rainfall $\geqslant$ 2 mm)	
	AND	$\neg(T_{\text{min}} \geqslant 6.7°\text{C})$	
	AND	$(T_{\text{max}} \geqslant 10.8°\text{C})$	
	AND	¬(Sunshine $\geqslant$ 6.2 hours)	This part of the rule
	AND	(Sunshine $\geqslant$ 4 hours)	corresponds to day 17.
OR			
		(Rainfall $\geqslant$ 2 mm)	
	AND	$(T_{\text{min}} \geqslant 6.7°\text{C})$	This part of the rule
	AND	(Sunshine $\geqslant$ 4.2 hours)	corresponds to day 29.
OR			
		¬(Rainfall $\geqslant$ 2 mm)	
	AND		This part of the rule
	⋮	⋮	covers the left half
			of the tree.
Then		(Nextday DRY)	

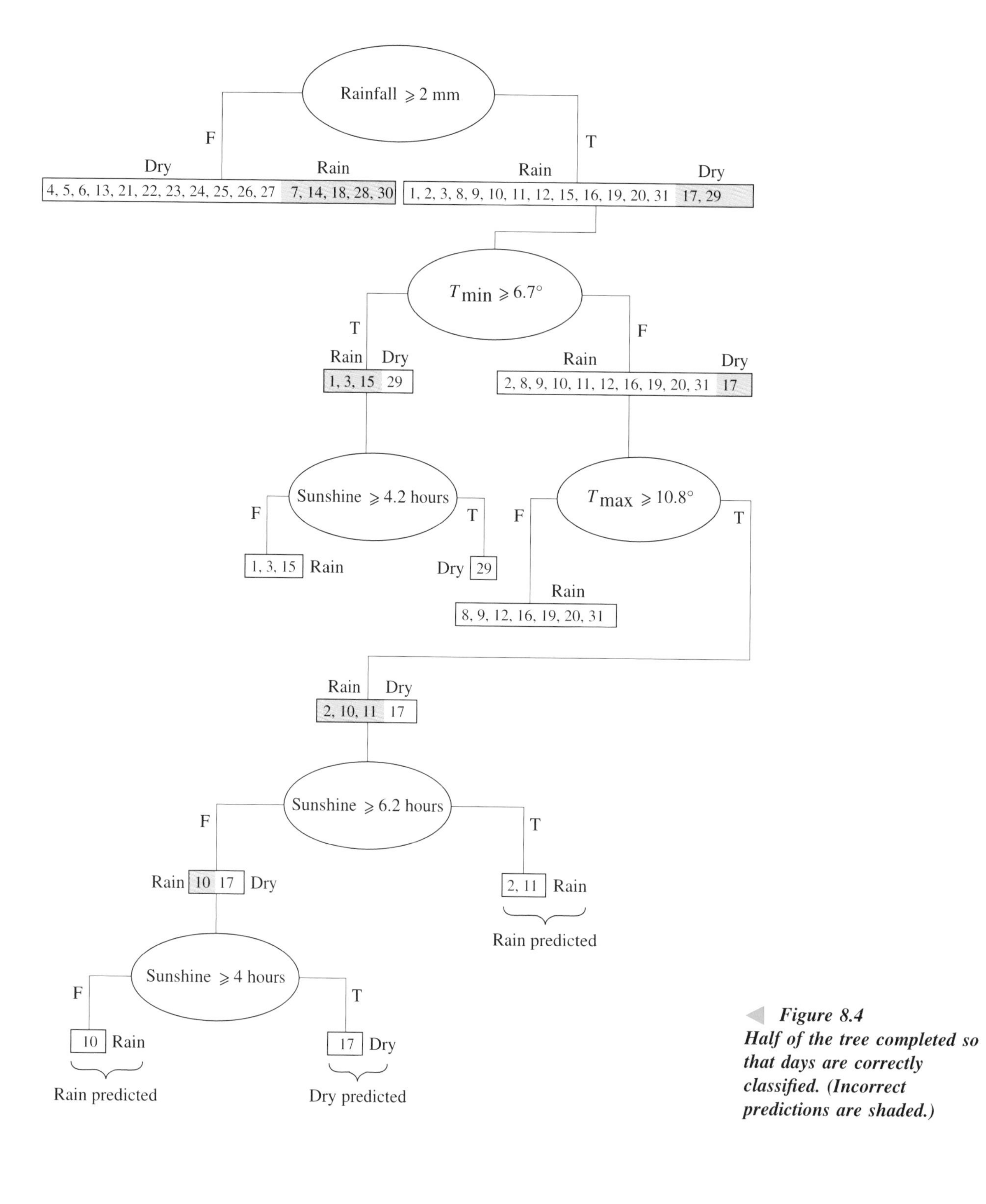

Figure 8.4
Half of the tree completed so that days are correctly classified. (Incorrect predictions are shaded.)

This example shows that even with relatively simple rules it is possible to classify the data. The way that the rules are produced can be described as learning by example, where the system learns from tabulated data. Whether or not the resulting rules are useful for classifying new data depends on how representative the training data are of the sort of data it can expect to have to classify.

This method of abstracting rules from data is not incremental. If the system is learning continuously, on making every new observation it would have to go through this process from the beginning to build a new decision tree and new rules. This is a case in which the system may log new data (learning by memory) and revise its rule base (learning from examples and experience) later when it is idle.

A disadvantage of this approach is that rogue data, i.e. abnormal or incorrect observations, may not be detected and may result in incorrect rules.

There are many more sophisticated ways of constructing these rules, but the basic principle is the same. Quinlan's ID3 (Interactive Dichotomizer 3) is basically the same, but instead of using the percentage of correct classification as the evaluation function it uses a measure called entropy. Other methods use standard statistical techniques to show that the classification is better than chance alone could do.

8.6 Learning by analogy

When confronted with a new problem, the history of the system can be searched to see if a similar problem has been seen (and solved) before. If one is found, the method of finding a solution can be applied again to the new problem to see if it works. If it does, then it is possible to say that the problem has been solved by analogy to the other problem.

This can be illustrated by the following problems. The first requires you to prove that RN = OY given that RO = NY in Figure 8.5(a).

A proof goes like this:

Step 1:	RO	= NY	(given)
Step 2:	RO + ON	= NY + ON	(add ON to both sides)
Step 3:	RN	= NY + ON	(RN = RO + ON)
Step 4:	RN	= OY	(NY + ON = OY)

The second problem requires a proof that, in Figure 8.5(b), angle $\angle$BAD = angle $\angle$CAE, given that $\angle$BAC = $\angle$DAE.

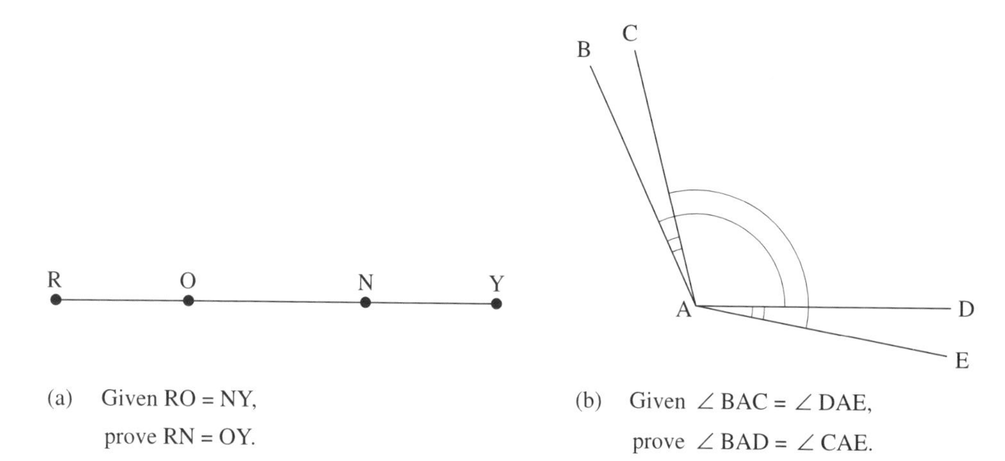

(a) Given RO = NY, prove RN = OY.

(b) Given ∠ BAC = ∠ DAE, prove ∠ BAD = ∠ CAE.

Figure 8.5 Analogical reasoning in geometry.

The analogical proof goes like this:

Step 1:	∠BAC	= ∠DAE	(given)
Step 2:	∠BAC + ∠CAD	= ∠DAE + ∠CAD	(add ∠CAD to each side)
Step 3:	∠BAD	= ∠DAE + ∠CAD	(∠BAD = ∠BAC + ∠CAD)
Step 4:	∠BAD	= ∠CAE	(∠CAE = ∠DAE + ∠CAD)

The analogy works by associating angles with lines, and addition of lengths with addition of angles.

Figure 8.6 shows another example of learning by analogy. In the first case, Figure 8.6(a), the vehicle learns how to solve the particular navigation problem. Although the environment is different in Figure 8.6(b), the analogy between the two problems is that the desired trajectory cannot be achieved if the vehicle goes exclusively in forward motion. In the first case, reversing and then going forward solves the problem. In the second case the analogy works if this tactic is applied twice.

In order to reason by analogy a machine has to establish the analogy between the parts of the problems, and the analogy between the relationships between those parts. This poses some difficult pattern recognition problems which, during learning, the machine must set up and solve by itself. In particular, the machine must be able to *represent* the problem and its solution in a way which allows it to make subsequent analogies. In principle, learning by analogy can be very powerful, but in practice analogies of any complexity can be hard to implement.

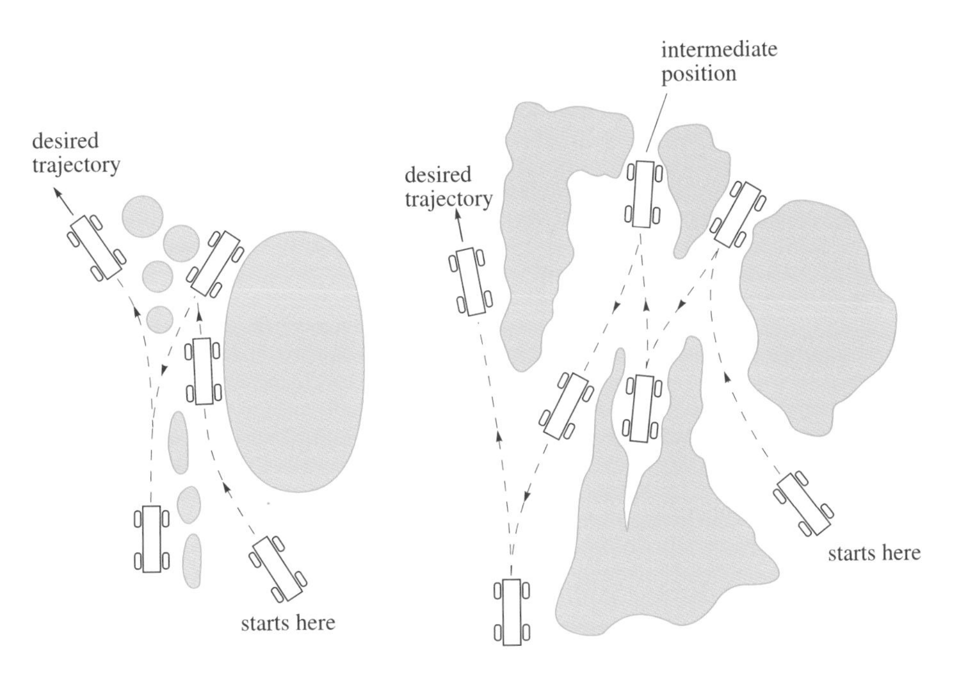

(a) The vehicle has to reverse in order to go forward on its desired trajectory.

(b) By analogy, the vehicle has to reverse and go forward to the intermediate position. From there it can reverse and go forward on its desired trajectory.

◀ ***Figure 8.6* *Learning by analogy.***

8.7 Learning by discovery

Learning by discovery allows the machine to form its own conclusions, based on the data it receives. An example of such a system is a self-organizing neural network. As input data arrive they start to form clusters in the pattern space. The clusters can be separated to give output classifications. How the data are organized is determined by the network alone, so the network 'discovers' relationships between its inputs and its outputs for itself.

Example

As another example, consider a battery-powered autonomous vehicle exploring a new and totally unknown environment in order to learn a new map. Suppose the vehicle has only touch sensors: in order to map out the environment the vehicle must move until its sensors respond to something. Since the shortest distance between two points is a straight line, let us suppose that the vehicle is programmed to move off in an arbitrary direction until its sensors are activated. Suppose that on encountering an object, the vehicle is programmed to back off, move to one side and seek another response. In this way the vehicle can build up a map of points at which it sensed an object. From this it can build a picture of its universe. As far as the machine is concerned the data stream through time appears as

```
...fffbbbaaafffcccfffbbbaaafffcccfff...
...001000000000000001000000000000001...
```

where f = forward, b = backward, c = clockwise, a = anticlockwise, 1 means an object sensed and 0 means no object is sensed. These data correspond to the positions shown in Figure 8.7.

Thus with respect to Figure 8.7, the vehicle is going forward at t_1 with no object sensed (f,0), forward at t_2 with no object sensed (f,0), and forward at t_3 when it senses an object (f,1). At t_4 the vehicle is going backwards with no object sensed (b,0), as it is at t_5 and t_6. At t_7, t_8, t_9 the vehicle rotates anticlockwise through a total of 90° with no object sensed (a,0). At t_{10}, t_{11}, t_{12} it goes forward with no object sensed (f,0). At t_{13}, t_{14}, t_{15} the vehicle rotates a total of 90° clockwise with no object sensed (c,0). At t_{16} and t_{17} it goes forward with no object sensed (f,0), and at t_{18} it is going forward when it senses an object (f,1). The vehicle repeats this and thus senses three colinear objects, which it may interpret as an edge.

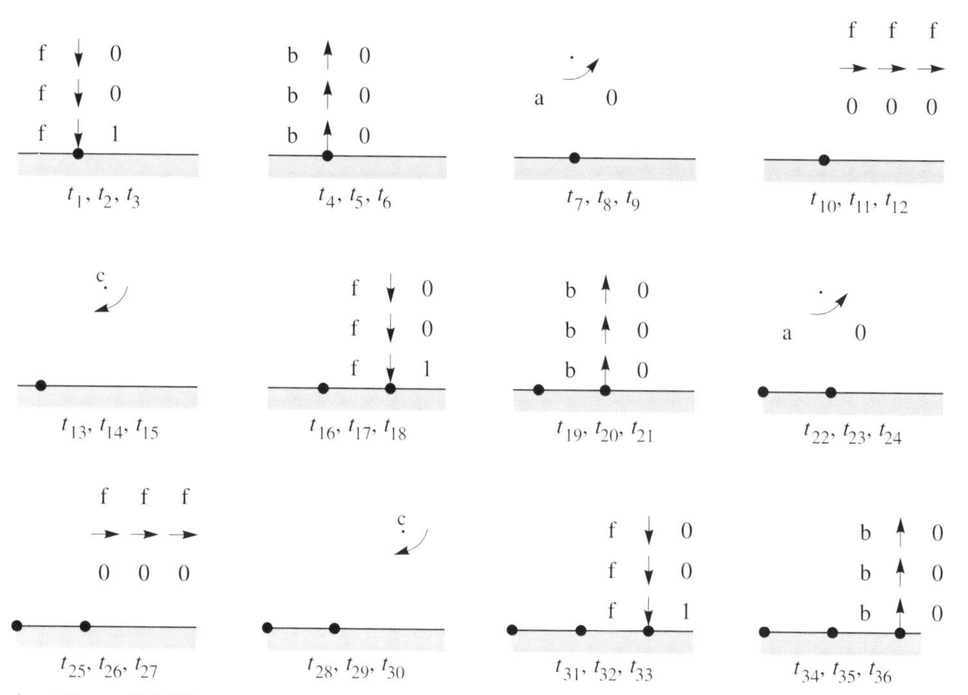

Figure 8.7 The sequence of data associated with an autonomous vehicle learning its environment.

If the vehicle uses coordinate geometry to represent the points at which it detects objects, the discovery of a sequence such as that shown in Figure 8.7 can be tested for its geometric properties. For example, one can try to fit a curve to the data. In this case linear regression would show that the points form a straight line, and give the equation of that line in terms of coordinate geometry. There are techniques for fitting sets of points to other geometric objects such as arcs of circles and other curves, and the vehicle can, in principle, take samples of objects and learn the shape of their edges.

So far we have considered a vehicle which represented its environment using coordinate geometry. This is called a *vector* representation, since objects are represented by sequences of numbers which represent points, lines, and other geometric objects.

There is another widely used representation for two-dimensional space, namely an array, or *grid*, of *cells* (usually squares) which represent areas of the actual space. For example, a square cell might represent a square metre. This is called the *resolution* of the grid. Any finite area will be represented by a finite number of cells. This representation is popular because it is easy to associate the cells with *pixels*, and display the representation of the environment on a computer screen.

Consider an autonomous vehicle which is planning its path in an environment about which it has imperfect knowledge. For example, it may have a map of the fixed part of the environment, but not know *a priori* if it will encounter other objects able to move in the space.

Suppose the vehicle senses an unexpected object. The new data about a previously unknown object fills in an area of that map that was previously blank, and that is *learning*. Using that data later to find a new path is *search*. Knowledge can be learnt even if it is never used. In an unpredictable world you cannot say in advance what is useful and what is not, so all knowledge that is acquired is potentially useful in some future circumstances.

The database may be a two-dimensional array in which a zero represents unoccupied space and a 1 represents an object, as illustrated in Figure 8.8.

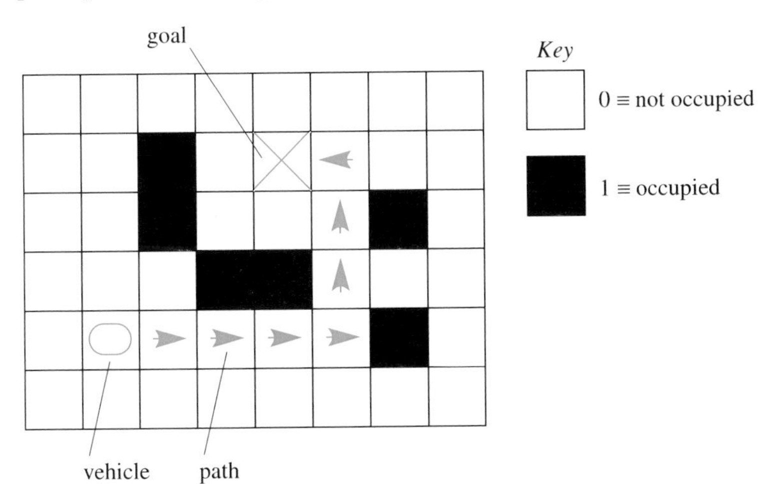

Figure 8.8
Binary array used to map the environment.

Let's assume that initially the map is full of zeros. Suppose that the vehicle moves around trying to reach the goal square and when it bumps into something it records a 1 in the appropriate cell of the array. Gradually it will build up a map of the whole environment. The raw input data are signals from sensors indicating the presence or absence of an object.

Some 'decoding' has to take place to work out the position of the object relative to the vehicle, so that a 1 can be placed in the correct position on the map. Filling in the map corresponds to learning about the environment and is not linked to any specific task that may be required to be performed in that environment.

8.8 Conclusion

In this chapter we have seen that learning involves many of the ideas discussed in this book, especially search. Learning describes some of the behaviour of a machine. A machine can be said to have learnt if its behaviour is altered by the input of new data. We have considered:

- *Learning by memory:* We described this as the accumulation of data in memory, e.g. data logging.
- *Learning by updating parameters:* This involves the machine using incoming data to update parameters.
- *Bayesian learning:* This method allows probabilities to be updated as more knowledge about cases is accumulated. We illustrated it by a machine monitoring itself for fault diagnosis.
- *Learning from examples:* This occurs when a machine generalizes from a set of examples. We illustrated this by showing how rules to predict tomorrow's weather can be abstracted from historical weather data.
- *Learning by analogy:* This happens when the machine makes an analogy between a problem it has solved before and a new problem. We illustrated this with a vehicle path-planning problem.
- *Learning by observation and discovery:* This occurs when a machine purposefully interacts with its environment in order to acquire data from which it can abstract useful knowledge. We illustrated this with an example of an autonomous vehicle mapping out its environment.

Although the techniques we have described make a useful start, they come nowhere close to the fantastic learning abilities of human beings. Learning remains an important and very active research area in artificial intelligence and the design of intelligent machines.

CHAPTER 9 INTELLIGENT CONTROL

9.1 Introduction

The subject of control was introduced in Volume 1, Chapter 8, by presenting methods for controlling systems which can be described by linear models using differential equations. As a reminder, a system is *linear* if it has the property that a sinewave with a frequency f applied at the input results in a sinewave with a frequency f appearing at the output once the system has settled down. The only differences between the two sinewaves are their amplitudes and their relative phase. This should be true for all relevant values of f, and should not vary with time.

This kind of model is a mathematical description of the behaviour of a system, and will always be an approximation to the actual behaviour of the system. When a linear model is used as the basis for the design of a controller, the performance of the controller will depend on the accuracy of the model. There are many systems for which a linear model is sufficiently accurate for linear control to be appropriate. However, there are many systems which are not – and a linear model would be so inaccurate that any controller designed using a linear model would perform badly. In these systems it is appropriate to use more complex models such as non-linear or time-varying models. Other complications that might arise are systems for which there are no known models, or systems where it is impossible to measure the output directly. Another difficulty arises in systems which are too complex to be represented or modelled by mathematical functions alone, and in which qualitative relational information must also be used. In these cases the conventional control methods described in Volume 1 may be inadequate.

There is at present a great deal of research effort being applied to the realm of 'intelligent' control. The term ***intelligent control*** reflects the fact that these techniques arise in the discipline of artificial intelligence, and does not infer that these techniques are 'cleverer' than classical methods. Intelligent control is most useful in situations where classical linear control is not suitable. There are basically three ways that intelligent control overcomes the limitations or the lack of a model. The first is that it may *learn* to control a system using methods such as neural networks and genetic algorithms which do not explicitly require a model. The second is that it can make do with very simple models, such as descriptions of a system in words, and takes this description to produce a controller using fuzzy logic. The third is that it can use incomplete and imprecise models and overcome the related uncertainty by using techniques from artificial intelligence.

In this chapter a distinction will be made between controlling systems which can be represented by mathematical formulae, and controlling systems which are too complex for this. Complex systems are usually hierarchical, and at the lowest levels it is not uncommon to find subsystems which can be modelled by formulae. For example, a motor car engine may be modelled by relationships between numerical variables, but a road system containing many motor cars cannot adequately be modelled by formulae alone.

To illustrate the application of different methods of intelligent control to a system which can be modelled using equations, a problem called the ***broom-balancer*** will be investigated. At some time you may have tried to keep a pole or a broom balanced on your hand, and you may have been quite good at it. If not, have a try (preferably somewhere safe where things will not get broken) and you will find that it is quite a tricky control problem. This problem, also called the *trolley and pole* or the *inverted pendulum*, has been widely used to investigate intelligent control strategies. It has the advantage that the system can be controlled using classical linear methods, and this can be used as a benchmark for intelligent techniques.

After some in-depth study of intelligent control applied to the broom-balancer, the more general problems of hierarchical control will be discussed.

9.2 The broom-balancer

Figure 9.1 shows the basic structure of the broom-balancer. A trolley runs on a track, like a railway, and the broom handle (or pole) is hinged to the trolley (or cart), pivoting in the same plane. The aim is to keep the broom balanced for as long as possible without moving the trolley beyond the ends of the track.

A similar problem arose when people first tried to launch rockets. After taking off, a rocket should be pointing upwards. If it is tilted over at an angle it will tend to rotate due to gravity, yielding a problem rather like trying to balance a broom. This led engineers to the solution of placing horizontal thrusters at the base of the rocket to supply horizontal forces to compensate for this rotation and keep the rocket pointing upwards.

In the two-dimensional problem of the broom-balancer, the trolley is moved by applying a horizontal force in either direction – left or right in Figure 9.1. When the broom starts to fall to the right, the trolley is moved to the right and the broom should move back to a vertical position. One more feature is added to the system, namely that the track has end-stops, so that not only has the broom got to be balanced but the trolley must stay near the centre of the track. (In a rocket this final limitation is not so critical.)

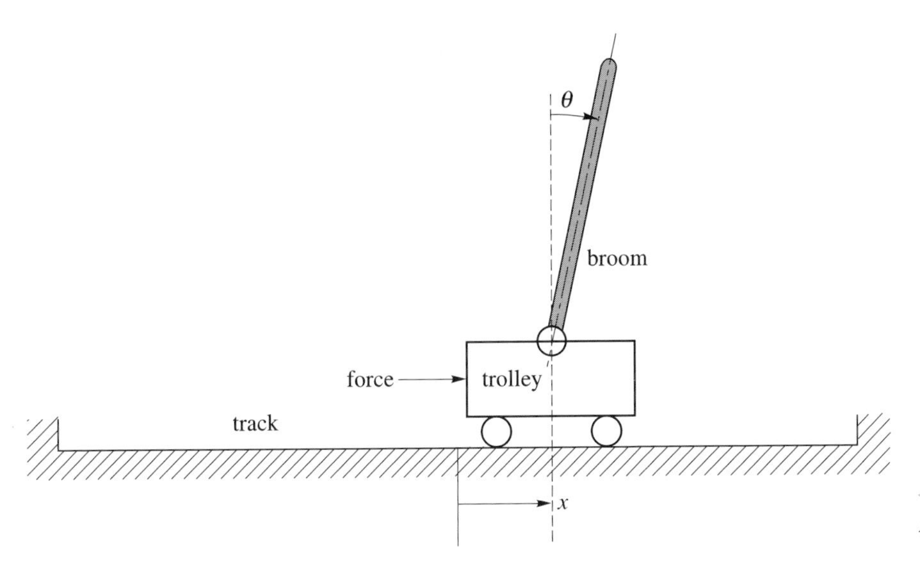

Figure 9.1 Broom-balancer.

The objective is to keep the broom upright and the trolley in the middle of the track. In other words, the angle θ must be kept as close as possible to zero and the position x from a fixed reference point must be kept as close as possible to zero. It will be assumed that the system has sensors which allow it to measure θ and x directly.

This system has been modelled mathematically. Barto *et al.* (1983) derived the following formulae, relating the angle θ and position x to the angular velocity of the broom about its pivot, $\dot{\theta}$, the angular acceleration of the broom about its pivot, $\ddot{\theta}$, the velocity of the trolley, $\dot{x}$, and the acceleration of the trolley, $\ddot{x}$.

Note that in the following analysis we use the dot notation for derivatives with respect to time, t. For example, velocity $\dot{x}$ is equivalent to $\mathrm{d}x/\mathrm{d}t$ or v, and angular acceleration $\ddot{\theta}$ is equivalent to $\mathrm{d}^2\theta/\mathrm{d}t^2$.

Model of the broom-balancer

$$\ddot{x} = \frac{F + ml[\dot{\theta}^2 \sin\theta - \ddot{\theta}\cos\theta] - \mu_c \operatorname{sgn}(\dot{x})}{M + m}$$

$$\ddot{\theta} = \frac{g\sin\theta + \cos\theta\left[\dfrac{-F - ml\dot{\theta}^2\sin\theta + \mu_c \operatorname{sgn}(\dot{x})}{M + m}\right] - \dfrac{\mu_p\dot{\theta}}{ml}}{l\left[\dfrac{4}{3} - \dfrac{m\cos^2\theta}{M + m}\right]}$$

where

θ is the angle of the broom, measured clockwise from the vertical (rad)

$\dot{\theta}$ is the angular velocity (rad s^{-1})

$\ddot{\theta}$ is the angular acceleration (rad s^{-2})

x is the horizontal position of the trolley, measured to the right from a reference point (m)

$\dot{x}$ is the velocity (m s^{-1})

$\ddot{x}$ is the acceleration (m s^{-2})

μ_p is a constant representing the frictional force between the broom and the trolley (=0.000002)

μ_c is a constant representing the frictional force between the trolley and the track (=0.0005)

M is the mass of the trolley (= 1.0 kg)

m is the mass of the broom (= 0.1 kg)

l is the half-length of the broom (the length of the broom is $2l = 1$ m)

F is the applied force (N)

g is the acceleration due to gravity (m s^{-2})

and $\operatorname{sgn}(\dot{x})$ is a function which takes the following values:

$+1$ when $\dot{x} > 0$,

-1 when $\dot{x} < 0$, and

0 when $\dot{x} = 0$.

Although it may be possible to measure $\dot{\theta}$ and $\dot{x}$ directly, the accelerations depend on the forces applied to the system. These forces are gravity, friction, and any control force F that is applied to the trolley.

The formulae in this broom-balancer model allow the unknown accelerations to be calculated if one knows the values of position and velocity. The difficulty with this system is that it is inherently unstable, and the model is non-linear. The obvious complexity means that the model needs to be simplified if classical linear control is to be attempted, and this will be done in the next section.

9.3 Classical solution

To illustrate how a non-linear model of a system can be approximated by a linear model, the formulae for the broom-balancer in the previous section can be simplified by making the following assumptions:

1. The broom is always nearly vertical. This means that θ will be less than 5°, so that $\sin\theta$ is approximately θ and $\cos\theta$ is almost 1.
2. Any terms with higher powers of $\dot{\theta}$, such as $\dot{\theta}^2$, will be small, and can be neglected.
3. Friction is negligible, so $\mu_c = 0$ and $\mu_p = 0$.
4. The mass of the broom is negligible, so $m = 0$.

The model then reduces to two relatively simple linear differential equations:

$$\ddot{x} = \frac{F}{M} \tag{9.1}$$

$$\ddot{\theta} = \frac{3g\theta}{4l} - \frac{3F}{4Ml} = \frac{3}{4Ml}(Mg\theta - F) \tag{9.2}$$

The next step uses Laplace transforms. If you are not familiar with the mathematics of Laplace transforms, it does not matter. In the following, just remember that in deriving transfer functions θ is replaced by Θ, $\dot{\theta}$ is replaced by $s\Theta$, and $\ddot{\theta}$ is replaced by $s^2\Theta$, where s is the Laplace operator. A *transfer function* can be derived from equation (9.2) as follows:

$$\frac{\Theta}{\boldsymbol{F}} = \frac{\frac{-3}{4Ml}}{s^2 - \frac{3g}{4l}}$$

After substituting, for example, the values $g = 9.8$, $l = 0.5$, and $M = 1$, we get:

$$\frac{\Theta}{\boldsymbol{F}} = \frac{-1.5}{s^2 - 14.7} \tag{9.3}$$

This is the transfer function of the system, which shows the relationship between the output, Θ, and the input, $\boldsymbol{F}$, in terms of Laplace transforms. It tells us mathematically what happens to θ when a force F is applied.

The first thing to note is the minus sign in the numerator of equation (9.3). This indicates that when the force applied is negative the angle, θ, becomes more positive and *vice versa*. This is intuitively what you would expect, since a positive force pushes the trolley to the right, which causes the broom to fall to the left.

The second thing to note is that since the transfer function is second order it has two poles in the s-plane. These are found by equating the denominator to zero:

$$s^2 - 14.7 = 0$$

so that

$$s = \pm 3.83$$

The fact that one of these poles lies in the right half of the s-plane indicates that the system is unstable, as explained in Chapter 8 of Volume 1.

9.3.1 Linear controller

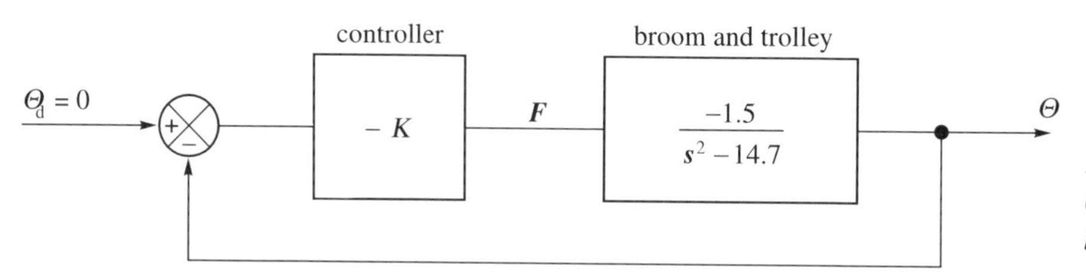

Figure 9.2 *Closed-loop control of the broom-balancer.*

Figure 9.2 shows a block diagram of a feedback control system with the linear model of the broom-balancer that we have just derived. The input, θ_d, is the desired angular position, which equals 0 since the aim is to keep the broom vertical. The value of the force that moves the trolley is produced as the output of the controller. Suppose the controller is simply proportional to the error $(\theta_d - \theta)$ with a gain $-K$, so that

$$F = -K(\theta_d - \theta) \quad \text{and} \quad \boldsymbol{F} = -K(\Theta_d - \Theta)$$

Then by equation (9.3) the closed-loop transfer function can be found to be:

$$\frac{\Theta}{\Theta_d} = \frac{1.5K}{s^2 - 14.7 + 1.5K}$$

This closed-loop transfer function shows that the closed-loop poles can be positioned by altering the value of K. A negative gain is used to counteract the previously mentioned minus sign in the transfer function. This has the effect that when the broom is falling to the left, the angle is negative, and now the force applied to counter this is also negative, which means the trolley gets pushed to the left. This proportional controller, however, is unable to stabilize the system.

A popular method used by control engineers to overcome the limitations of proportional control by itself is to include derivative action in the controller, i.e. to design what is called a *proportional plus derivative* (P+D) *controller.* This helps to balance the broom by producing a force that is proportional to the angular velocity of the broom as well as its angle. With proportional control alone, the force applied depended on the angle, but it is clear that when the broom is falling quickly, the force applied has to be larger to return it to a steady upright position.

The effect of introducing derivative action can be seen by examining the transfer function. A P+D controller has the form

$$\frac{\boldsymbol{F}}{\boldsymbol{E}} = -K_p(1 + \boldsymbol{T}_d\boldsymbol{s})$$

The term $\boldsymbol{E}$ is the Laplace transform of the error, e, between the desired output and the actual output, i.e. $e = \theta_d - \theta$. The other two terms, K_p and T_d, are the gain and derivative time constant respectively. Substituting $\boldsymbol{F} = -\boldsymbol{E}K_p(1 + \boldsymbol{T}_d\boldsymbol{s}) = -(\Theta_d - \Theta)K_p(1 + \boldsymbol{T}_d\boldsymbol{s})$ into equation (9.3) gives the following closed-loop transfer function:

$$\frac{\Theta}{\Theta_d} = \frac{1.5K_p(1 + \boldsymbol{T}_d\boldsymbol{s})}{s^2 + 1.5K_p\boldsymbol{T}_d s - 14.7 + 1.5K_p}$$

With this transfer function there are many possible ways to determine the position of the closed-loop poles. For example, if the value of T_d is set to 0.26 the resulting transfer function can be simplified, using a process called pole–zero cancellation, to

$$\frac{\Theta}{\Theta_d} = \frac{0.39K_p}{s - 3.83 + 0.39K_p}$$

We can now position the remaining pole anywhere along the horizontal axis. For example, if we make K_p equal to 40, the pole is at

$$s = -11.77$$

In theory we have produced a good design – the broom can be balanced. In practice, however, the system fails. When the system is steady there will be a small steady-state error in the angle of the broom. Because of this error the trolley accelerates in one or other direction and hits the end of the track after a short period of time. This behaviour is plotted in Figure 9.3. The system starts with the trolley at $x = -0.5$ m and the broom at an angle of $\theta = 0.1$ rad. The broom reaches the vertical within 2 seconds, but the trolley accelerates until it hits the end of the track at $x = 2.0$ m after 6.72 seconds.

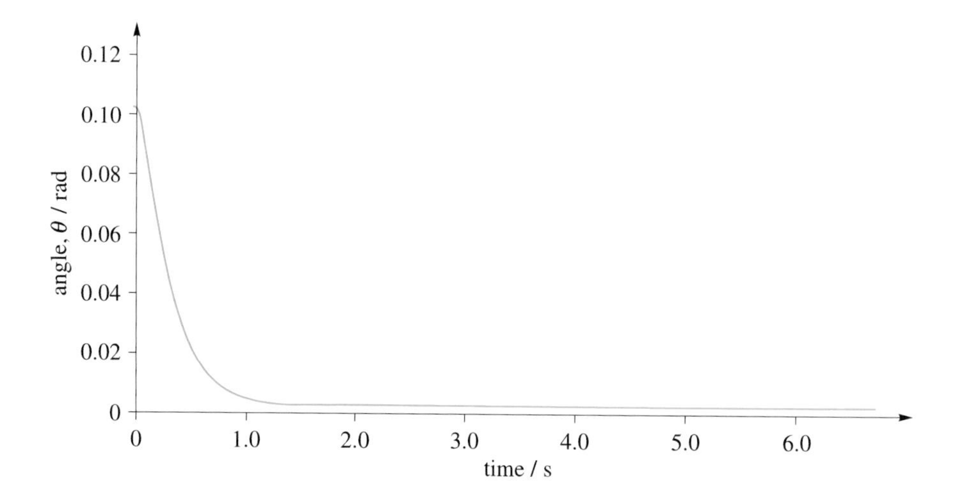

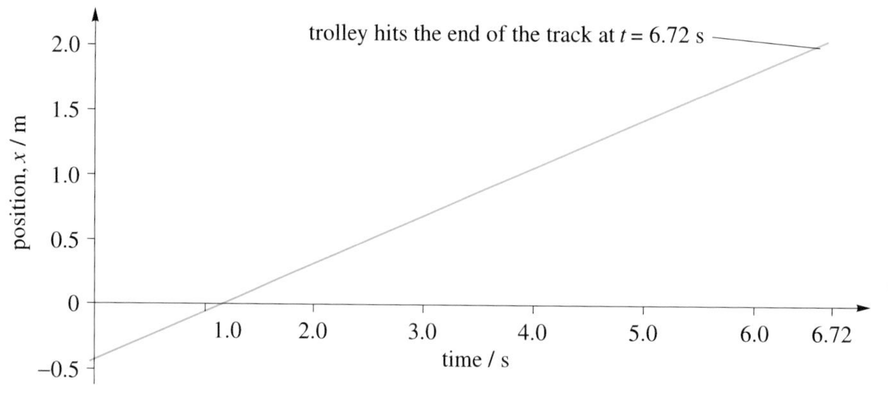

Figure 9.3
Graphs of θ and x against time for the broom-balancer with P+D control.

We can see why this happens by going back to the original linear equations. Equation (9.2) was:

$$\ddot{\theta} = \frac{3g\theta}{4l} - \frac{3F}{4Ml}$$

If we assume that the broom can be balanced, then when the system reaches a steady state the broom will not be moving, and any derivatives of θ will be zero. Also, when the broom is balanced the angle θ is not quite zero due to the steady-state error but is a constant, denoted by θ_{ss}. Therefore,

$$0 = \frac{3g\theta_{ss}}{4l} - \frac{3F}{4Ml}$$

and $F = Mg\theta_{ss}$

This means that the force being applied is also constant, which will be denoted by F_{ss}:

$$F_{ss} = Mg\theta_{ss}$$

Substituting into equation (9.1):

$$\ddot{x} = \frac{F_{ss}}{M}$$

This shows the acceleration of the trolley when a force, F_{ss}, is applied. As the force is constant the trolley has a constant acceleration, which means that the trolley will move and finally collide with the end of the track. For rocket launchers, this solution may be adequate, as the rocket can move horizontally indefinitely. For this reason the P+D solution has been used for rockets.

It is possible to design a more complicated controller that takes into account the interaction between the trolley and the broom. There isn't space in this book to go into this, but we can show a modification that corrects for the interaction. The problem appears to be that the broom is balanced but no account is taken of the position of the trolley. If the input to the controller, which is currently the error between the desired angle and the actual angle, was to include a small fraction of the position of the trolley, even when the broom is vertical there would be some error unless the trolley was in the centre of the track. One possibility is therefore to add a fraction of the velocity and position of the trolley.

Using the P+D model, $\boldsymbol{F} = -K_p(1 + \boldsymbol{T}_d\boldsymbol{s})\boldsymbol{E}$, and substituting the parameters $T_d = 0.26$ seconds and $K_p = 40$ gives

$$\boldsymbol{F} = -40(1 + 0.26\boldsymbol{s})\boldsymbol{E}$$

and in Laplace form

$$\boldsymbol{F} = -10.4\boldsymbol{Es} - 40\boldsymbol{E}$$

On converting this, $\boldsymbol{F}$ becomes F, $\boldsymbol{Es}$ becomes $\dot{e}$, and $\boldsymbol{E}$ becomes e. Therefore we get

$$F = -10.4\dot{e} - 40e$$

But since $\theta_d = 0$, $e = \theta_d - \theta = -\theta$, it follows that

$$F = 10.4\dot{\theta} + 40\theta \tag{9.4}$$

The effect of x can be introduced in the following way:

$$F = 10.4\dot{\theta} + 40\theta + \dot{x} + x \tag{9.5}$$

The small contribution of the position and velocity is just enough to ensure that the trolley stays in the middle of the track. Initially the term $\dot{x} + x$ contributes very little to the value of the force. However, when the broom has reached its steady state the value of $10.4\dot{\theta} + 40\theta$ is small, so that the terms $\dot{x}$ and x now have an influence. The system now reaches a steady state when all of the terms are zero, which means that the broom is upright and stationary and the trolley is in the centre of the track and stationary. This behaviour is shown in Figure 9.4, where initially $x = -0.5$ m and $\theta = 0.1$ rad as before. The broom takes a bit longer to settle in a vertical position, but the trolley is prevented from hitting the end of the track, so that the system is successfully controlled.

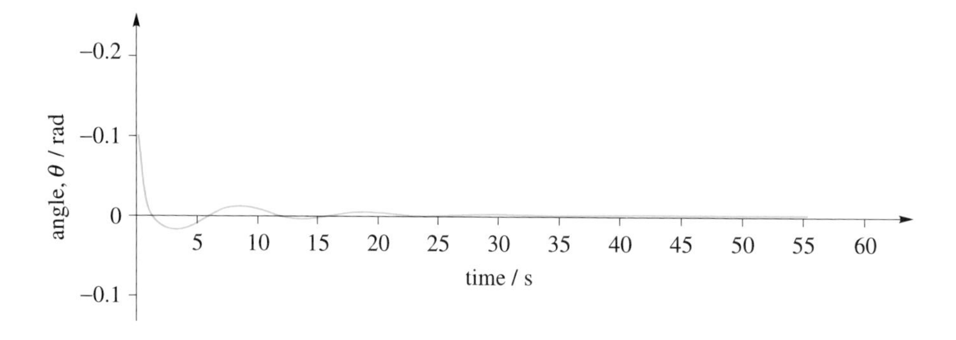

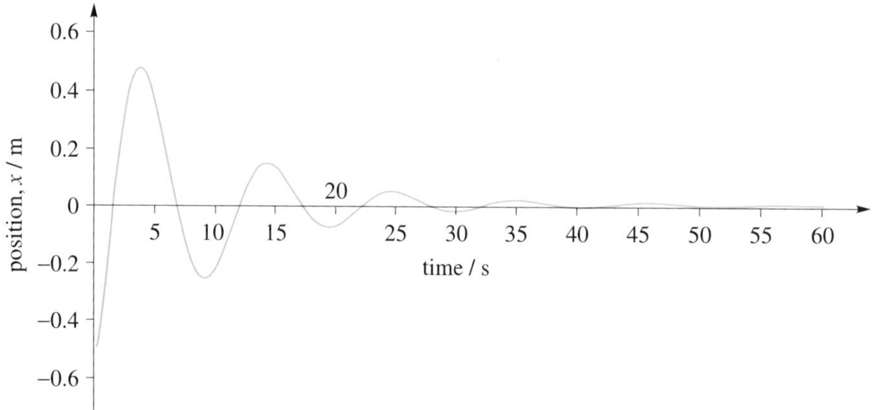

Figure 9.4
Graphs of θ and x against time with a modified P+D controller.

This solution works by good fortune rather than intentional design. Generally it would not be acceptable to simply add new terms to the controller without analysing their effect on the stability of the system at the very least.

9.3.2 Non-linear and bang-bang control

A further modification that is sometimes made by engineers is to make the controller non-linear. This may appear to be quite contrary to what we said earlier about trying to keep the system linear, but there are good reasons for doing it.

The non-linearity that is introduced is a hard-limiter. This gives an output of $+F_{max}$ when its input is positive and an output of $-F_{max}$ when its input is negative. It is introduced into the system as part of the controller, so that the controller calculates its output using the P+D equation, but that output is then passed through the hard-limiter. The resulting controller produces what is called ***bang-bang control***. It is called 'bang-bang' because the controller only has two possible output values, which are as hard as possible in one direction or the other. In some controllers which produce a mechanical force on the output you can hear the system banging as it switches from one extreme to the other.

According to Pontryagin, a Russian control engineer, under certain conditions bang-bang control produces optimal control. This is known as ***Pontryagin's maximum principle***. Now there are a variety of different meanings to optimal control, and the one that is meant here is minimum-time optimal. This means that if you want a system to go from state A to state B in the shortest time, this can be achieved using bang-bang control.

An example of this is driving a car. The fastest way of getting from point A to point B with the car at rest at both points is to accelerate away from A as hard as possible and then to brake as hard as possible to come to rest at B. Figure 9.5 shows a graph of velocity against time when this is done. The area under the graph equals the distance travelled.

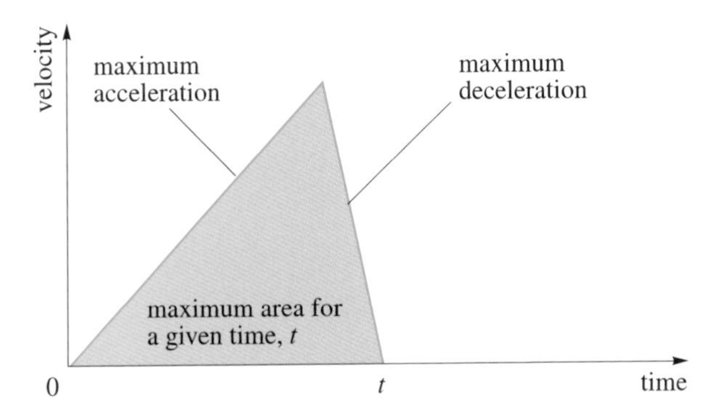

Figure 9.5 Velocity profile of a car under bang-bang control.

The penalties paid for this control strategy are that the accelerator or the brakes are used constantly, so there is a high fuel consumption, brake linings don't last long and there is severe mechanical stress placed on the system. But, barring accidents, you reach the destination more quickly than using any other strategy.

The constraints under which bang-bang control is time-optimal are that the model of the system being controlled is linear, where the controller is linear prior to passing through a hard-limiter, and where only the controller output is constrained by having, say, a maximum value. The broom-balancer fits these constraints. It can approximate to a linear model, and there is almost certainly an upper limit to the force that can be applied. So, if we would like the broom-balancer to be time-optimal, which means that we want the broom to be balanced as quickly as possible, bang-bang control seems to be a good choice. However, the way that we've arrived at the parameter settings is largely by good luck. Methods do exist that enable designers to find a bang-bang solution for a limited number of problems, but usually it is very difficult. Figure 9.6 shows graphs of the broom-balancer with bang-bang control, plotted against time. The system again starts with $x = -0.5$ m and $\theta = 0.1$ rad. Here θ reaches the vertical position almost straight away, showing the time-optimal behaviour. The position of the trolley, which isn't the variable that is optimally controlled, eventually reaches the centre of the track but takes over a minute to get there.

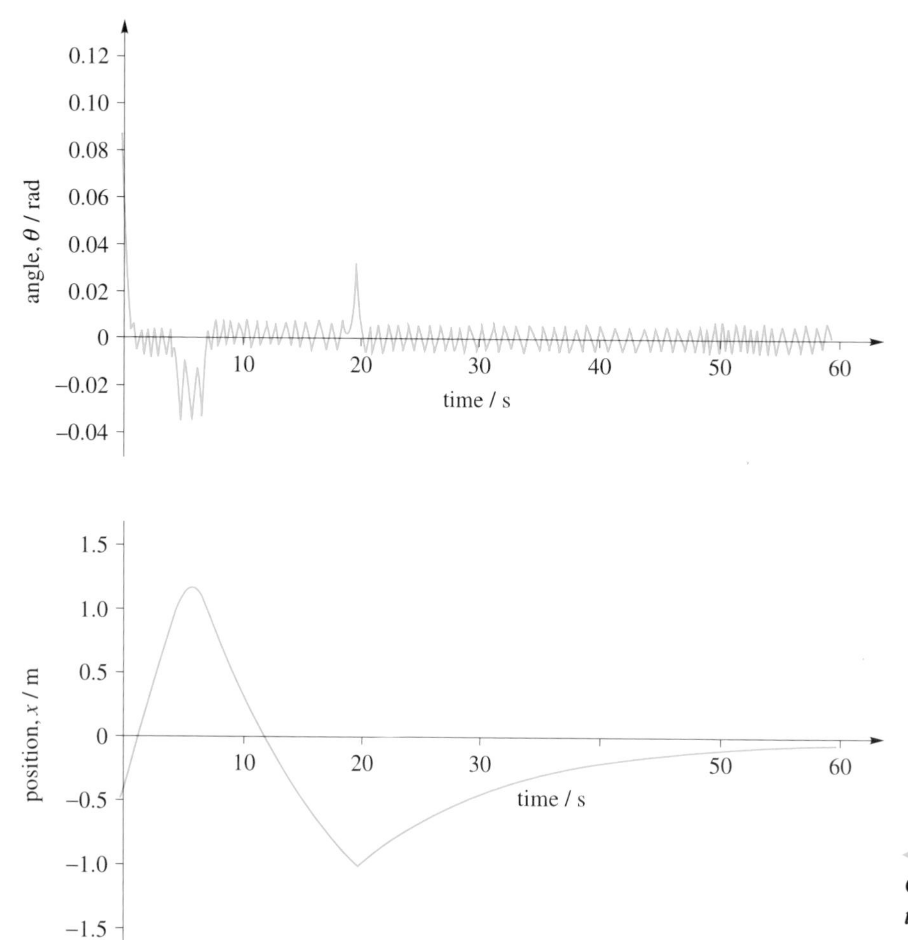

Figure 9.6
Graphs of θ and x against time for broom-balancer with bang-bang control.

9.3.3 Summary of classical control

We have seen how classical control makes use of differential equations to model a system. In many cases it is possible to simplify these differential equations to make them linear, and then apply fairly standard design techniques to produce a feedback system with the desired characteristics. In the case of the broom-balancer this was possible, which is why we chose this example to illustrate the methods.

We have also seen how the interaction between the broom and the trolley could be controlled by adding terms to the controller. The method used in this chapter was rather *ad hoc*, but produced the desired result. Then finally, Pontryagin's maximum principle was introduced to show how the response of the system could be made time-optimal by including a hard-limiter and thus producing bang-bang control.

In the following sections we shall look at some of the more recent developments in the field of intelligent control which can also be applied to systems which are more complex than the broom-balancer. We will continue to apply them to the broom-balancer for comparison.

9.4 Neural network solution

9.4.1 Single neuron

In the previous sections we have described the broom-balancer and shown how classical control theory can be used to derive solutions to the problem. To do this we used Barto *et al.*'s mathematical model of the system given in Section 9.2. But what if no such model existed? For the remainder of this chapter we will assume that no model is available in the form of a differential equation, and show how the AI techniques explained elsewhere in this book can be used to develop new control strategies.

In this section we will show how a neural network can be used to control a system. The advantage of this approach is that it is not necessary to have a model of the system in the form of a differential equation, since the network *learns* how to control the system.

In the case of the broom-balancer the *inputs* to the network will be the data which describe the state of the broom. These could include the position and velocity of the trolley, and the angle and angular velocity of the broom. So it is assumed that the system has, at least, a stream of data (x_t, θ_t) which give the values of position and the angle of the broom at time t.

Suppose the samples arrive every T seconds. Then the velocity at time t is approximated by $(x_t - x_{t-1})/T$, and the angular velocity is approximated by $(\theta_t - \theta_{t-1})/T$. Given these inputs, the desired output of the network is the force which restores x to zero and θ to zero.

The training data for the network will usually have input values x_t and θ_t, and the calculated values $(x_t - x_{t-1})/T$ and $(\theta_t - \theta_{t-1})/T$, together with the force that was applied. The output of the network will depend on the kind of transfer function the network will learn, as will be discussed in subsequent sections. The training data may come from observing a person controlling the system, or they may come from simulations in which the results of many trials are recorded.

The expression for the P+D controller was effectively a weighted sum of the angular velocity, angle, velocity and position. With bang-bang control, the controller could be described as follows:

$$\text{output} = +F_{max} \quad \text{when } w_1\dot{\theta} + w_2\theta + w_3\dot{x} + w_4x \geqslant 0$$

$$\text{output} = -F_{max} \quad \text{when } w_1\dot{\theta} + w_2\theta + w_3\dot{x} + w_4x < 0$$

You should recognize this from Chapter 4 on neural networks as being in the form of a neuron where the neuron fires (+1 output) when the weighted sum of its input is greater than some threshold (in this case 0), otherwise it doesn't fire (−1 output). This implies that we can use a single neuron to control the broom-balancer. This was precisely what Bernard Widrow thought, and went on to demonstrate it with his ADALINE network (Widrow and Smith, 1964).

The main difficulty of this solution is the parameter setting – just how do we get the appropriate values? Widrow suggested training with a teacher in which a human tries to control the trolley. The ADALINE monitors the system outputs (such as the angle of the broom and position of the trolley) and the corresponding control action taken by the human. By pairing the system outputs with the controller output a training set can be constructed where the present set of system outputs are the inputs to the ADALINE, and the controller output is the desired output of the ADALINE.

Clearly then, the best that this solution will ever achieve is to be able to learn to control a system as well as an existing controller (human or otherwise). The only situation where this is advantageous is where a system is currently controlled manually and has to be automated, but little is known about the system. The only way to automate it is to mimic the human controller, and the ADALINE does just this.

A major handicap of the single neuron is its inability to emulate all input–output relationships. This means that it might not be able to mimic a controller over the full range of circumstances. The *multilayer perceptron* was shown in Chapter 4 to have an advantage over the ADALINE, which is that it can mimic any consistent input–output relationship. In principle, the multilayer perceptron can be used to emulate any existing controller.

9.4.2 Multilayer network

The property of a multilayer network that is most interesting is its ability to emulate any consistent input–output relationship. In a linear system this relationship is the transfer function, and the goal in many control strategies is to find an approximation to the inverse of the system's transfer function.

For example, in feedback control such as that shown in Figure 9.7 the closed-loop transfer function of a system is

$$\frac{Y}{X} = \frac{CG}{1 + CGH}$$

where G is the transfer function of the system being controlled, H is the transfer function of the feedback path, C is the transfer function of the controller, Y is the system output and X is the system input.

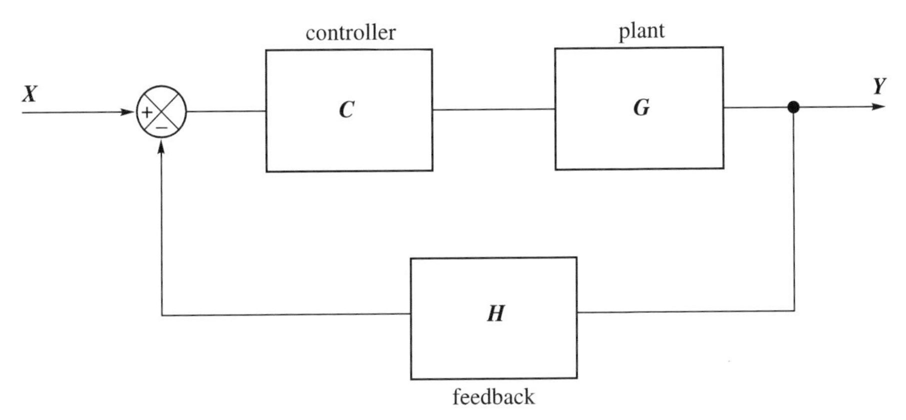

Figure 9.7
Generalized linear feedback control system.

Ideally, designers would like the transfer function Y/X to equal 1, so that the system responds immediately to any change in input with no error between input and output. For example, with the broom-balancer one wants to input values of $x = 0$ and $\theta = 0$, and for the system to hold these values. Any deviation from this is fed back through H.

Making $Y/X = 1$ can be achieved, for example, if $C = \frac{2}{G}$ and $H = 0.5$:

$$\frac{Y}{X} = \frac{\frac{2G}{G}}{1 + \frac{2G \times 0.5}{G}} = \frac{2}{1 + 1} = 1$$

This shows that from a control engineer's point of view, having the transfer function $1/G$ or G^{-1} (called the ***inverse*** of G) would be very useful. So as a general principle, if the inverse of G can be approximated, good control can be achieved using feedback. Since a multilayered neural network can approximate any consistent relationship, it should be able to train to G^{-1}.

There are a number of ways that this can be done, but we will look at just one, called the ***generalized learning architecture***, shown in Figure 9.8.

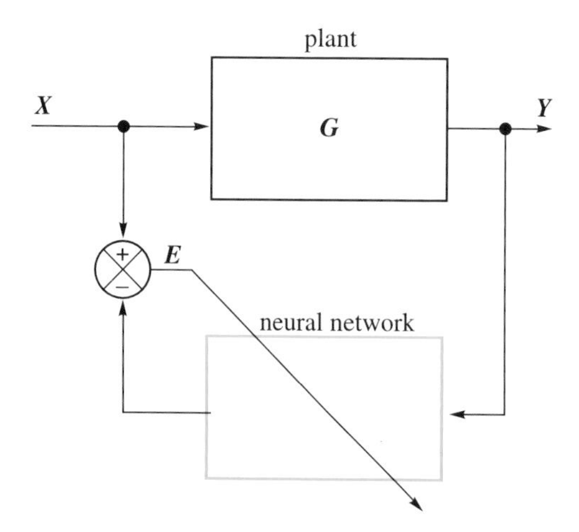

Figure 9.8 Generalized learning architecture.

A set of input–output training data is produced by sending inputs, x, into the system (plant) and recording those values of x with their corresponding output values, y. By setting the inputs of the neural network to y and making the desired output x, the network is trained to do the opposite of $\boldsymbol{G}$, and the network will approximate the inverse $\boldsymbol{G}^{-1}$. Once it has trained it effectively represents the inverse transfer function of the system being controlled, and it can be used as part of a controller.

One problem with this method is in deciding how to select the training data. The inputs and outputs should be representative of the states that the system will enter, but it can be difficult to know in advance how to do this.

For the broom-balancer, the inputs to the plant would be the force, F_t, applied to the trolley. The output would be the values of x_{t+1} and θ_{t+1}, and possibly their derivatives. For the neural network, the inputs would be x_{t+1} and θ_{t+1}, and the desired output would be F_t. The actual output of the neural network, let us call it G_t, can then be compared to the desired value and an error calculated, shown as $\boldsymbol{E}$ in Figure 9.8. The error can be used to adjust the weights in the network using back-propagation, as indicated by the diagonal arrow through the network.

9.4.3 Recurrent networks

A ***recurrent network*** is one where feedback is allowed. A special case of a recurrent network is the ***fully-connected*** one, where the output of each neuron is connected to the input of every other neuron and its own input. Figure 9.9 shows a fully-connected recurrent network with three neurons and two inputs.

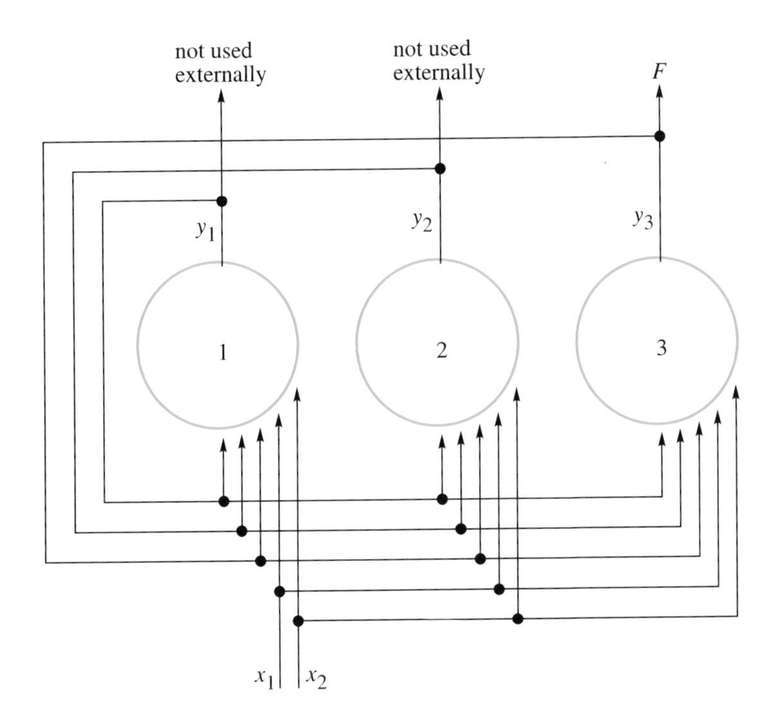

Figure 9.9
Fully-connected recurrent neural network.

An important feature of recurrent networks is that it becomes unnecessary to enter the terms $\dot{x}$ and $\dot{\theta}$ explicitly. This is because, for example, at any time the system has both x_{t-1} and x_t as inputs and so implicitly the term $x_t - x_{t-1}$ can enter the system at time t. $\dot{x}$ is approximated by $(x_t - x_{t-1})/T$, where T is the (constant) time interval between samples. In principle, the network weights will adjust the value of $(x_t - x_{t-1})$ as appropriate, and it can be assumed that $(x_t - x_{t-1})/T \approx \dot{x}$ is implicitly in the system. By a similar argument, $\dot{\theta}$ is also implicitly in the system.

This sort of network is of particular interest because it is comparable with the final P+D controller that we used earlier. The expression for the controller output of equation (9.5) was:

$$F = 10.4\dot{\theta} + 40\theta + \dot{x} + x$$

To convert this to a discrete approximation, the derivatives $\dot{x}$ and $\dot{\theta}$ are approximated by the difference between the current value and the previous value divided by the sampling period, T:

$$F_k = \frac{10.4(\theta_k - \theta_{k-1})}{T} + 40\theta_k + \frac{x_k - x_{k-1}}{T} + x_k$$

Let's assume for illustration that $T = 0.02$ seconds, then:

$$F_k = 520(\theta_k - \theta_{k-1}) + 40\theta_k + 50(x_k - x_{k-1}) + x_k$$

so that

$$F_k = 560\theta_k - 520\theta_{k-1} + 51x_k - 50x_{k-1}$$

Here we have an expression containing the present values of angle and position and the previous values of angle and position. We therefore only have two ***state variables***, x and θ, although they also have to be stored for one sampling period.

Figure 9.10 shows a network which produces the value of F_k according to the expression above. It requires three neurons and has the interesting feature that most of the weights are forced to be zero by the absence of a connection. The final output neuron can have a linear output or could be hard-limited to produce bang-bang control.

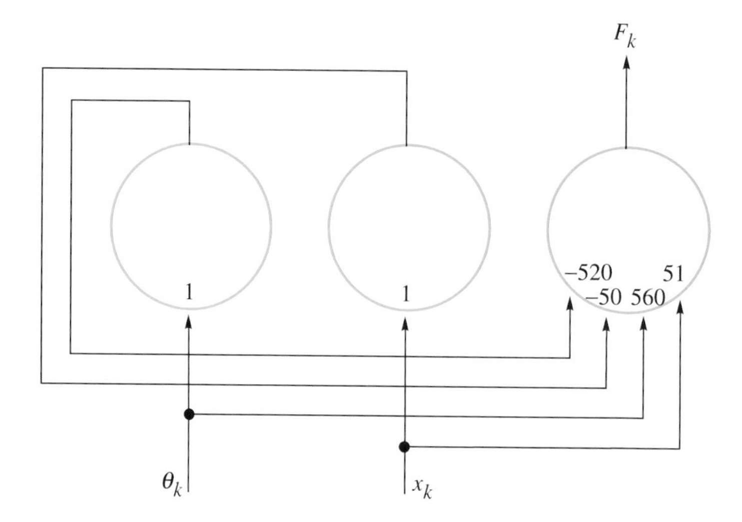

Figure 9.10
Recurrent network emulation of the P+D controller of equation (9.5).

One aspect of the network which differs from conventional networks is that the output of each neuron is ***semi-linear***, which means that it is only linear over a range of values. In theory, when the network is controlling the system correctly, there is no reason why the outputs cannot be linear. However, during training the weights will not be ideal and the output can grow because of positive feedback. It is therefore necessary to limit the output as shown in Figure 9.11. The exact value at which the output flattens doesn't matter since it is only a safety measure and not part of the control action.

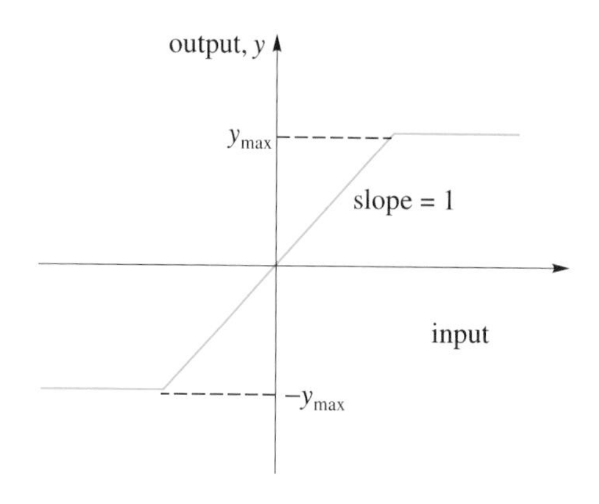

Figure 9.11
Semi-linear output function limited by $-y_{max}$ and $+y_{max}$

Here we have calculated the network weights using values from an earlier section. In principle, it should be possible to build and train a network which provides appropriate control. However, there are practical difficulties due to the way the problem has been defined. First, it is difficult if not impossible to train a network by back-propagation because of the semi-linear output functions (and possibly a hard limiter on the output neuron). Also, back-propagation requires an appropriate error function to be defined. The desired behaviour is to balance the broom as long as possible, but this consideration has been omitted, which makes the network behaviour ill-defined.

An alternative is to use the ***time-to-failure*** as a measurement of how well the network is performing, where failure is defined as the broom falling over or the trolley hitting the end-stops. In the next section genetic algorithms will be used to find the weights using such a measurement.

9.5 Genetic algorithms

The back-propagation technique for training network weights described in Chapter 4 on Neural networks assumed that the network was a multilayer perceptron. Recurrent networks which have feedback from higher levels to lower levels do not have this architecture, and so back propagation cannot be used as a training method.

Finding the weights in the recurrent neural network can be viewed as a search problem, and the characteristics of genetic algorithms can be used to find the weights for the broom balancing problem. Genetic algorithms were introduced as a search technique in Chapter 3, and in Chapter 5 on Scheduling they were illustrated in an application to the travelling salesman problem. This section illustrates how the techniques described in this book may be combined to find a solution to control problems.

Let each chromosome consist of all the weights for all the neurons. At first the weights are chosen randomly, and the system is run until the broom falls over or the end of the track is reached. The time-to-failure is recorded and its reciprocal used as the *fitness function*. After all the members of the population have been tried, a new population is created by breeding from the old population.

Breeding consists of an *elitist strategy* with *single-point crossover* and *mutation*. Elitism means that a fraction of the population that consists of the fittest individuals is preserved from one generation to the next (we used 10% in our experiments). Single-point crossover consists of splitting the two parent genes at a random point and swapping the genetic material to produce two different offspring (as described in Section 3.3.5 of Chapter 3).

For example, in Figure 9.9 there are three neurons each with five inputs. Each chromosome therefore has 15 entries. Let the two chromosomes be

$$(w_1, w_2, w_3, w_4, w_5, w_6, w_7, w_8, w_9, w_{10}, w_{11}, w_{12}, w_{13}, w_{14}, w_{15})$$

and

$$(w_1', w_2', w_3', w_4', w_5', w_6', w_7', w_8', w_9', w_{10}', w_{11}', w_{12}', w_{13}', w_{14}', w_{15}')$$

First the crossover point is found at random; let us suppose the crossover point is 3. Then all the genetic material to the left of point 3 will be swapped, resulting in

$$(w_1, w_2, w_3', w_4', w_5', w_6', w_7', w_8', w_9', w_{10}', w_{11}', w_{12}', w_{13}', w_{14}', w_{15}')$$

and

$$(w_1', w_2', w_3, w_4, w_5, w_6, w_7, w_8, w_9, w_{10}, w_{11}, w_{12}, w_{13}, w_{14}, w_{15})$$

The weights are all real numbers, and if crossover alone was used the weights would simply be shuffled about. Mutation is therefore necessary to alter the values of the weights. The form of mutation used in this application is called ***real number creep***, and consists of changing the value of a weight by some random value of between −10% and +10% of the weight. If a small population is used the choice of weights is limited, so a high mutation rate is needed, including the possibility of a mutation rate of 1.0 in which all values are mutated during breeding.

The output of the controller is hard-limited, so a value of $\pm F_{max}$ is produced. The exact value of F_{max} is set in advance, in which case the genetic algorithm finds the best set of weights for that pre-selected value of F_{max}. Alternatively, the value of F_{max} could be included in the chromosome and would then be subject to crossover and mutation itself.

For the broom-balancing problem we found that the genetic algorithm quickly converged to a solution in which some members of the population could balance the broom for hundreds of seconds. In our attempts, a working solution was found in less than 100 generations.

In this example genetic algorithms were used as a method to train a recurrent network, but one might ask if this technique could be applied to training neural networks in general. This depends on the nature of the problem and the search space. Genetic algorithms provide a powerful search mechanism, but their use might be considered to be a sledgehammer to crack a nut for the problem of training multilayer perceptrons. Since back propagation is designed for training multilayer perceptrons, in many instances it will converge more quickly to a solution when training than genetic algorithms. However, in principle genetic algorithms can be used to train multilayer perceptrons and, like simulated annealing, they offer the possibility of exploring more of the search space because they do not just use hill climbing or gradient descent.

Although the P+D controller gives a better solution than any we have managed to generate from scratch using a genetic algorithm, these experiments show that intelligent control techniques can produce working solutions. The P+D controller has been used as a benchmark reference in this chapter because it can be modelled by a formula. However, many control problems do not start with a formula and sometimes it is not practical to try to find a formula. In such systems intelligent control may provide the *only* way of controlling them. The P+D solution is better because the broom gets to the upright position more quickly and then doesn't 'wobble about' very much. The solutions found with the genetic algorithm are much less steady, and never really settle down. However, they all balance the broom indefinitely. This should be expected as the fitness function used doesn't take into account any performance criteria other than the length of time that the broom is balanced. A more sophisticated fitness function that took into account values such as the time to settle would produce better solutions in this respect.

Recurrent networks provide a sufficiently flexible architecture to emulate many conventional control strategies, and to create some new ones. Finding the weights using a genetic algorithm seems to be one of the best ways to adjust the network to give the best performance.

9.6 Fuzzy rules

Earlier we showed that the problem of controlling the broom-balancer could be solved using a P+D controller with a hard-limiter on the output. The reasoning behind this was that we could use an approximate linear model of the broom-balancer to design the P+D controller, but then we wanted the advantage of bang-bang control to speed up the response. In effect, the controller is obeying a rule of the form:

If $(10.4\dot{\theta} + 40\theta + \dot{x} + x) \geqslant 0$

Then output is $+F_{max}$

Else output is $-F_{max}$

As stated, this rule is Boolean with outputs 'Yes' (apply $+F_{max}$) and 'No' (apply $-F_{max}$).

The values used in the antecedent predicates of the *If* statement may be inaccurate due to measurement error, and perhaps more importantly, the model used was only an approximation to the actual system. This means that our confidence in the control action is not 100%, and that having such a dramatic switch from $+F_{max}$ to $-F_{max}$ may be bad. As it stands, there will be some situations where the value of $(10.4\dot{\theta} + 40\theta + \dot{x} + x)$ is slightly greater than 0 which causes a force of $+F_{max}$ to be produced. If one of the variables changes by a tiny amount such that $(10.4\dot{\theta} + 40\theta + \dot{x} + x)$ becomes slightly less than 0, the force suddenly switches to $-F_{max}$. Sensitivity to such small changes is not desirable when there are so many inaccuracies in the model and the measurements.

A way around this is to use ***fuzzy control*** to make that transition more fuzzy by smoothing the control action from one side to the other. One can still have bang-bang control, but the decision about the output of the controller can be less 'crisp'.

Clearly, if the model is accurate, then fuzzy control is not going to improve matters. But in situations where the model is more complex, or where the model perhaps changes slightly over time, or where the states are difficult to measure accurately, fuzzy control is useful.

Fuzzy logic was described in Chapter 6 on Reasoning. Essentially, we describe the system linguistically, using phrases like 'the broom is falling to the left' rather than having an accurate measurement of the states. These phrases refer to the fuzzy sets that describe the system, and at any time the outputs are measured and used to calculate the set memberships of these fuzzy sets. The main difference between fuzzy sets and the more usual 'crisp' sets is that a variable can be a member of more than one set, with a degree of membership that is a figure between 0 and 1. In crisp sets, a variable can only be a member of one set and then it has a membership of 1 of that set; all other crisp sets have a membership of 0.

Figure 9.12 shows an example of the membership functions, χ, of the four state variables, $\dot{\theta}$, θ, x, $\dot{x}$ and the output force, F. The membership functions were selected to be triangular and overlapping in this way so that the total confidence at any point is 1. There are many more possibilities for the membership functions, but this is one of the simplest. All the variables are divided into NEGATIVE, SMALL or POSITIVE. Again they could be further divided, but these three divisions are the simplest. Note that the membership functions for the output variable, the force F, are truncated as there is a maximum value for the force.

In any state, the membership of each set is found. As we've already said, a variable can belong to more than one set. The output is also defined by membership of sets as shown. These sets, together with a set of rules, are all that are needed. The rules are in the form:

If $\dot{\theta}$ is POSITIVE

Then force is POSITIVE

If $\dot{\theta}$ is NEGATIVE

Then force is NEGATIVE

This rule is found by observing and qualitatively understanding the system rather than by examining the differential equation model. Intuitively, if the broom is accelerating to the right (positive), then to slow it down you have to move the trolley to the right, which is done by applying a positive force. So applying this rule controls the velocity of the broom. Similar rules can be found for controlling the angle of the broom and the velocity and position of the trolley.

In the following it is assumed that values of x, $\dot{x}$, θ and $\dot{\theta}$ are available to the system. In practice this means that x and θ are available from the sensors, and the $\dot{x}$ and $\dot{\theta}$ are calculated from them as discussed previously. Given these data, the associated fuzzy set memberships are calculated using the functions given in Figure 9.12. These fuzzy set membership values are then available for testing the rules.

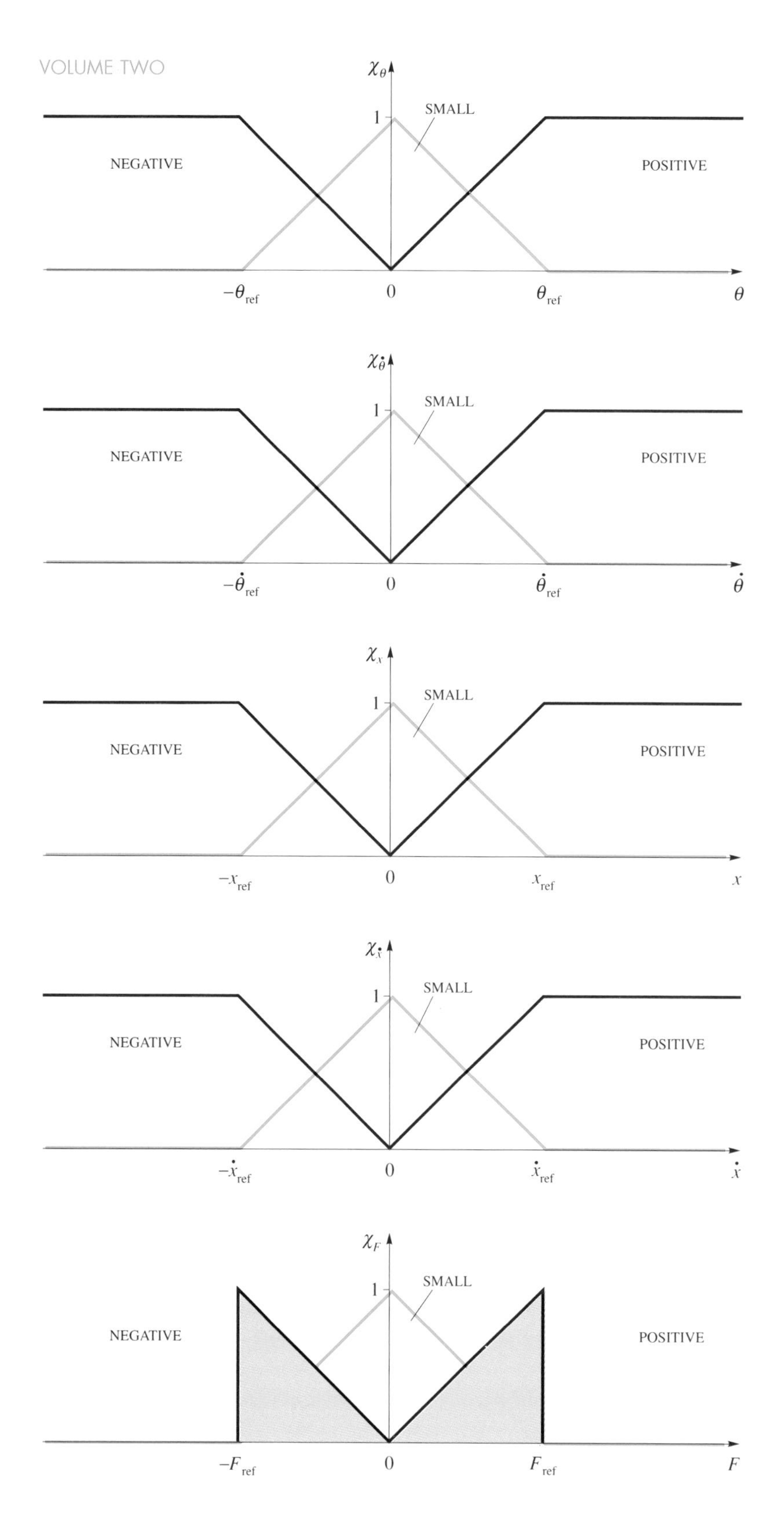

Figure 9.12
Membership functions of the state variables and the output force.

We started from the following rules, assuming a first-come, first-served conflict resolution strategy:

If $\dot{\theta}$ is POSITIVE
Then force is POSITIVE
If $\dot{\theta}$ is NEGATIVE
Then force is NEGATIVE
If θ is POSITIVE
Then force is POSITIVE
If θ is NEGATIVE
Then force is NEGATIVE
If $\dot{x}$ is POSITIVE
Then force is POSITIVE
If $\dot{x}$ is NEGATIVE
Then force is NEGATIVE
If x is POSITIVE
Then force is POSITIVE
If x is NEGATIVE
Then force is NEGATIVE
Else force is SMALL

It is important to notice the order in which the rules are tested. The angular velocity $\dot{\theta}$ is tested first, and if it is POSITIVE or NEGATIVE a force is applied. If the angular velocity of the broom is SMALL only then will any of the other rules be examined. What this means is that the rules have been prioritized. The angular velocity is controlled first. When this is SMALL, the angle itself is controlled. When this is SMALL the velocity of the trolley is controlled, and finally when this is SMALL the position of the trolley is controlled.

A decision has to be made about the reference values for the variables so that the terms NEGATIVE, POSITIVE and SMALL can be defined. These have to be guessed, but guided by some knowledge of the range of system parameters. For example, the reference angle of the broom is chosen to be 0.1 radians or about 6° and the reference position of the trolley is set to 0.1 m. These figures suggest that we want the broom-balancer to end up very close to the centre of the track with the broom almost vertical. The reference values for the angular velocity of the broom and the velocity of the trolley are chosen to be 1 rad s^{-1} (about 60° s^{-1}) and 1 m s^{-1} respectively. In all cases the reference values are symmetrical, which means, for example, that the variable x is NEGATIVE if it is less than –1 m, POSITIVE if it is greater than 1 m, and SMALL if it is between these values.

The controller works as follows. First, calculate the fuzzy membership of each of the input sets. We will use the letters P for POSITIVE, S for SMALL and N for NEGATIVE.

In Figure 9.13, the current state of the system is that $\theta = 8°$, $\dot{\theta} = 53°\ \mathrm{s}^{-1}$, $x = 1.2$ m and $\dot{x} = -0.8\ \mathrm{m\,s}^{-1}$, so the 12 membership values would be:

$\chi_{\dot{\theta}\mathrm{N}} = 0.0$ $\quad\chi_{\dot{\theta}\mathrm{S}} = 0.1$ $\quad\chi_{\dot{\theta}\mathrm{P}} = 0.9$
$\chi_{\theta\mathrm{N}} = 0.0$ $\quad\chi_{\theta\mathrm{S}} = 0.0$ $\quad\chi_{\theta\mathrm{P}} = 1.0$
$\chi_{\dot{x}\mathrm{N}} = 0.8$ $\quad\chi_{\dot{x}\mathrm{S}} = 0.2$ $\quad\chi_{\dot{x}\mathrm{P}} = 0.0$
$\chi_{x\mathrm{N}} = 0.0$ $\quad\chi_{x\mathrm{S}} = 0.0$ $\quad\chi_{x\mathrm{P}} = 1.0$

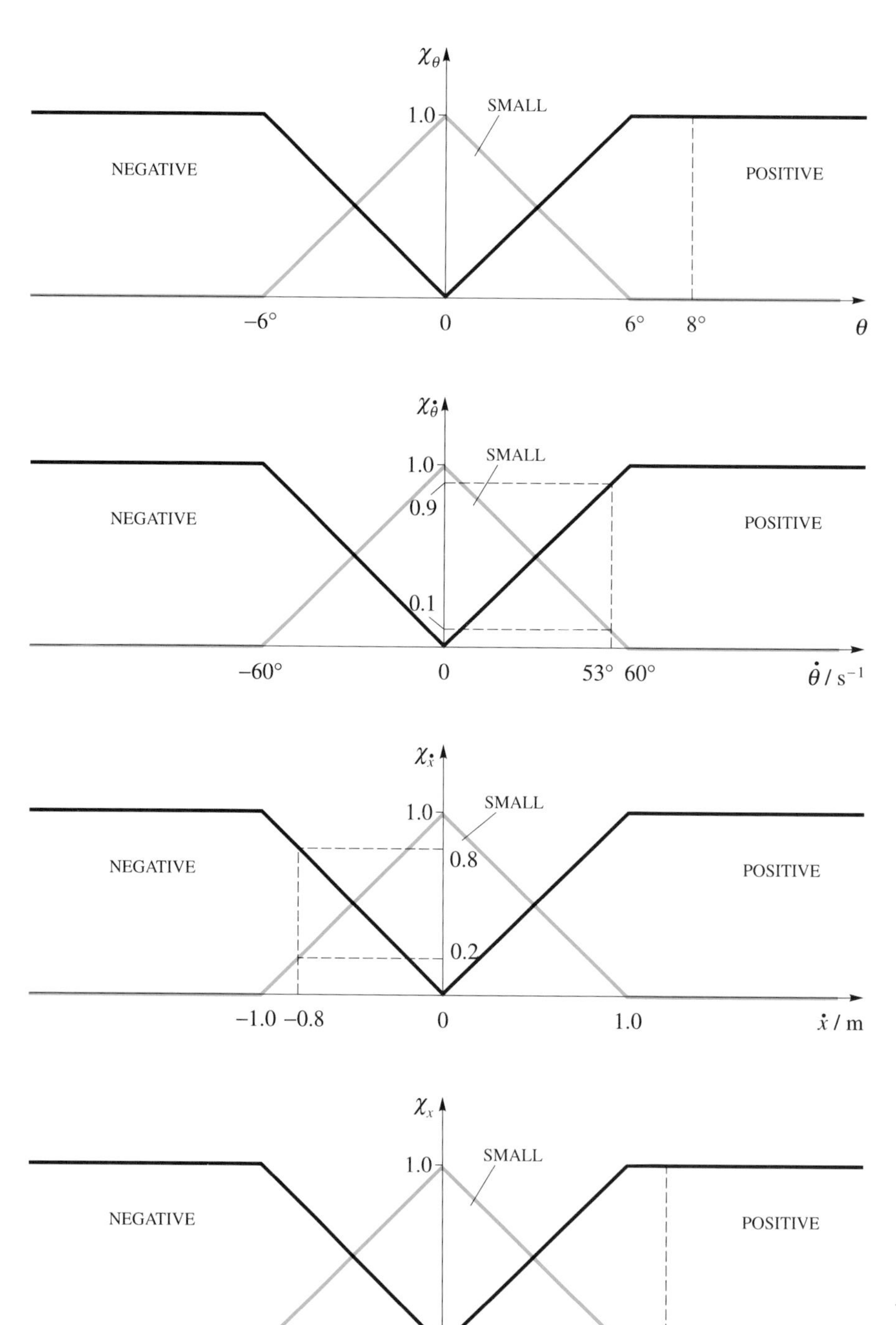

Figure 9.13 Membership values for particular instance.

The confidence in the membership of the output set is found by converting the rules that we already have into a fuzzy form. The first step is to rearrange the rules into a logical form. This is done by looking at what combinations of rules produce a force of POSITIVE (P), NEGATIVE (N) and SMALL (S) in turn.

$F = \text{P}$

If $\dot{\theta} = \text{P}$
OR $\dot{\theta} = \text{N}$ AND $\theta = \text{P}$
OR $\dot{\theta} = \text{N}$ AND $\theta = \text{N}$ AND $\dot{x} = \text{P}$
OR $\dot{\theta} = \text{N}$ AND $\theta = \text{N}$ AND $\dot{x} = \text{N}$ AND $x = \text{P}$
Then $F = \text{P}$

$F = \text{N}$

If $\dot{\theta} = \text{N}$
OR $\dot{\theta} = \text{P}$ AND $\theta = \text{N}$
OR $\dot{\theta} = \text{P}$ AND $\theta = \text{P}$ AND $\dot{x} = \text{N}$
OR $\dot{\theta} = \text{P}$ AND $\theta = \text{P}$ AND $\dot{x} = \text{P}$ AND $x = \text{N}$
Then $F = \text{N}$

$F = \text{S}$

If $\dot{\theta} = \text{S}$ AND $\theta = \text{S}$ AND $\dot{x} = \text{S}$ AND $x = \text{S}$
Then $F = \text{S}$

You should recall that the fuzzy equivalent of AND is the MIN operator, and that the fuzzy equivalent of OR is the MAX operator. We can therefore convert the rules into a fuzzy form as follows:

$$\chi_{FP} = \text{MAX}(\chi_{\dot{\theta}\text{P}}, \text{MIN}(\chi_{\dot{\theta}\text{N}}, \chi_{\theta\text{P}}), \text{MIN}(\chi_{\dot{\theta}\text{N}}, \chi_{\theta\text{N}}, \chi_{\dot{x}\text{P}}), \text{MIN}(\chi_{\dot{\theta}\text{N}}, \chi_{\theta\text{N}}, \chi_{\dot{x}\text{N}}, \chi_{x\text{P}}))$$

$$\chi_{FN} = \text{MAX}(\chi_{\dot{\theta}\text{N}}, \text{MIN}(\chi_{\dot{\theta}\text{P}}, \chi_{\theta\text{N}}), \text{MIN}(\chi_{\dot{\theta}\text{P}}, \chi_{\theta\text{P}}, \chi_{\dot{x}\text{N}}), \text{MIN}(\chi_{\dot{\theta}\text{P}}, \chi_{\theta\text{P}}, \chi_{\dot{x}\text{P}}, \chi_{x\text{N}}))$$

$$\chi_{FS} = \text{MIN}(\chi_{\dot{\theta}\text{S}}, \chi_{\theta\text{S}}, \chi_{\dot{x}\text{S}}, \chi_{x\text{S}})$$

Using the same membership values as before, in Figure 9.12, and applying the fuzzy rules, the membership of the force fuzzy sets can be calculated as before:

$\chi_{\dot{\theta}\mathrm{N}} = 0.0$ $\chi_{\dot{\theta}\mathrm{S}} = 0.1$ $\chi_{\dot{\theta}\mathrm{P}} = 0.9$

$\chi_{\theta\mathrm{N}} = 0.0$ $\chi_{\theta\mathrm{S}} = 0.0$ $\chi_{\theta\mathrm{P}} = 1.0$

$\chi_{\dot{x}\mathrm{N}} = 0.8$ $\chi_{\dot{x}\mathrm{S}} = 0.2$ $\chi_{\dot{x}\mathrm{P}} = 0.0$

$\chi_{x\mathrm{N}} = 0.0$ $\chi_{x\mathrm{S}} = 0.0$ $\chi_{x\mathrm{P}} = 1.0$

$\chi_{F\mathrm{P}} = \mathrm{MAX}(0.9,\ \mathrm{MIN}(0.0,\ 1.0),\ \mathrm{MIN}(0.0,\ 0.0,\ 0.0),\ \mathrm{MIN}(0.0,\ 0.0,\ 0.8,\ 1.0)) = 0.9$

$\chi_{F\mathrm{N}} = \mathrm{MAX}(0.0,\ \mathrm{MIN}(0.9,\ 0.0),\ \mathrm{MIN}(0.9,\ 1.0,\ 0.8),\ \mathrm{MIN}(0.9,\ 1.0,\ 0.0,\ 0.0)) = 0.8$

$\chi_{F\mathrm{S}} = \mathrm{MIN}(0.1,\ 0.0,\ 0.2,\ 0.0) = 0.0$

The membership values are shown in Figure 9.14 as shaded areas.

Where sets overlap the maximum membership value is chosen. The force is finally calculated by finding the 'centre of gravity' of the shaded areas (this method of defuzzification is slightly different to that given in Chapter 6). The centre of gravity is the point at which the total shaded area to its left equals the total shaded area to its right. In Figure 9.15 the shaded area looks like Figure 9.14, but the area has been quantized into forces 1 newton apart.

The resulting force will be that at which the difference between the sum of the samples to the left and the sum of samples to the right is a minimum.

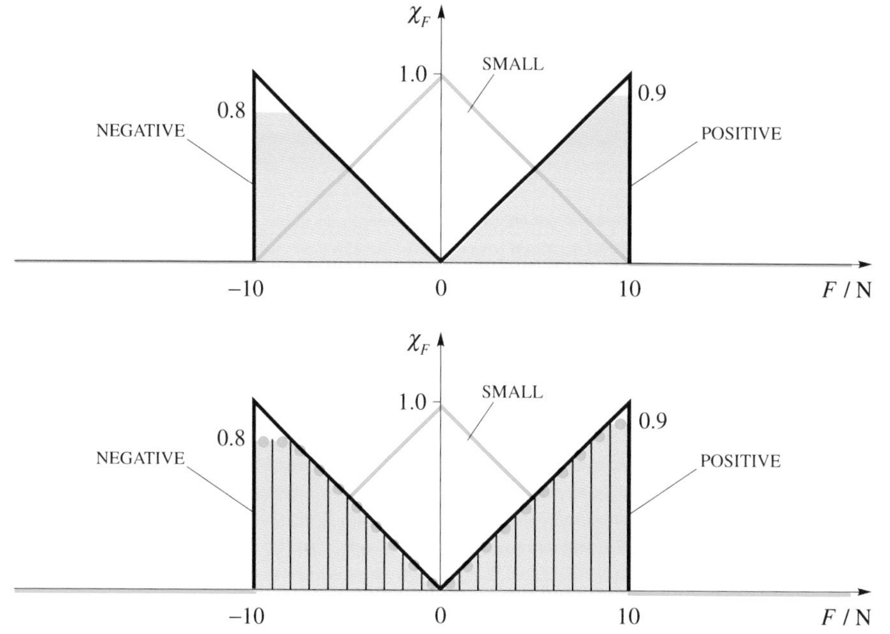

Figure 9.14
Membership values of the output variable, F.

Figure 9.15
Defuzzification of the outp

The formula used here for finding the centre of area, F_c, is:

$$F_c = \frac{\sum_{i=m}^{n} \chi_{max}(F_i) \cdot F_i}{\sum_{i=m}^{n} \chi_{max}(F_i)} \qquad (9.6)$$

where F_i is the centre of the ith discrete interval, in the range m to n, and $\chi_{max}(F_i)$ is the largest fuzzy set membership value associated with F_i when the membership functions intersect. In Figure 9.15 the membership functions of NEGATIVE and POSITIVE do not intersect, and that for SMALL is zero. In this case $m = -10$ and $n = +10$, and the centre of area can be found by evaluating the two summations in equation (9.6):

$$\begin{aligned}\sum_{i=-10}^{10} \chi_{max}(F_i) \cdot F_i &= -9.5 \times 0.8 - 8.5 \times 0.8 - 7.5 \times 0.75 - 6.5 \times 0.65 \\ &\quad - 5.5 \times 0.55 - 4.5 \times 0.45 - 3.5 \times 0.35 - 2.5 \times 0.25 \\ &\quad - 1.5 \times 0.15 - 0.5 \times 0.05 + 0.5 \times 0.05 + 1.5 \times 0.15 \\ &\quad + 2.5 \times 0.25 + 3.5 \times 0.35 + 4.5 \times 0.45 + 5.5 \times 0.55 \\ &\quad + 6.5 \times 0.65 + 7.5 \times 0.75 + 8.5 \times 0.85 + 9.5 \times 0.9 \\ &= 1.375\end{aligned}$$

$$\begin{aligned}\sum_{i=-10}^{10} \chi_{max}(F_i) &= 0.8 + 0.8 + 0.75 + 0.65 + 0.55 + 0.45 + 0.35 + 0.25 \\ &\quad + 0.15 + 0.05 + 0.05 + 0.15 + 0.25 + 0.35 + 0.45 + 0.55 \\ &\quad + 0.65 + 0.75 + 0.85 + 0.9 \\ &= 9.75\end{aligned}$$

Substituting into (9.6), the force in newtons (N) is

$$F_c = \frac{1.375}{9.75} = 0.141 \text{ N}$$

These fuzzy rules are capable of controlling the broom-balancer. Again, the performance of the system is not as good as the P+D controller, but the broom can be balanced indefinitely. The graphs for the broom-balancer under fuzzy control are shown in Figure 9.16.

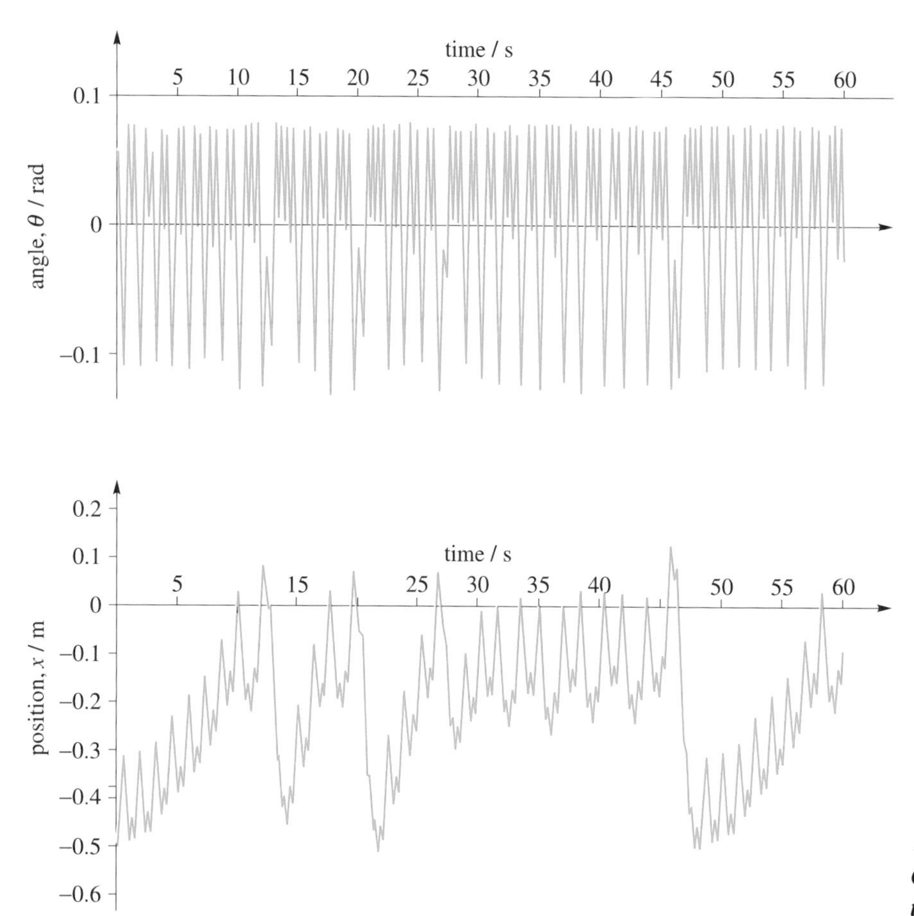

Figure 9.16
Graphs of θ and x against time for the fuzzy controller

The advantage of using fuzzy rules is that the models and measurement do not have to be precise. The designer can use loosely defined terms like large, medium and small, and the membership functions themselves can be defined very loosely. However, this does require a certain amount of good guesswork.

Research in this area has now focused on setting some of the parameters such as finding the membership functions of the fuzzy sets using adaptive methods such as neural networks. First a rough guess is made of the fuzzy rules. Then the fuzzy rules are transformed into an equivalent neural network. The network is then shown examples and the weights adjusted to improve the overall performance. Then the network is transformed back into a fuzzy rule-based system.

The advantage of the neural network is that it provides an adjusting mechanism, whereas the advantages of the fuzzy rule-based system are that it can be efficiently coded and is robust.

9.7 Hierarchical control of complex systems

9.7.1 Complex control problems

So far in this chapter we have examined how the techniques of artificial intelligence can be applied to an example of a difficult control problem. In the case of the broom-balancer the problem is difficult because the system is inherently unstable. However, this problem had the important features of (1) being modelled by a formula and (2) there being measurable control variables which allowed feedback loops to be identified, and these formed the basis of the various control strategies. But what if one is trying to control a system where there is no known formula and the controlled variables are difficult to define, let alone measure?

For example, consider a hypothetical colony of mechanical ants working together on a mining project. The goal of this mechatronic system is to extract as much mineral as possible in the shortest possible time at the least possible expense. Suppose that the ants have no *a priori* map of the area in which they are working. The intelligent control strategies just studied cannot be invoked for this system because it is too complex to define a single meaningful error, and the prospect of it being controlled by a few continuous control variables is remote. How could one begin to control a system of such complexity?

Complex systems will have the following general features:

- They will have emergent behaviour: parts will form wholes in which the whole may have properties and performance not possessed by any of the parts; for example, the ants may be specialized and form teams.
- There will be a hierarchy (more precisely a heterarchy) of parts and wholes; for example, there may be divisions made up of teams responsible for mining certain areas giving a three-level hierarchy of ants / teams / divisions. Note that it is possible that some ants will belong to more than one team, or that some teams will belong to more than one division. In this case the part/whole structure is a *heterarchy*.
- They must function with uncertain information about their environment; for example, the ants have no map and must learn the environment.
- The system and many of its subsystems cannot be represented by numerical information alone. Geometrical, topological and abstract relational information at many hierarchical levels may all be needed to represent the system adequately in order to control it.
- The performance measures which may be applied to the whole system cannot be disaggregated to give measures allowing top-down closed-loop control; for example, the measure of quantities mined in unit time emerges from the interacting behaviour of the ants.

- Some parts of the system may be controllable by closed-loop techniques, even though the whole system may not; for example, the electro-mechanical parts of an individual ant may be controlled by closed-loop techniques allowing an ant to move its limbs with precision.

9.7.2 Control of a simple vehicle

In this book there is not enough space to discuss all the possible problems involved in controlling complex hierarchical systems. However, the following simple system illustrates a number of points.

Example

Consider a vehicle on a grid which in one time interval is capable of making one move North, East, South or West (Figure 9.17). Suppose this vehicle has sensors which allow it to detect fixed obstacles up to two squares away in any direction. The vehicle's goal is to move 10 squares to the East. What control strategy could or should be used?

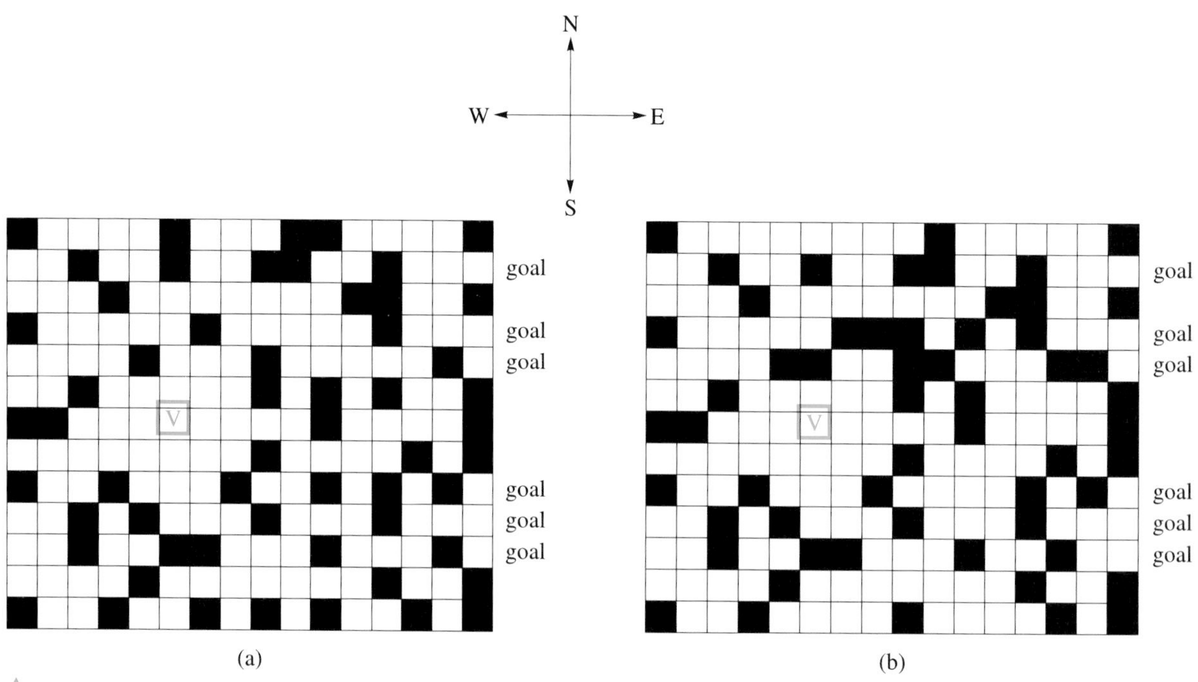

Figure 9.17
Vehicle V can move one square at a time to the North, East, South or West. It has to reach one of the goal squares without occupying the black obstacle squares.

First let us consider some *ad hoc* suggestions and see where this leads.

Strategy 1:

(1) Go East until an obstacle is encountered to the East.

This is a poor strategy because it fails if an obstacle is encountered.

Strategy 2:

(1) Go East until an obstacle in encountered.

(2) When an obstacle is encountered to the East, go North until it is possible to go East again.

This strategy will work if there is not a long vertical wall of obstacles, as in Figure 9.18(a), and if there are no traps, as there are in Figure 9.18(b). So it could be modified to give strategy 3.

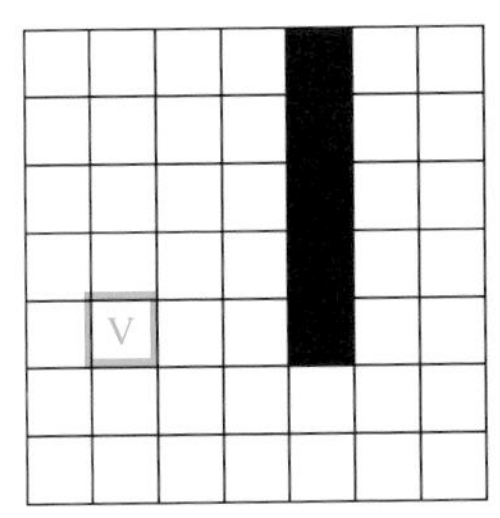

(a) A vertical wall of obstacles

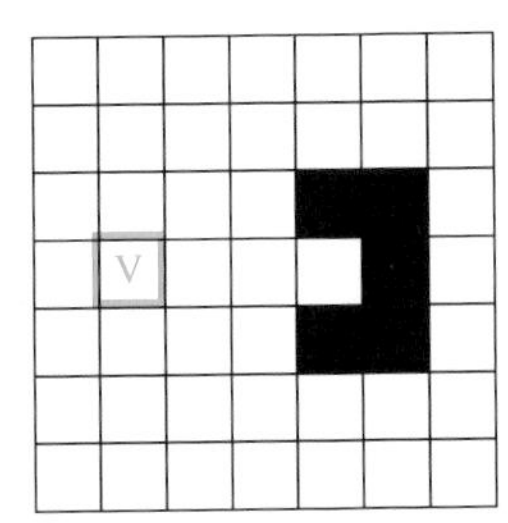

(b) A trap formed by obstacles

Figure 9.18
Special configurations of obstacles.

Strategy 3:

(1) Go East until an obstacle is encountered

(2) When an obstacle is encountered to the East, go North unless that obstacle is part of a wall.

(3) If an obstacle to the East is encountered and that obstacle is part of a wall, go South.

(4) If an obstacle is encountered, and that obstacle is part of a trap, if possible go North, else go South, else go West.

The use of terms such as 'wall' and 'trap' is a quantum leap for the representation of this system. What is a 'wall'? What is a 'trap'? Are the objects shown in Figure 9.19 walls or traps? Does the machine need to 'see' more than two squares ahead in order to perceive such objects?

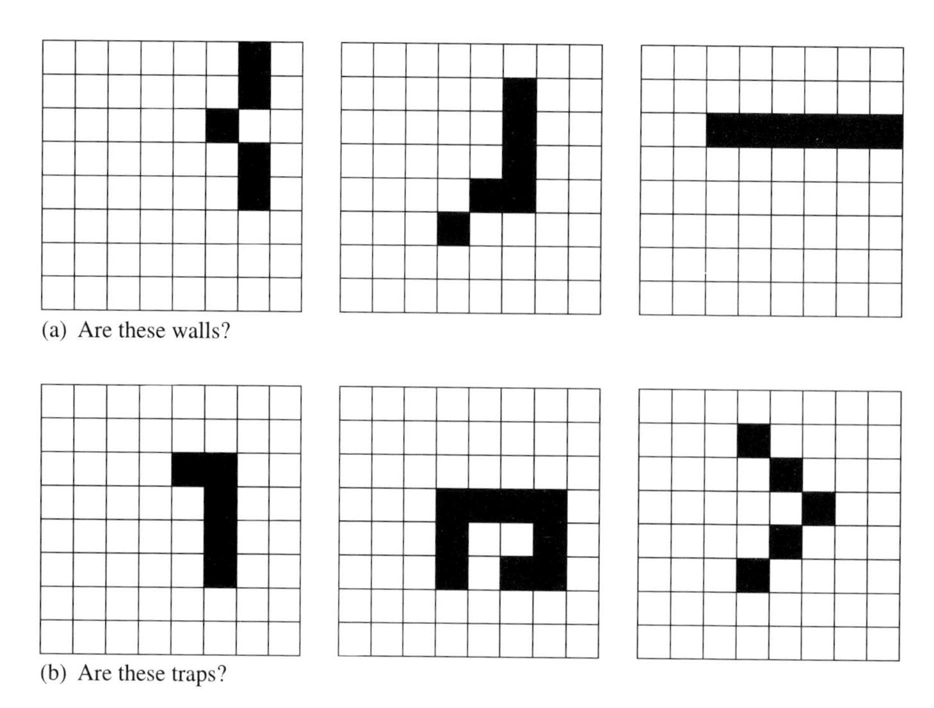
(a) Are these walls?
(b) Are these traps?

Figure 9.19
Problems of classification and pattern recognition in control.

This illustrates that strategies to solve this kind of problem depend on pattern recognition, as discussed in Chapter 2. When more complex spatial information is available from the sensors, techniques of computer vision may be required, as discussed in Chapter 10.

9.7.3 Cognition for control

The control decisions for this hypothetical vehicle must be made by its cognition subsystem, as discussed in Chapter 5 of Volume 1. Even though the environment is structured by the grid and blocks on it, as shown in Figure 9.17 it can get quite complicated. Recall the following from Section 5.2.4 of Volume 1:

> A temptation for the mechatronics system designer is to try and cover all cases. The target is a machine cognition subsystem that will create as accurate and as complete a model of objects as possible in the world of interest and use this description to plan actions. However, a complete model may contain much information that is of little or no use to the task in hand, and the system cannot know that it has a good model, except by performing 'experiments' and examining the consequences of actions in the world. This suggests that *learning* about actions and their consequences is an important part of the cognition function.

Certainly there are combinatorially many possible configurations for the vehicle to negotiate, and the mechatronics designer cannot anticipate them all. Compare the universe of Figure 9.17(a) with that of Figure 9.17(b). Which is easiest to negotiate and why? You can probably see at a glance that (a) is 'simpler' than (b). What does this mean? If you count them, you will find that both environments have exactly the same number of blocks in them. The difference between the two is that (b) is more 'connected' than (a). For example, the longest 'chain' of black squares (connected by an edge or a corner) in (a) has 8 members, while the longest in (b) has 27 members. In fact the long chain in (b) acts as a kind of curtain which makes the task hard to complete.

Once a vocabulary such as 'connected' and 'chains' has been elicited for the problem, one can formulate knowledge using it. For example, if one knew that the longest chain in the environment had length 2, one could reason that there can be no traps in the environment, and a relatively simple control strategy is appropriate.

What kind of experiments might our vehicle perform? In fact, the only control strategy available given the lack of an overview is 'move and see what you find'. In this case the machine can find one of three outcomes: that it has encountered the goal, that it has encountered an obstacle, or that it can keep on moving.

Examples of tactical rules for getting round local obstacles were given above. The worst case is that the robot gets stuck in a nasty trap from which it is difficult to escape. A machine that cannot remember where it has been, i.e. cannot *learn* the environment, may get stuck in a trap forever, rather like a lobster in a lobster pot. However, a machine that remembers its moves should be able to extricate itself, by analogy with 'Ariadne's thread' as mentioned in Chapter 8.

This type of control must involve experiments of the 'try and see' variety, because this is the only way that the machine can get feedback on what is in the environment, and what are the consequences of its actions. In this context we can see these experiments as the machine learning more about its environment, and so adding to its knowledge base.

9.7.4 Scheduling and path planning for control

Suppose the vehicle introduced in Section 9.7.2 has a vision system which allows it to see all of the environment. Then the vehicle does not need to engage itself in exploratory experiments to find out what is in the environment. The vehicle now has a different problem: given its position at V, what is the optimal path to get to one of the goal squares?

This can be viewed as a *search* problem, and treated in the manner of Section 3.4.3 in Chapter 3. For example, every square can be given a number according to its distance from the nearest goal square and a best-first search will result in a goal being found.

Alternatively it can be viewed as a *path planning* problem and treated in the manner of Chapter 5. This can be achieved by constructing a network with a link of length one between the centres of vertically or horizontally adjacent white squares.

9.7.5 Control as search

When trying to control a system one is effectively searching the space of all possible control actions which give the desired outcome. This means that the kind of considerations given to selecting a search technique may be relevant in designing the control strategy. For example, at each stage that one makes a choice, should it be best-first? When would breadth-first be better than depth-first for a vehicle seeking a way through its environment? When should one consider using random search?

Although there is no definitive answer to these questions, in general best-first is to be preferred if this is known to give a solution in the particular case. Otherwise breadth-first search might be appropriate for a vehicle which is establishing a base from which it expects to operate, while depth-first search might be best for a vehicle navigating a landscape just once. A random element may be appropriate when the vehicle has many choices, or to facilitate 'jumping' to what might be a better starting point when it gets stuck.

9.7.6 Controlling complex systems

In his classic book on Cybernetics first published in 1956, W. Ross Ashby writes:

> Science stands today on something of a divide. For two centuries it has been exploring systems that are either intrinsically simple or that are capable of being analysed into simple components. The fact that such a dogma as 'vary the factors one at a time' could be accepted for a century, shows that scientists were largely concerned in investigating such systems as *allowed* this method; for this method is often fundamentally impossible in the complex systems.
>
> *(Ashby, 1956, p.5)*

It may be difficult to accept that the mathematical science of the last two hundred years is simple, especially if one has just struggled through undergraduate courses on calculus and the like. Nonetheless, engineers attempting to control systems should understand the divide that comes between those systems which can be represented by formulae in which one can 'vary the factors one at a time', and those in which one cannot.

Classical control and the intelligent *control techniques* applied to the trolley and broom problem are *low-level* control strategies. This does not mean that they are trivial or more humble than control applied at higher levels of hierarchical aggregation. It does mean that in any system there is a limit to the degree of precise control that can be achieved at more aggregate levels.

For example, a large aeroplane is a relatively complex system. The engines and other mechanical subsystems are controlled with great precision by what we have called low-level control strategies. These increasingly include the techniques that we discussed in the first part of this chapter.

These low-level techniques cannot be applied to controlling the whole aircraft. For example, route selection depends on scheduling algorithms. Air traffic controllers in the wider system could not possibly depend on low-level control techniques since they must handle tremendous quantities of discrete relational information. They do this using a mixture of geometric representation and a model of the system expressed in natural language. This model is distributed over the many maps, manuals and handbooks that human beings have used to impose structure on the world's air space.

In general we would like to answer the questions: what exactly is a complex system, and what systematic approach can be taken to controlling such systems?

Poul Anderson gave the following profound insight into the nature of complex systems:

> I have yet to see any problem, however complicated, which, when you looked at it the right way, did not become still more complicated.

Anderson's comment leads us to reflect on the nature of human understanding versus the objective complexity of the universe. Perhaps the best definition is that a system is complex if you do not understand it. To some people the motor car engine is a total mystery and they regard it as something very complex; to others the engine is a relatively simple system. But then again, automotive engineers who do know engines very well may consider them to be very complex in the interaction of their geometry, chemistry and physics.

Engineers attempting to control complex systems should be aware that science is not a body of neutral objective truth: *science is a belief system* and every bit as likely to be wrong as a political belief system. If engineering control systems are based on flawed beliefs then anything can happen. The more complex the system the more unpredictable the outcome, as illustrated by epidemics, wars and economics in social systems.

For many complex systems that people create there is no *a priori* scientific knowledge. New systems evolve, or the research engineer invents a new system about which nothing is known. Then engineering and science go hand in hand. This is clearly illustrated by the development of information technology in which fundamental research has been done in the attic or the garage by anyone sufficiently interested or motivated, not just by professional scientists.

The question as to how to control complex systems should be seen in terms of a closely related question: how can one analyse and understand complex systems?

This is a research question which is beyond the scope of the book, but it involves the following:

1. The system must be observed systematically and the data recorded according to rigorous scientific standards.
2. The system must be explicitly represented in an appropriate vocabulary.
3. The relationship between time and the system's dynamics must be understood.
4. Predictions about the system must be expressed in terms of the vocabulary and time constructs.
5. A theory or model of a system is as good as the last successful prediction: one incorrect prediction makes part or all of the model wrong.

The first of these requires no justification: one cannot control a poorly understood complex system. The last of these is the scientific principle laid down by Karl Popper: a scientific theory can never be proved to be correct, it can only be demonstrated to be consistent with the observations made to date. One cannot know if a new observation will contradict the theory, and there are many examples in science of this happening. One day scientists believe one thing is true, the next they must believe that it is false. However, science is relatively stable due to its ideally high standards of evidence, and its insistence on replicability in which scientists must report their experiments in a way which enables others to repeat and validate them.

Prediction is the way that one tests theories, and the aim of control is to make machines and complex systems behave in predictable ways. It is important to realize that time is a social construct with many interpretations. Although atomic clocks behave in a very regular way, human beings do not and sometimes complain that time drags or that time flies: social time is different to clock time. Many complex systems are mixtures of machines and human beings, e.g. road systems, factories, mines and space stations. Often the system's own 'heartbeat' defines an appropriate time in which to make predictions, rather than saying 'such and such will happen at a precise clock time on a precise day'. Complex systems are controlled more strategically. For example, time is defined in terms of a sequence of events so that Phase II can be correctly predicted to follow the completion of Phase I, but the precise clock time for either cannot be predicted.

Representation is a major theme in artificial intelligence, and the second of the points listed above is very important. All the information and knowledge built into machines must be explicit, and this requires an appropriate symbolic and numerical vocabulary to be constructed. Sometimes this is *system-specific*, which means that the representation has to be custom-built for that system. For example, in biological systems one refers to cells and cytoplasm, while in social systems one refers to committees and regulations. However, there are many constructs which are common to many systems. In particular, much of the vocabulary developed in this book can be applied to a wide diversity of complex systems and their subsystems. Furthermore, all the science and mathematics that one learns contributes to this general vocabulary, as indeed do the arts, humanities and social sciences. The main contribution that this book can make in enabling you to control complex systems is to give you the basic building blocks, and to illustrate how they can be applied in some particular cases.

9.8 Conclusion: principles for intelligent control design

The field of intelligent control is expanding rapidly. The use of neural networks in control has produced a flurry of research activity all over the world, but perhaps it is fair to say that at this time (1995) there are relatively few commercial applications of neural networks in control. Similarly, genetic algorithms have yet to have a major impact but hold so much potential that it is difficult to imagine an area of research where they won't find any use. In control, they could open up new approaches to adaptive control and in particular self-tuning control where parameters have to be adjusted in order to meet a performance requirement.

Fuzzy control is currently being introduced into all sorts of control areas – from self-focusing cameras to cement works. Of particular interest is the hardware that is being developed in Japan, where the FC110 Digital Fuzzy Processor has been built. Clearly, when investment is made in designing new forms of hardware there is almost certainly going to be a market for these devices. So the future looks promising for fuzzy control.

As we've seen, in cases where conventional control can be applied there is no reason to use intelligent control. In the example of the trolley and pole, conventional control performed better than the recurrent neural network or the fuzzy controller.

In this book we have presented some of the techniques from artificial intelligence that can be used to design better machines. Many of these techniques can be applied in the control of complex systems, but it is difficult to give a recipe for how they can be applied to any particular engineering design problem. However, there are some features that will be common to most situations, and it is possible to lay out some principles which can guide the design process:

- Construct an explicit vocabulary of the features, objects and properties of the system. Any concept which is to be used will have to be well defined in an operational sense, and making the vocabulary explicit will help to discourage wishy-washy generalities which cannot be usefully implemented.
- Observe the system and try to record its behaviour in terms of the vocabulary. Express this in a way that can be entered into the machine, e.g. tables of data, rules and facts in a rule-based system.

- Try to quantize observations so that they can be weighted in terms of probability and/or fuzzy set membership.
- Identify subsystems. Attempt to devise relatively independent control strategies for them.
- Try to understand how subsystem variables aggregate into variables at higher levels.
- Construct deterministic tactics to cope with well understood local problems. If conventional control techniques are applicable, then use them.
- Optimize where possible in subsystems, but beware of hill climbing into a local optimum which is not a global optimum solution for the whole system.
- Beware of being absolutely deterministic – have a random element to allow you to jump out of a bad situation.
- Decide whether to expect to implement a top-down control strategy, or whether you intend to let the system behaviour emerge from relatively independent subsystems.

The importance of intelligent control is that it becomes possible to control systems where conventional control is not possible, rather than to compete with conventional control. We can therefore expect to see many new developments in intelligent control applications in the future.

Further reading

In this chapter we come to the frontiers of knowledge on how to model and control complex systems. This has been a very active area of research over the last fifty years and many penetrating insights have been gained. John Casti has put together a highly accessible account of these theories, ranging from the basic principles of modelling and beliefs, to catastrophe theory, cellular autonoma, chaos and discrete dynamics, game theory, brains and minds, classical control, computation, and complex systems.

If you want to know more, the following books are highly recommended:

Casti, J.L. (1992) *Reality Rules I: Picturing the world in mathematics – the fundamentals*, John Wiley & Sons, Chichester.

Casti, J.L. (1992) *Reality Rules II: Picturing the world in mathematics – the frontiers*, John Wiley & Sons, Chichester.

Acknowledgement

The control strategies and results reported in this chapter are based on work done by our colleagues Nick Hallam and Neil Woodcock.

References

Ashby, W.R. (1956) *An Introduction to Cybernetics*, Methuen & Co., New York.

Barto, G., Sutton, R.S. and Anderson, C.W. (1983) 'Neuronlike adaptive elements that can solve difficult learning control problems', *IEEE Trans. on Systems, Man and Cybernetics*, Vol. SMC-13, No. 5, September/October, pp. 834–846.

Widrow, B. and Smith, F.W. (1964) 'Pattern recognising control systems', In: Ton, J.T. and Wilcox, R.H. (eds) *Computer and Information Sciences*, Spartan Books, Cleaver Hume Press, pp. 288–317.

CHAPTER 10
COMPUTER VISION

10.1 Introduction

Figure 10.1 shows a *digital image*. It is an array of numbers called *greyscales* associated with an image. Each of the cells in a digital image is called a *picture element*, or *pixel*. Usually the greyscales are interpreted in terms of brightness: pixels with large numbers are bright, pixels with smaller numbers are darker.

Computer vision attempts to answer the following questions:

(1) Can any objects be recognized in the digital image?

- (1.1) Where is each object?
- (1.2) How big is each object?
- (1.3) How is each object oriented?

(2) Do the recognizable objects make up other objects?

- (2.1) Can a *scene* be recognized?
- (2.2) Can we recognize objects and scenes:

  when bits are missing through *occlusion* (part of an object is hidden behind another object),

  when bits are missing through poor image quality,

  when spurious bits are added through poor image quality?

Before you read on further, take a few minutes to try to answer questions (1) and (2) for the image in Figure 10.1.

153	160	158	160	156	164	163	162	152	153	151	157	154	157	154	160	161	158	154	151	152	151	152	150	153	153	167	154	162	158	160
159	164	160	160	156	155	152	161	155	152	151	156	155	157	156	156	150	154	153	150	150	155	161	160	160	157	159	156	157	152	156
149	156	158	160	157	160	158	156	151	156	153	153	151	148	147	153	156	150	153	155	154	157	155	153	156	154	156	153	168	156	158
155	159	156	152	148	155	152	159	152	155	154	157	152	154	147	153	148	142	156	164	157	153	157	154	155	154	152	153	160	154	156
153	159	158	161	154	159	155	156	151	161	164	145	133	133	136	141	136	142	144	153	159	157	155	151	155	155	159	155	159	158	167
153	161	160	161	157	161	153	156	157	169	165	142	126	124	115	112	111	117	130	148	153	153	157	156	157	161	159	154	160	162	163
146	156	153	159	156	156	145	161	180	187	154	128	116	105	95	96	89	101	109	126	137	143	152	143	151	163	161	155	157	154	156
149	155	152	160	163	164	150	168	172	169	149	112	99	98	85	76	81	90	83	93	110	119	123	119	133	143	154	155	154	152	157
148	159	158	163	161	157	137	138	146	129	87	80	85	87	84	91	96	95	87	91	97	101	113	112	122	137	153	157	163	157	159
151	160	163	168	157	139	112	110	105	94	78	80	84	94	92	94	99	100	93	89	94	100	104	97	106	118	145	155	162	158	156
149	159	163	160	136	111	84	86	84	83	71	78	92	102	97	98	100	103	91	96	86	80	86	94	95	86	123	152	158	152	157
156	164	169	156	120	90	72	84	75	71	77	93	108	122	123	126	125	119	109	105	85	78	82	87	90	80	111	141	157	152	153
157	164	168	150	104	72	77	87	76	80	91	117	138	147	153	159	155	153	142	128	108	88	83	92	92	76	103	136	155	156	162
157	161	164	139	92	70	79	87	72	90	108	134	153	159	157	161	163	162	151	149	126	101	90	85	89	83	95	123	151	157	159
155	160	166	137	74	54	81	94	70	73	120	151	148	153	154	159	153	151	155	157	138	114	86	85	90	83	87	110	142	151	155
158	159	163	136	83	68	83	91	74	83	122	150	152	155	151	154	158	161	157	163	146	121	94	84	88	84	90	111	142	153	153
152	158	159	129	85	73	78	86	70	85	126	155	157	157	160	158	156	162	162	162	146	119	89	91	99	89	86	113	144	150	153
156	161	161	135	82	64	79	91	74	82	122	156	159	156	158	158	161	167	164	162	139	109	86	88	93	84	89	119	147	153	159
152	153	156	136	104	82	68	81	77	86	103	132	148	155	160	162	161	168	164	155	130	99	89	95	86	61	81	122	142	145	153
150	153	155	141	110	78	57	68	73	85	96	122	141	151	155	160	165	171	156	134	106	95	91	84	82	68	92	131	148	144	151
155	157	154	147	118	82	51	58	68	83	94	110	129	141	147	151	152	155	141	126	99	79	84	88	76	63	107	138	148	149	157
155	159	155	145	125	99	60	53	69	83	86	98	106	116	126	135	142	147	120	105	89	85	80	74	73	88	125	147	156	152	159
150	155	150	148	137	118	87	72	63	79	83	84	81	88	92	102	111	116	108	104	87	74	75	74	80	106	134	143	150	151	158
152	158	150	146	143	131	112	93	74	72	72	82	71	65	68	74	83	90	81	78	82	79	77	83	97	120	142	144	151	152	156
150	157	158	157	152	149	128	113	88	81	79	77	71	74	70	79	71	78	81	84	78	77	91	97	121	140	153	157	164	161	158
153	157	155	155	150	160	151	138	114	101	88	84	75	71	68	78	80	85	81	85	94	102	118	126	143	151	155	155	159	155	162
151	161	160	159	154	159	157	156	139	121	107	95	81	75	61	63	80	101	104	114	126	131	141	146	155	152	153	154	159	155	161
155	162	157	154	150	162	158	159	161	154	139	130	118	110	103	116	125	139	152	155	158	157	154	149	151	149	153	152	155	151	155
158	161	154	155	148	153	151	157	156	163	162	163	153	152	147	152	164	175	174	168	164	158	158	156	160	156	157	161	161	155	159
157	160	154	155	153	156	153	163	157	157	157	164	161	166	163	164	164	163	159	158	152	151	153	149	153	156	159	157	162	158	163
153	155	152	156	147	153	151	157	157	160	156	161	161	160	157	161	156	155	153	150	149	150	151	149	151	152	157	159	158	155	161

Figure 10.1
A digital image.

10.2 Abstracting information from digital images

Faced with the question 'Can any objects be recognized in the digital image?', an understandable response could be a peeved 'How do I know, it could be anything!' Fortunately, in most applications of computer vision we know quite a lot about the image already. For example, if the image comes from a satellite we would expect it to contain objects which differ from those in a medical image. As it happens, the object in Figure 10.1 is common in an industrial manufacturing context. Does this, and the hint that there is just one object, help you to decide what it is?

Figure 10.2 illustrates the kinds of problems you may have encountered when trying to interpret Figure 10.1. Inspection of the greyscale numbers in Figure 10.2 shows there is a marked vertical column of low values in the centre of the image. This suggests there is a dark object against a lighter background.

One approach to finding out what is in the image is to reduce the complexity of 256 different grey levels to just two, i.e. to make the image *binary*. This simplification to black-and-white is known as ***binarization***. Figure 10.3(a) shows the central part of the image in Figure 10.2: all those pixels which have greyscales less than or equal to 140 are shown in black. Binarization thus reveals part of a long thin object here.

Suppose it is known that the image may contain one or more of the following objects: washers, pins, nuts, bolts, screws, wire, bar codes and ball bearings. Can it be decided which of these objects is present in Figure 10.3(a)? Intuitively, the choice can be narrowed down to pins, bolts and screws, because they are all long thin objects with a head. Of these, the screw is an unlikely candidate because the sides of the object in the image are almost parallel. A bolt is less likely than a pin because of the lack of serration at the sides (which would be manifest as an irregularity of edge at this level of resolution of about 100 pixels to the inch). So it can be concluded that the object is a pin. Even so, it does not look much like a pin.

Suppose we try to improve the image by changing the *threshold* from 140 to 160 greyscale units, as shown in Figure 10.3(b). Then the head of the pin becomes more blob-like, and the shaft has parallel sides. But how is the best threshold chosen? Unfortunately there is no single answer to this. If the threshold is chosen to be just ten greyscale units above 160, the information in the image begins to disintegrate as shown in Figure 10.3(c).

In bench experiments like this, one can adjust parameters such as the threshold and observe what happens. In engineering practice we cannot afford the luxury of machines whose performance in any particular case is very sensitive to the setting of parameters. In general, one seeks vision techniques that are *invariant* to variables such as ambient light, which can change the greyscale levels in an image considerably. Even the signal from a video camera changes through time

172	169	173	168	172	171	175	172	172	169	172	166	166	166	166	167	168	169	166	169	175	171	166	169	168	168	166	171	165	172	171
170	176	173	162	166	170	176	170	168	165	173	172	172	164	170	177	174	171	172	170	166	171	173	174	170	177	169	172	170	172	169
177	169	166	167	176	165	170	167	168	169	171	166	170	171	168	169	173	171	170	171	165	168	167	168	165	173	176	180	169	173	175
166	169	170	161	168	171	171	168	169	163	171	168	166	163	168	169	170	170	165	163	165	166	166	170	166	171	166	169	166	174	171
172	167	170	166	170	168	170	165	170	168	168	165	168	163	165	167	168	169	172	170	164	168	168	169	164	176	175	172	166	168	165
170	169	169	167	168	172	177	166	166	171	177	169	166	165	168	169	178	173	166	163	164	166	163	169	170	171	167	172	166	172	173
171	170	166	166	172	168	171	170	176	171	172	170	174	170	171	174	172	168	166	170	170	169	162	163	168	174	166	168	164	170	168
169	164	163	166	169	167	175	166	165	168	172	167	170	166	164	165	168	171	170	166	164	167	162	164	165	172	165	168	166	170	169
176	171	172	166	172	170	174	169	166	166	172	170	172	168	161	154	160	171	172	173	170	168	161	160	163	168	166	170	168	171	176
174	167	167	165	171	170	174	169	172	169	172	176	172	154	146	142	142	156	174	176	169	170	163	166	169	174	161	169	171	174	170
176	166	168	169	172	166	175	168	167	171	183	179	160	133	115	112	114	131	160	173	169	170	167	168	170	174	172	175	171	174	170
174	170	170	165	167	163	170	166	168	168	175	173	165	124	78	71	83	114	144	163	165	161	161	167	170	175	169	171	167	172	170
169	163	167	163	167	164	170	165	164	166	170	167	156	114	59	45	65	100	139	162	159	163	165	165	167	174	169	171	169	171	166
172	168	169	167	173	168	172	166	167	167	175	168	165	134	66	1	69	130	150	158	165	170	167	168	171	176	173	172	167	173	175
170	167	172	166	174	169	175	172	167	163	176	172	174	150	77	0	74	150	165	163	171	174	168	166	167	173	167	169	171	173	174
171	168	169	166	173	171	178	173	166	166	177	174	173	159	88	0	61	151	168	159	164	172	166	168	166	172	167	166	164	170	168
178	172	172	168	167	163	173	170	170	169	178	181	183	162	88	2	53	152	170	162	167	172	167	165	166	173	170	172	167	169	168
169	170	176	171	177	174	175	171	169	168	177	175	183	171	102	2	51	155	176	166	164	172	173	172	167	175	175	173	171	178	174
174	170	172	168	173	168	177	172	165	170	181	175	177	170	95	1	58	154	172	164	166	172	167	166	165	170	172	173	167	172	173
167	164	167	165	174	169	175	172	170	169	178	174	180	168	99	4	33	150	172	159	162	168	169	170	170	175	176	178	169	167	170
176	170	169	168	176	169	176	171	165	169	178	173	177	166	97	2	43	151	176	164	164	173	173	173	174	176	171	170	167	170	168
172	166	174	171	180	175	181	175	168	168	177	177	183	169	106	0	38	152	175	164	168	178	174	173	171	174	171	173	169	175	172
173	168	167	171	177	168	175	171	171	168	178	173	178	165	98	1	39	150	177	163	166	173	175	173	171	178	171	173	171	175	171
173	167	174	168	170	170	176	166	164	170	176	177	182	168	111	1	27	147	175	159	158	170	171	167	165	175	173	174	171	177	172
171	166	168	168	173	164	173	169	167	169	178	177	183	167	105	3	29	145	173	162	165	173	169	166	163	171	169	171	169	173	172
175	171	173	172	174	168	176	172	168	164	173	175	177	173	112	2	29	141	170	161	160	168	169	167	168	176	171	173	170	176	174
175	171	173	171	178	169	172	174	170	167	175	176	180	170	120	7	18	141	176	165	164	172	168	168	165	173	171	175	171	178	174
171	166	168	167	172	165	175	173	170	171	177	175	178	173	116	0	19	140	172	159	158	170	168	168	167	173	169	170	167	170	166
169	171	172	163	170	168	171	169	169	167	173	176	179	172	125	2	12	137	176	160	157	168	167	168	165	175	167	171	173	172	167
177	168	172	173	176	170	178	173	163	169	180	176	179	177	128	3	9	136	174	163	164	174	172	173	170	173	170	173	168	172	170
172	169	172	164	169	170	171	169	170	168	176	175	182	175	130	1	8	137	179	165	161	174	169	165	170	174	165	170	169	172	168

Figure 10.2

Another digital image. Is it a washer, a pin, a nut, a bolt, a screw, a bar code, a ball bearing, or a piece of wire?

due to noise created by its internal circuitry: no two digital images are exactly the same. Vision techniques which cannot cope with this and other uncertainties will not be robust in practice.

One way to understand how to abstract information from a given class of digital images is for the vision engineer to study displays of them on a monitor.

(a) greyscales ⩽ 140

169	172	166	166	166	166	167	168	169	166	169	175	171
165	173	172	172	164	170	177	174	171	172	170	166	171
169	171	166	170	171	168	169	173	171	170	171	165	168
163	171	168	166	163	168	169	170	170	165	163	165	166
168	168	165	168	163	165	167	168	169	172	170	164	168
171	177	169	166	165	168	169	178	173	166	163	164	166
171	172	170	174	170	171	174	172	168	166	170	170	169
168	172	167	170	166	164	165	168	171	170	166	164	167
166	172	170	172	168	161	154	160	171	172	173	170	168
169	172	176	172	154	146	142	142	156	174	176	169	170
171	183	179	160						160	173	169	170
168	175	173	165						144	163	165	161
166	170	167	156							162	159	163
167	175	168	165						150	158	165	170
163	176	172	174	150				150	165	163	171	174
166	177	174	173	159				151	168	159	164	172
169	178	181	183	162				152	170	162	167	172
168	177	175	183	171				155	176	166	164	172
170	181	175	177	170				154	172	164	166	172
169	178	174	180	168				150	172	159	162	168
169	178	173	177	166				151	176	164	164	173
168	177	177	183	169				152	175	164	168	178
168	178	173	178	165				150	177	163	166	173
170	176	177	182	168				147	175	159	158	170
169	178	177	183	167				145	173	162	165	173
164	173	175	177	173				141	170	161	160	168
167	175	176	180	170				141	176	165	164	172
171	177	175	178	173					172	159	158	170
167	173	176	179	172					176	160	157	168
169	180	176	179	177					174	163	164	174
168	176	175	182	175					179	165	161	174

(b) greyscales ⩽ 160

169	172	166	166	166	166	167	168	169	166	169	175	171
165	173	172	172	164	170	177	174	171	172	170	166	171
169	171	166	170	171	168	169	173	171	170	171	165	168
163	171	168	166	163	168	169	170	170	165	163	165	166
168	168	165	168	163	165	167	168	169	172	170	164	168
171	177	169	166	165	168	169	178	173	166	163	164	166
171	172	170	174	170	171	174	172	168	166	170	170	169
168	172	167	170	166	164	165	168	171	170	166	164	167
166	172	170	172	168	161			171	172	173	170	168
169	172	176	172						174	176	169	170
171	183	179								173	169	170
168	175	173	165							163	165	161
166	170	167								162		163
167	175	168	165								165	170
163	176	172	174						165	163	171	174
166	177	174	173						168		164	172
169	178	181	183	162					170	162	167	172
168	177	175	183	171					176	166	164	172
170	181	175	177	170					172	164	166	172
169	178	174	180	168					172		162	168
169	178	173	177	166					176	164	164	173
168	177	177	183	169					175	164	168	178
168	178	173	178	165					177	163	166	173
170	176	177	182	168					175			170
169	178	177	183	167					173	162	165	173
164	173	175	177	173					170	161		168
167	175	176	180	170					176	165	164	172
171	177	175	178	173					172			170
167	173	176	179	172					176			168
169	180	176	179	177					174	163	164	174
168	176	175	182	175					179	165	161	174

(c) greyscales ⩽ 170

	172										175	171
	173	172	172			177	174	171	172			171
	171			171			173	171		171		
	171											
									172			
171	177						178	173				
171	172		174		171	174	172					
	172							171				
	172		172					171	172	173		
	172	176	172						174	176		
171	183	179								173		
	175	173										
	175											
	176	172	174								171	174
	177	174	173									172
	178	181	183									172
	177	175	183	171					176			172
	181	175	177						172			172
	178	174	180						172			
	178	173	177						176			173
	177	177	183						175			178
	178	173	178						177			173
	176	177	182						175			
	178	177	183						173			173
	173	175	177	173								
	175	176	180						176			172
171	177	175	178	173					172			
	173	176	179	172					176			
	180	176	179	177					174			174
	176	175	182	175					179			174

Figure 10.3 ***Binarizing the image of a pin with different threshold levels: (a) pixels with greyscales ⩽ 140 are black, (b) pixels with greyscales ⩽ 160 are black, (c) pixels with greyscales ⩽ 170 are black.***

Figure 10.4 shows a computer screen display of a whole image of the pin, and Figure 10.5(a) shows an enlarged display of the head of the pin. Note a common problem in computer vision: part of the head of the pin is missing due to a reflected highlight. Of course, you know that the head of a pin is solid, but the computer does not unless given this information. Similarly, in Figure 10.5(c) the 'point' of the pin does not look very sharp in the image.

Although binarization and greyscale techniques can be used to create displays of digital images, in computer vision one is not primarily interested in creating pictures. The task is to *abstract usable information from digital pictures*. In practice this usually means going from the greyscale array to a string of *symbols*. These symbols might be alphabetical, such as the word 'pin', or they might be numerical, reflecting an encoding of the class of the object or the value of some parameter associated with it. Converting greyscale images to binary images is a stage in simplifying the image by *classifying* the pixels as either black or white. In subsequent algorithms one can use tests such as: 'if the pixel is black then do something, if it is white then do something else'. Binarizing images loses information, of course, and the design engineer should ensure that essential information is not likely to be lost before adopting this approach.

Returning to the fundamental questions of computer vision posed in Section 10.1, question (1.1) can be answered by saying that the object is located at a certain position in the image, for example at the pixel in the 369th column of the 290th row. But what does this mean? The object typically occupies quite a large number of pixels. If you want to know where the pin is, you must first define the concept

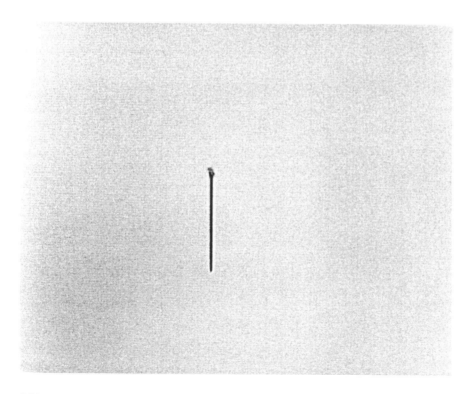

Figure 10.4
Digital image of a pin.

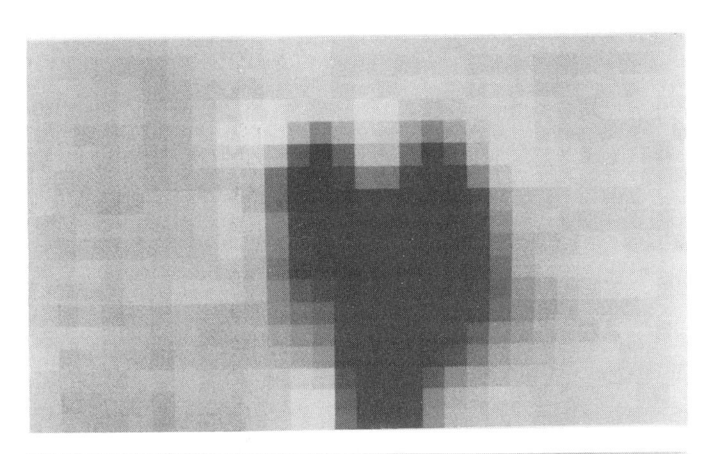
(a) The head

(b) The shaft

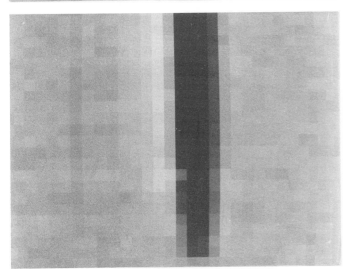
(c) The point

Figure 10.5
Enlargement of the pixels in the image of the pin in Figure 10.4.

of position in an operational way. For example, you might define the pin to be in a position determined by the centre of its head and its point. Then four numbers determine its position: $(x_{head}, y_{head}, x_{point}, y_{point})$. This is a simple but fundamental fact: *you* must determine the representation for the information that computer vision will produce. Given our definitions, the position of the pin's head can be established to within about one or two pixels. This may be sufficiently accurate for some purposes, in which case this would be a good method to use (it is simple and computationally inexpensive). However, it may not be accurate enough for other purposes. Simple thresholding as in Figure 10.3 is usually useful for only the most simple applications of computer vision.

One can begin to answer question (1.2) of Section 10.1 by putting dimensions on the pin: it is about three or four pixels wide, i.e. about 3.5 pixels divided by 100 pixels per inch = 0.035 inches. Using a micrometer it is found that the pin has a diameter of about 0.67 mm, which is about 0.026 inches. This is an error of about 35%, which suggests that the threshold was set too high. Nevertheless, this shows that computer vision can be used to make measurements. This approach could be used to *calibrate* a vision system with appropriate thresholds, but this usually means having to invest a considerable effort in keeping illumination and other factors constant. Another approach is shown in Section 10.5.2 which can give much more accurate measurements in a wide variety of external conditions without the need to calibrate to find appropriate thresholds.

Having established a geometric representation for the pin through the positions of its head and point, to answer (1.3) one can *define* the orientation, for example, in terms of the angle the line between those points makes with the horizon.

In answer to question (2), in this case the pin can be considered to be made up of three sub-objects – its *head, point* and *shaft* – but at the moment we have no way of discriminating 'head' pixels from 'shaft' pixels.

10.3 The nature of digital images

There are many sources of digital images, including some of scenes that cannot normally be seen with the unaided eye. Television cameras provide one of the most widely used sources of images, but currently (1995) there is a mis-match between analogue television technology and digital computers. In order to obtain a digital image from a television camera one must *digitize* the analogue signals from the camera using special hardware.

10.3.1 Images from television cameras

Figure 10.6 shows how one line of a television signal has been digitized to form one row of pixels for a digital image. In general, one needs special *analogue-to-digital conversion* hardware to convert the output of a camera into a form that can be accessed by a computer. Typically, the signals from the camera are plugged into a special graphics board inside the computer which converts the signals to digital form, stores the data in its on-board memory, and enables them to be displayed on a monitor.

The signal from a television camera may be encoded as a PAL video signal on a single sheathed wire such as that used to carry a signal to a television set. Some cameras output separate red, green and blue signals on one wire each, often with another wire carrying timing signals for synchronization. Such an output is called RGB, and it gives a better quality image than video. In fact, the video signal is a combination of the RGB data carried on a single waveform which can be sent along a single sheathed wire cable. The PAL signal has to have the RGB information components separated out (decoded) in order to drive the separate red, green and blue electron guns of the colour cathode ray tube (CRT).

Britain uses the PAL system, in which a TV camera produces 25 frames per second. It does this by producing two *interlaced* scans of alternate lines, each scan taking 1/50 second, with a frequency of 50 Hz. Contemporary television technology satisfies millions of domestic viewers but is not ideal for scientific and industrial applications. For example, interlacing can cause jitter which is uncomfortable for operators. This television technology is the result of incremental changes in standards over some fifty years and may be coming to the end of its life. Although it may take some years to become the domestic standard, high-quality digital television is set to take over.

When an image from an RGB camera is digitized, we either (1) take an 'average' of the red, green, blue values to produce a monochrome image with one number per pixel, or (2) take one value each for red, green and blue to produce a digital image with three greyscale numbers per pixel.

10.3.2 Simon's Three-Pixel Principle

Figure 10.6 illustrates a very important limitation on television technology due to sampling considerations, as discussed in Chapter 3 of Volume 1. In practice it is never possible for a camera to respond to a perfect edge with a drop from the maximum greyscale to the minimum greyscale. As shown in Figures 10.4 and 10.5, the edges of the pin get blurred, and instead of the idealized vertical-sided waveform shown in Figure 10.7 the camera delivers a 'V' shaped wave. This is due to the way that the camera samples the greyscales at points, and the inevitability that light from neighbouring pixels will enter the camera when a

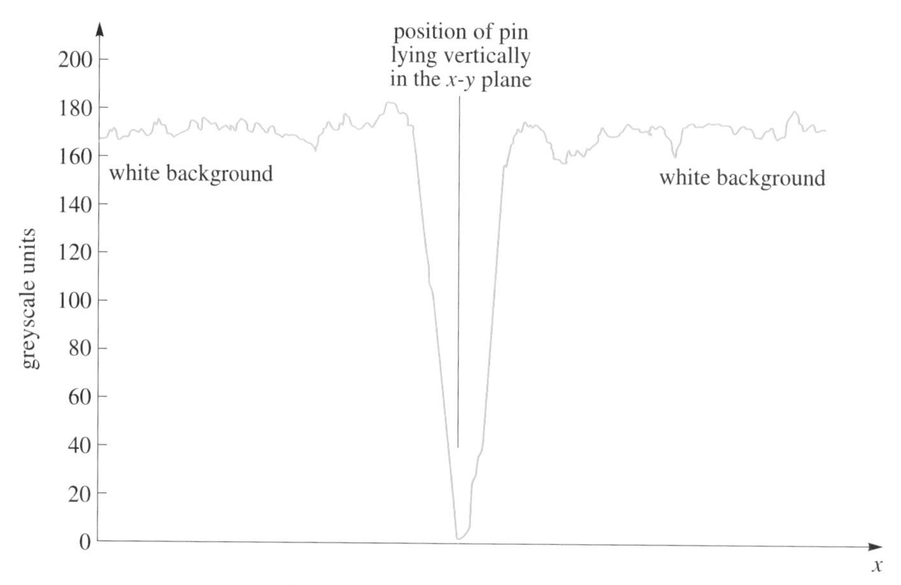

(a) Analogue signal delivered by the camera

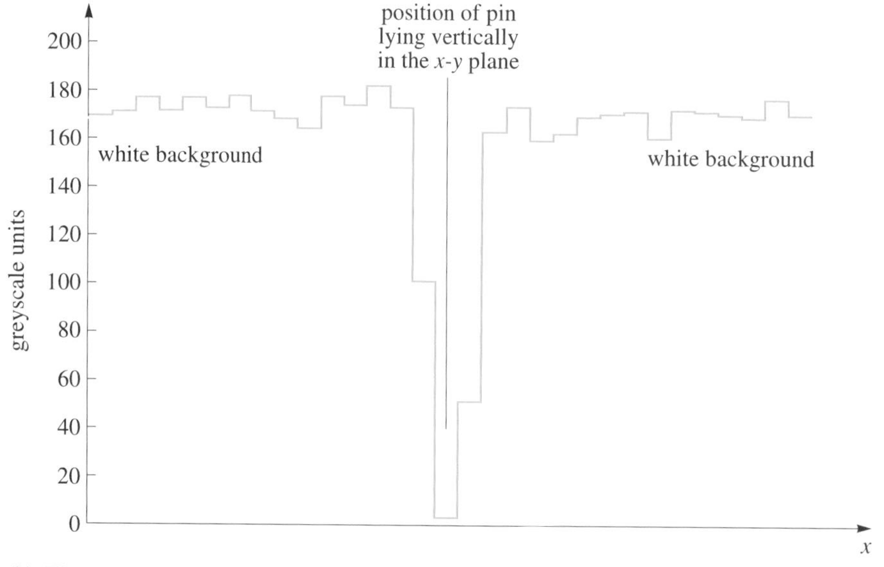

(b) The signal converted to discrete digital form

Figure 10.6 ***Converting continuous signals from the camera to discrete digital form as the greyscales of a row of pixels in a digital image.***

given pixel's greyscale is being sampled. We call this ***Simon's Three-Pixel Principle*** after Jean-Claude Simon (pronounced 'Seemon'), a French pioneer in computer vision and pattern recognition (Simon, 1986).

Simon's Principle says that 'if an object can be detected in an image, then the response to that object must be reflected in the greyscale values of at least three pixels'. So, if a satellite image showed a dark tarmac road in the light sand of the desert, no matter how high the satellite or how narrow the road, you would either detect the road over at least three pixels in any direction, or you would not detect it at all.

Theoretical suggestions that the road might be imaged as exactly one pixel wide by greyscales such as ... 100, 100, 99, 100, 100, ... are confounded by the problem that cameras cannot produce such precise images. The *signal-to-noise ratio* of a camera is a measure of the degradation of the image within the instrument itself. Typically a high-quality RGB camera will produce a signal with ± 4 greyscale units out of 255. This represents an error of 1.6%, which corresponds to a signal-to-noise ratio of 36 dB. Human eyes cannot usually detect a difference of one greyscale between two adjacent pixels.

Simon's Three-Pixel Principle raises the important question as to what theoretical limitations there are to locating objects in images. Is it three pixels? The answer to this question is emphatically no. In an industrial context the authors have developed a method of computer vision which detects objects to sub-pixel accuracy. How can this be possible in the light of Simon's Three-Pixel Principle? The answer is that, while the information that defines an object must be spread over three or more pixels, the object itself is located within some particular pixel. One speaks of the *support* of the object being detected. Usually the support consists of many pixels in the region of the object. Each of them contributes some *information* to the object-detection process. For example, the method described later in Section 10.5.2 effectively integrates and distils the information from many pixels to give very precise, sub-pixel-accurate, positioning of the edges of objects.

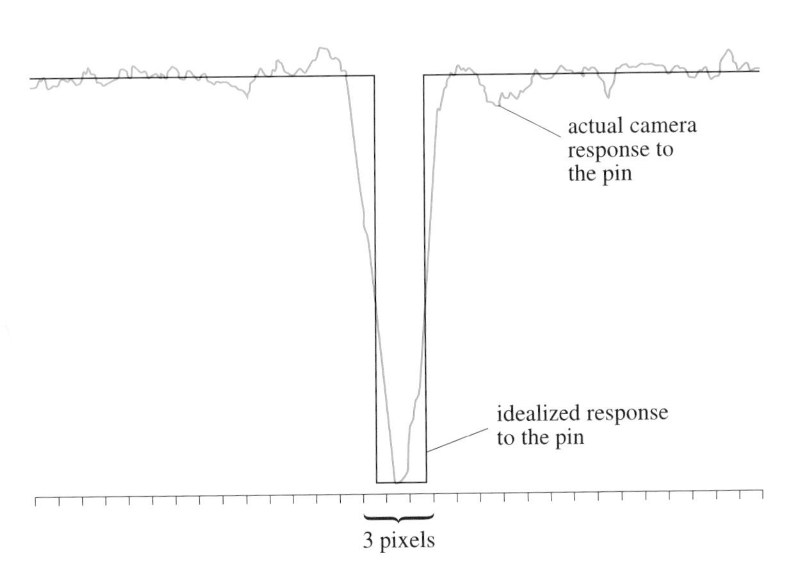

Figure 10.7 Illustration of Simon's Three-Pixel Principle: any object that can be detected in an image must be at least three pixels wide.

10.3.3 Humans' astonishing ability to read images

Computer vision has turned out to be an extraordinarily difficult problem. One reason for this is that humans constantly underestimate the difficulty of the problem because our own vision system is so spectacularly good at abstracting information from images. For example, Figure 10.8(a) shows a digitized image of part of a portrait of Pope Paul III, painted in the sixteenth century by Titian (1488/90–1576). As you look at the Pope's eyes in the digital image in Figure 10.8(a) there appears to be an abrupt change from the black pupils to the whites. The enlargement of his right eye in the digital image shown in Figure 10.8(b) shows that the transition from black to white is much more messy than one might have imagined.

Figure 10.9(a) shows a digital image of a selection of screws, bolts, washers, and tags. If you look at the long bolt in the top-left of the picture you will probably be able to see the serrated edge of the screw thread quite clearly. However, if you look at the enlargement in Figure 10.9(b) of the part in the white rectangle you will see that the sharp edges of the thread appear over three or more pixels at this level of resolution (*cf.* Simon's Three-Pixel Principle). It can also be noted that the bolt does not create a set of pixels of homogeneous greyscales. In fact there is a highlight along its length, and it would be easy to mistake the 'half' of the bolt in Figure 10.9(b) for an entire but thinner bolt. You are unlikely to make such a mistake, but then you have a vision system which involves a large part of your

(a) Digitized image of Titian's portrait of Pope Paul III (768 × 576 pixels) brain and which has adapted and been perfected over millions of years of evolution.

For many centuries artists have been fascinated by our human ability to 'read' things into pictures which, on closer examination, are not as explicit as we think. At the end of the last century the 'Impressionist' school of painting emerged which exploits to the full our ability to read things into pictures. Figure 10.10(a) shows a digitization of part of Renoir's painting of a boating party. Figure 10.10(b) shows an enlargement of the pixels of the right eye of the girl in the centre of the picture. At this level of detail it is very difficult to read the expression in the girl's eye, compared to the beauty and the emotions it conveys when put in the context of the whole face and the whole scene.

Human beings get tremendous enjoyment out of the miraculous behaviour of their vision systems, and we use our vision so effortlessly that it is tempting to think that vision is a simple process. It is not.

The point is further illustrated by the images of eight British postage stamps in Figure 10.11(a). Below these is an enlargement of the pixels making up Queen Elizabeth's eye and nose. It is unlikely that you would have recognized this outside the context of the whole portrait, which itself is highly stylized. You might find it easier to 'see' the Queen if you screw up your eyes or look at Figure 10.11(b) from a distance.

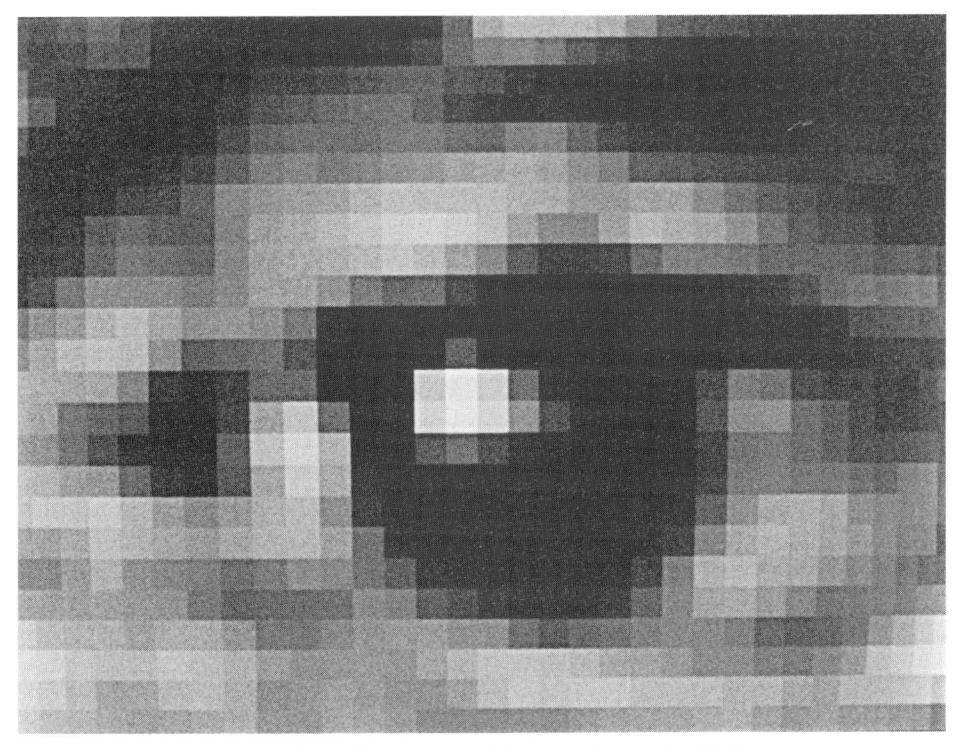

(b) Enlargement of the pixels making up the right eye of Pope Paul III

***Figure 10.8* Enlargement of the pixels making up the right eye in Titian's portrait of Pope Paul III.**

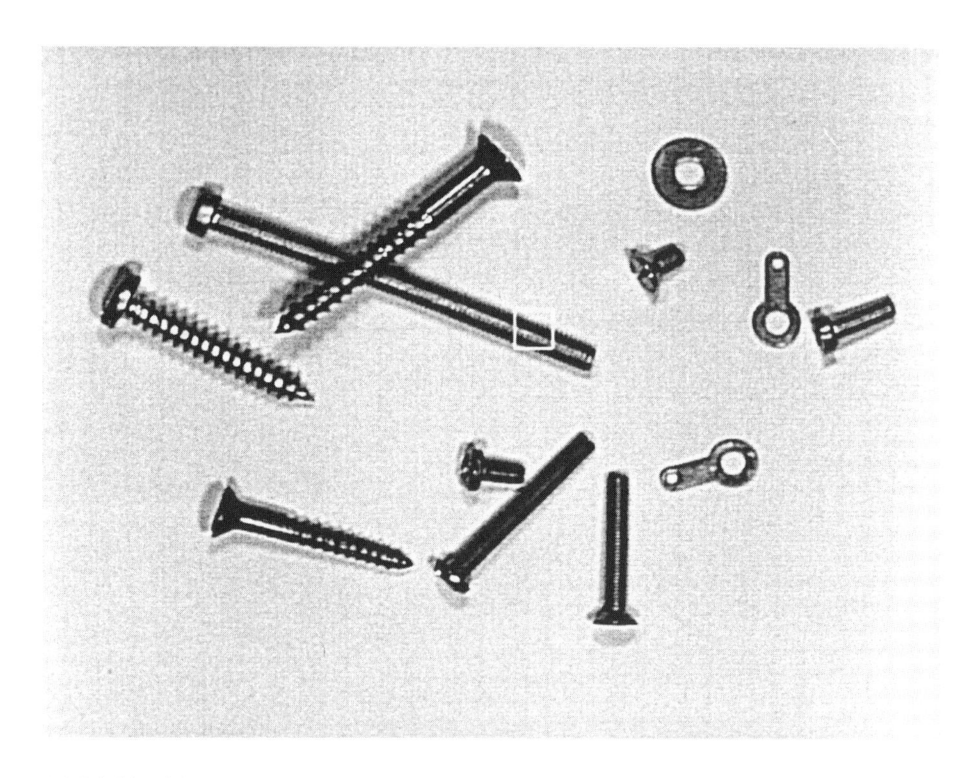

(a) Digitized image

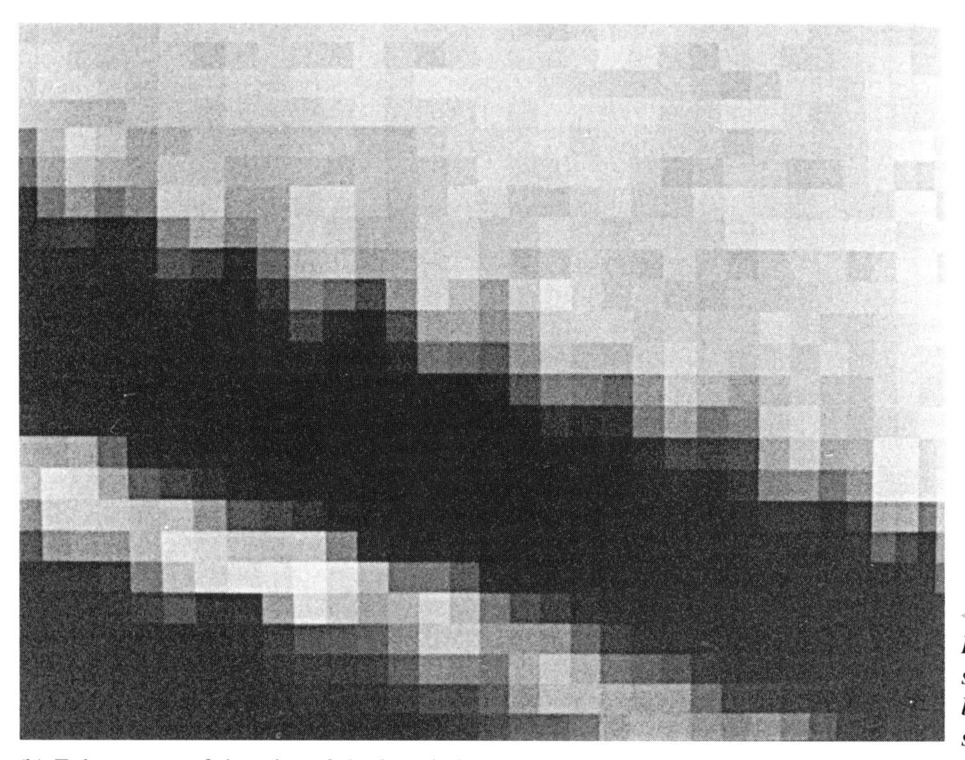

(b) Enlargement of the edge of the long bolt

Figure 10.9 Digitized image of small objects (screws, bolts, washers and tags) showing the serrated edge of a bolt.

(a) Digitized image

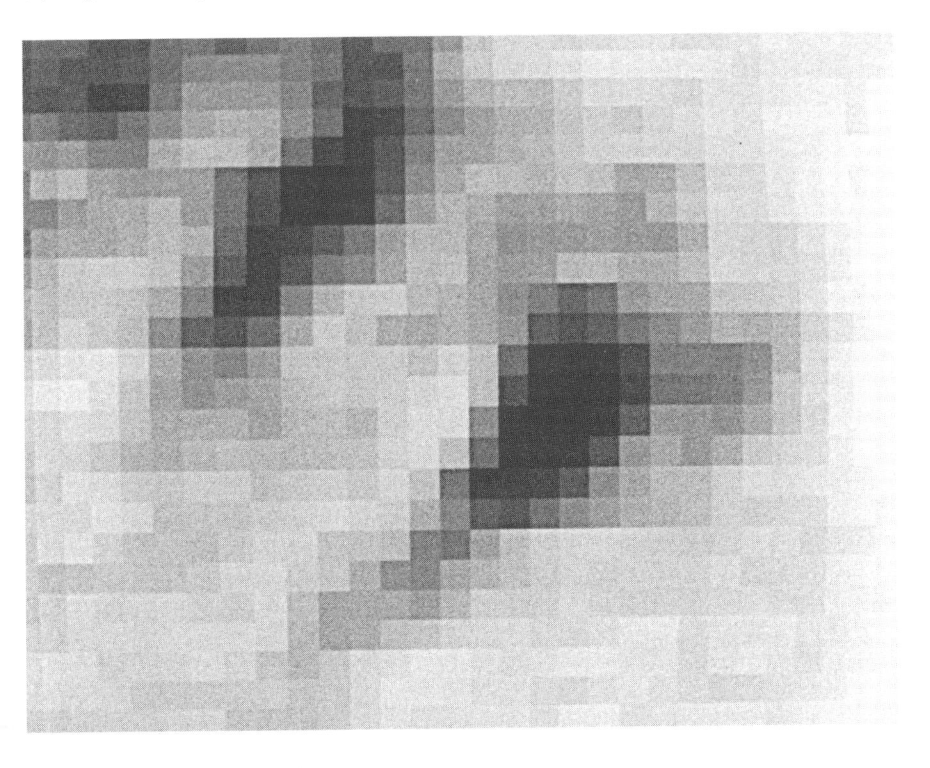

(b) Enlargement of the girl's eye

Figure 10.10
A close-up of the pixels making up the eye in the Impressionist painter Renoir's 'The Luncheon of the Boating Party'. Reproduced with the permission of The Phillips Collection, Washington, D.C.

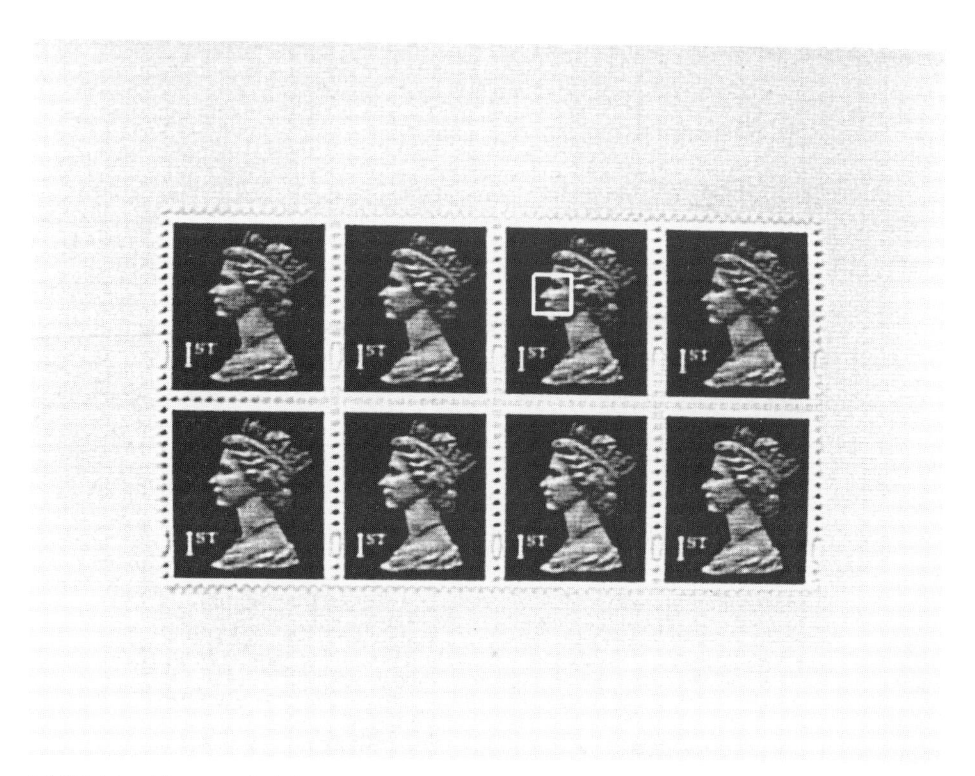

(a) Digitized image of eight postage stamps

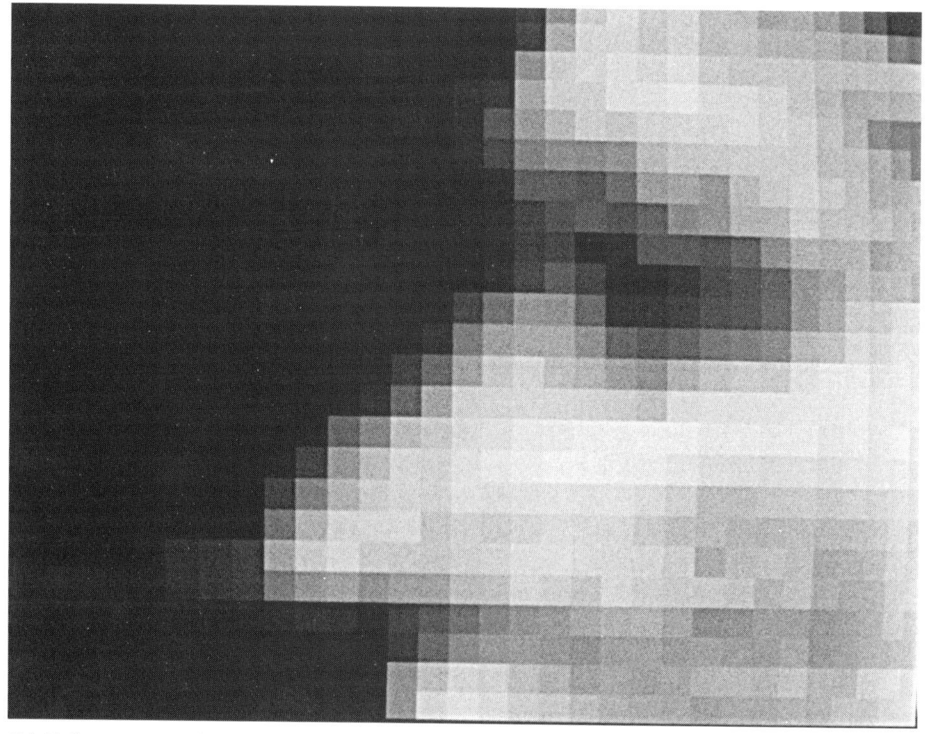

(b) Enlargement of the pixels making up Queen Elizabeth's eye and nose

Figure 10.11 Enlargement of the pixels in a postage stamp. Reproduced by permission of Royal Mail.

The digital images in this chapter suggest that it may be difficult to get a computer to do a fraction of what our eyes and brains do so easily. Nevertheless, computer vision is not an overwhelming challenge provided one adopts a rigorous scientific approach.

10.3.4 The generality of digital images

There are many sources of images, such as sensor arrays or scanning sensors. The sensors and transducers detecting one pixel's worth of information may be detecting light, as in the case of the television camera, or they may be detecting other phenomena. For example, sensors detecting the strength of magnetic fields may deliver magnetic images, sensors detecting pressure may deliver a pressure image, sensors detecting acid/alkali pH values may produce a pH image, and so on. In fact, it is remarkably easy to create digital images from simple sensors, as illustrated in Chapter 4 of Volume 1 by the example of a 'scanner' which uses eight pieces of wire.

Some of the non-light-based sources of images will deliver values on a scale which, by an abuse of language, we call greyscales. Others may deliver yes/no information and so give us a binary image. Once the image data is inside a computer's memory, the problem of abstracting useful information from a mosaic of 'coloured' dots is the same irrespective of the origins of the images.

Document scanners are becoming a major source of digital images. These work at a variety of resolutions, with 200, 300 and 400 dots per inch (d.p.i.) being common. Colour scanners are available which typically allocate one byte of data storage for each of the red, green and blue components of the image. A major problem with this kind of image scanning is the huge amount of data it creates. A typical page, without data compression, will generate some 8 inches × 300 d.p.i. × 10 inches × 300 d.p.i. × 3 bytes per pixel = 21.6 megabytes.

Although most people encounter these devices in the context of capturing images or text for desktop publishing, they have a wide range of applications in information processing systems. For example, commercial devices can be purchased for a few thousand dollars which 'read' credit card slips in a fraction of a second. Typically these machines read at 200 d.p.i. and one byte per pixel. The great problem is then to abstract useful information using computer vision, as will be discussed later in Section 10.5.3.

In many applications the speed of the available imaging devices can determine the feasibility of whole systems. Television cameras and associated frame-grabbing hardware can produce images at a maximum 'real time' rate of 25 per second. Against this high imaging rate is the disadvantage that domestic TV technology limits images to about 768 × 576 pixels. Although this pixel resolution is relatively low, combined with 24 bits per pixel (8 bits each for the red, green and blue image information) this technology delivers very high-quality images according to subjective judgement of our eyes. Colour document scanners also deliver very high-quality images, but they can be slow. The general principle is that it will take longer to produce higher quality images which have more dots or colours.

10.4 Computer vision versus computer graphics

Computer graphics and computer vision are highly inter-related but are different disciplines. Lay people often confuse the two because both involve digital images which can be displayed as pictures on computer screens. The important difference between the two is summarized in Figure 10.12.

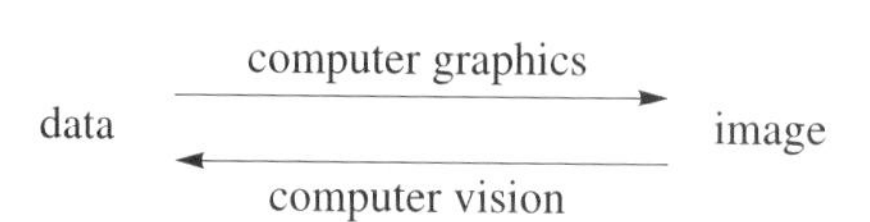

***Figure 10.12* Computer vision and computer graphics are complementary but different disciplines.**

10.4.1 Computer vision and computer graphics as complementary disciplines

In computer graphics data are used to create pictures as exemplified by, for example, the case of computer-aided design (CAD). In computer vision one starts with a picture and attempts to abstract data from it. Computer vision is orders of magnitude more difficult than computer graphics. Usually experts in computer vision are also experts in computer graphics and use this expertise to create graphic user interfaces (GUIs) which make their vision products easier to use and more attractive.

10.4.2 Representation and CAD data structures

Disciplines such as CAD have pioneered some important ideas for computer vision related to *representation*. It is easy to say that one wants a robot to 'see' a mechanical piece, but this begs the question of the robot's internal representation of that piece. For some applications the representations developed for CAD make an appropriate *target language* for computer vision. For example, a curved object might be recorded as a small number of x-y points which can be used to create a parameterized curve called a B-spline, which is an approximation to the edge of the object. This representation is different to one which simply records a lot of short lines sufficiently close together to give a piecewise linear approximation to the original curve.

10.4.3 2-D vision and 3-D stereo vision

Almost all digital images are two-dimensional, reflecting the geometry of the sensing devices and the way scenes are imaged. In many applications it is possible to abstract the required information from the two-dimensional image, but in some applications it is necessary to reconstruct the three-dimensional scene from the image to obtain symbolic and parametric information in three-dimensional coordinates.

Given various hypotheses about the nature of the scene, one can abstract 3-D information from a single image. In fact, the principles of reconstruction from perspective date back to Leonardo da Vinci (1452–1519). Geometric considerations put a limit on how much 3-D data can be reconstructed from a single 2-D image, especially when there is limited perspective information available in the image. However, *stereo-imaging* allows powerful 3-D imaging, as our own 3-D binocular vision illustrates. Computer-based stereo imaging involves, for example, two cameras with known vision and geometric properties arranged so that their 2-D pixels can be correlated. Various algorithms are then used to recognize objects and reconstruct some 3-D data. The details of this are beyond the scope of this book.

10.5 Object recognition and measurement

In computer vision one must be clear which data are actually required in any particular case. Once this is known the engineer can choose the least expensive or most effective way of delivering that information (assuming it is possible).

The simplest information one can demand of a vision system is whether an object is present or not. Domestic security lights triggered by infra-red radiation illustrate a very simple present/not-present system. The 'image' has a single pixel which can have two states: 'red' or 'not-red' according to the single infra-red sensor. The purpose of such a system is to detect and discourage 'intruders' and provide illumination for legitimate visitors. These systems have very poor discrimination, and they are frequently triggered in error by cats or other animals. Since in most cases there is a low cost to erroneous recognition, these systems are regarded as satisfactory for their purpose.

10.5.1 Detecting insects in a digital image using neural networks

In some cases the cost of erroneous recognition may be unacceptably high, and the vision system must be more discriminating than the simple infra-red sensor.

To illustrate this point consider a machine whose purpose is to kill some undesirable insects by ultraviolet irradiation but not to kill other benign insects. Let us suppose that the insects of interest are the following, as illustrated in Figure 10.13:

W	wasp	G	greenbottle
F	fly	f	blowfly
B	bee	L	ladybird

How can we begin the job of recognizing these objects in an image? What information can we use? Of many possibilities, we might immediately think of characteristics such as colour, shape, size, movement pattern, speed, and so on. As always in computer vision, these things are easy to say but much more difficult to pin down in an explicit representation. For example, how would you set about representing the concept of insect shape within a computer? It can be done, but, of the various possibilities, colour is one of the easiest characteristics to represent and we will see how far one can get using colour information alone.

One of the first problems is distinguishing the objects of interest (the insects) from objects of no interest (the background). As Figure 10.14 shows, the background can be very complex, and in the case of flowers it can move around considerably. Thus the problem of deciding if the image contains a bee is compounded by the problem of knowing what else the image contains. A simple solution to the problem of *background clutter* is to constrain the system so that the

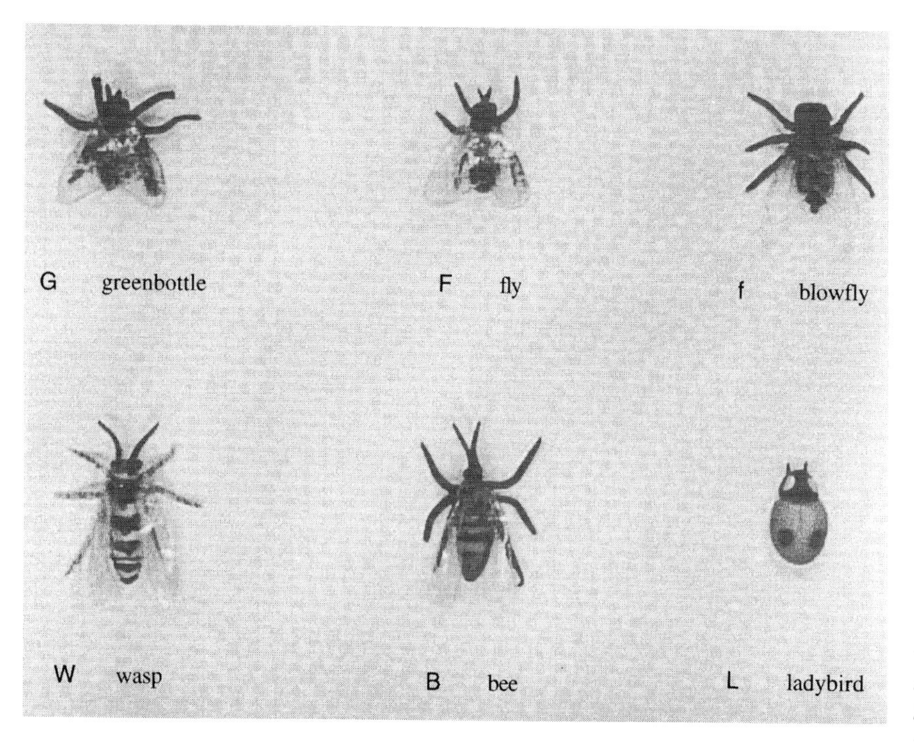

Figure 10.13
Six types of model insects to be recognized by computer vision.

Figure 10.14
The first problem in computer vision: discriminating objects of interest from the background clutter.

background is fixed and simple. So suppose that the machine will have a platform on which the insects will walk, and suppose that this platform will be of a constant light blue colour which allows the pixels to be classified as either background or insect with reasonable fidelity. In practice, of course, some pixels are misclassified so that in the following experiments some background pixels have their data included in the 'insect' statistics, while some genuine insect pixels have got lost in the background. Such 'filtering' of pixels into object and background immediately degrades the quality of the information available, but in this case the degradation is not critical.

Having abstracted a set of 'insect pixels' from the image, suppose that the greyscale values from a colour camera can be used to classify the pixels into one of the following colours:

Red	Green
Yellow	Ochre
Black	White

so that each insect will have its pixels assigned to these six classes. Table 10.1 shows the pixel frequencies for six pairs of model test insects for each of these colours. Not surprisingly, the ladybirds have the highest numbers of red pixels, the wasps have the highest numbers of yellow pixels, the greenbottles have the highest numbers of green pixels, and so on. However, not all the insects can be classified by having predominance in one colour. For example, honey bees are mostly black but have a yellowy-brown 'ochre' colour. Even though this colour characterizes the bees, their ochre count is less than that of the wasps whose bright yellow becomes this ochre colour in certain lights and shadows.

TABLE 10.1 PIXEL FREQUENCIES BY COLOUR FOR SIX PAIRS OF INSECTS

Object	Red	Green	Yellow	Ochre	Black	White	Unclassified
Ladybird 1	12872	554	423	291	9107	223	145
	13009	483	243	314	8728	364	403
Ladybird 2	18579	842	228	28	16879	917	283
	18402	938	229	25	16880	969	218
Fly 1	2775	3383	1845	294	16566	368	1966
	2831	3417	1878	334	16675	285	1482
Fly 2	1399	2322	3875	375	13271	3445	5924
	1414	2378	3797	384	13398	3316	5685
Bee 1	3669	4821	1281	2576	21829	693	779
	3594	4542	1336	2762	21442	632	762
Bee 2	3695	3037	535	2021	18198	344	266
	3623	3173	621	2102	18086	333	256
Wasp 1	2314	3590	12753	1686	8386	2846	2330
	2269	3834	12675	1608	8540	2724	2155
Wasp 2	3097	5034	15124	3626	9599	4406	845
	2756	3814	15018	3159	8936	6701	1556
Greenbottle 1	1678	6797	1207	106	15737	1402	1891
	1691	6734	1152	98	15831	1371	1356
Greenbottle 2	842	7931	3192	63	14925	3287	3113
	873	7900	3183	57	14708	3254	3142
Blowfly 1	1801	4033	432	231	24722	24	1185
	2161	3994	412	240	25271	27	1143
Blowfly 2	3958	3578	212	153	29592	11	458
	3893	3670	210	128	29997	17	437

Thus every insect is represented by six numbers: the number of its pixels classified as red, the number classified as green, the number classified as yellow, the number classified as ochre, the number classified as black, and the number classified as white. For example, for the first ladybird the numbers can be arranged as a sequence, or *vector*:

red	green	yellow	ochre	black	white
(12872,	554,	423,	291,	9107,	223)

while for the second ladybird the statistics are:

red	green	yellow	ochre	black	white
(18579,	842,	228,	28,	16879,	917)

Inspection of these data for the two ladybirds shows the entries in the vectors to be similar but not identical. It is intended to use a pattern recognition approach which will effectively classify insects according to their similarity to the test vectors for the six pairs of insects.

As explained in Chapter 4 of Volume 1, and Chapter 2 of this volume, each of these colours can be considered to define an axis in a multidimensional space. So one wants to classify the insects given their position in this six-dimensional colour/pixel frequency space.

Although it is very difficult to show this 6-D space on 2-D paper, let us choose just the yellow and green dimensions to get a feel for how the insects group together in this space. Four examples of each insect are used in what follows. Thus Figure 10.15 shows that the wasps form a cluster high in the yellow part of the space while the greenbottles form a cluster far to the right in the green part of the space. The ladybirds too form a distinct cluster near the origin because they have very few yellow pixels and very few green pixels.

Unfortunately the flies and bees are both very close in this 2-D subspace, which suggests that they will be difficult to classify. As it happens, it could be very important to separate bees from flies in this application. Fortunately, the other colour dimensions allow the flies and the bees to be separated.

Having established an operational pre-processing procedure which maps the various insects into this colour/frequency space, the computer vision task can be completed in a number of ways. This case is particularly well suited to the application of neural networks. The training data for the network are the vectors of colour frequencies as inputs, and the insect classes as outputs. So we might use a six-input and six-output network with six nodes in the hidden layer. In fact this is what we did, and the pattern recognition was very successful in correctly assigning new insects to their class.

In this application of neural networks we exploit some of their useful features. The first is that no two insects have exactly the same pixel colour frequencies and so the generalization of the network to 'similar' data is essential. Indeed,

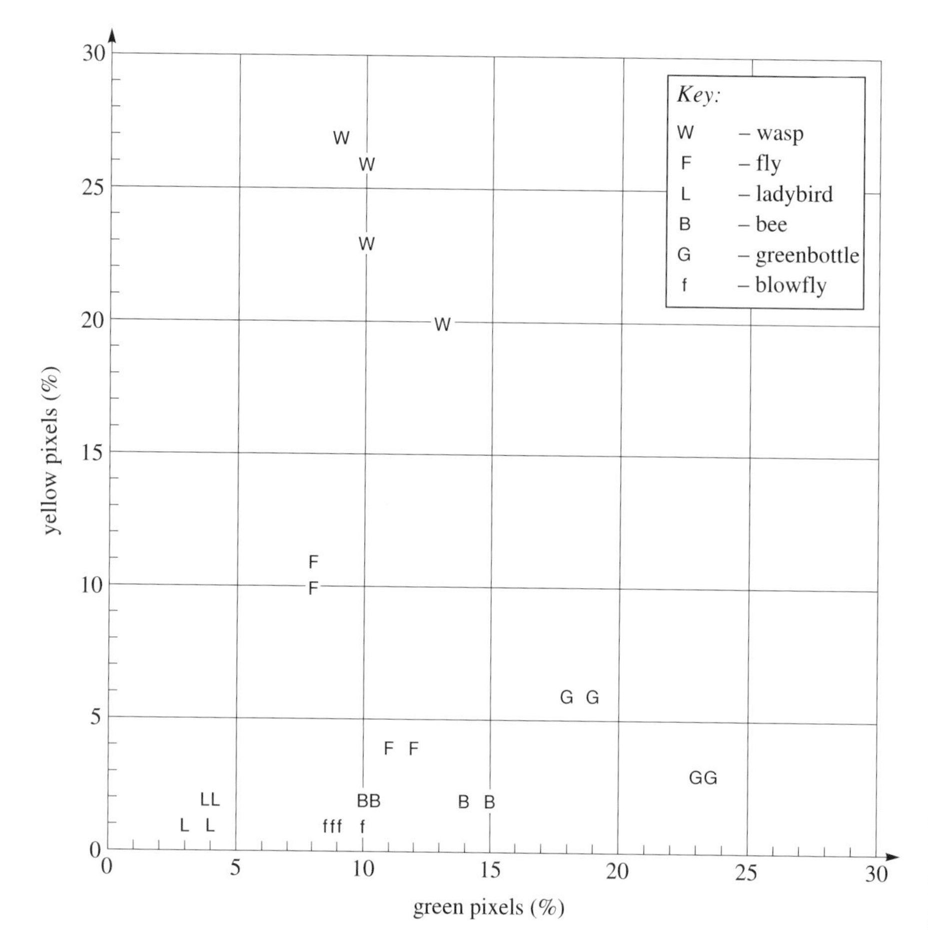

Figure 10.15
The insects clustered in the two-dimensional green–yellow subspace.

sometimes there is considerable variability in the pixel colour data, but nevertheless the neural network can cope with this.

Another useful feature of the neural network is its ability to cope with redundancy in the data. It happens that there is very little useful information in the frequency of white pixels. This is because the wings of the insects, although transparent in some lights, are highly reflective and can produce quite large 'white' responses in rather a random way. Neural networks cope with this kind of thing very well; as they train, the weights given to this error-causing redundant data are lowered and they automatically play less of a role in the classification. The exception might be the ladybirds, which do not have reflective transparent wings, and for them the white frequency could remain useful data. Thus the neural network paradigm automatically adjusts to exploit the relevant discriminatory information.

Once the insect has been recognized – or more precisely, classified – the system can take whatever action is appropriate according to its specification.

10.5.2 Measuring the diameter of a pin using sub-pixel edge detection

The kind of object recognition in the last section is of the 'yes/no', 'it-is-here-or-not' kind. In many applications of computer vision much more than this is required. Apart from knowing that an image contains an object, more information about that object may be needed.

Measurement usually requires a *geometric model* of the environment, which can be 2-D or 3-D. Recognition can use a simpler model with symbolic elements such as 'wasp is present is true', 'bee is present is false', and 'ladybird is present is false'.

The example of measuring the pin examined earlier in this chapter implicitly used a model of the pin in which it has 'sides' which can be represented by lines in Cartesian coordinate space. Thus we can define the 'diameter' of the pin to be the perpendicular distance between the two lines which make up the edges of the pin.

As explained in Chapter 4 of Volume 1, conventional attempts at edge detection have not been very successful because they depend on thresholding, as illustrated by the example in Section 10.2 of this chapter. Figure 10.16 shows the results of a new method of edge detection that has been developed by the authors in commercial applications of vision for scientific measurement. Using this method, the geometric 'edges' of the pin have been detected to sub-pixel accuracy approaching one-tenth of a pixel. By this method the measurement is that the pin is 2.5 pixels across. At 100 pixels per inch this means that the diameter measurement of the pin is 0.025 inches. This compares to a micrometer measurement of about 0.026 inches, and the system has delivered the measurement to an accuracy of about one-tenth of a pixel. This 4% error could of course be reduced by digitizing the pin at a higher magnification. This method is almost contrast-independent, i.e. it is highly tolerant to changes in the level of illumination.

10.5.3 Optical character and handwriting recognition

It is a remarkable fact that human beings now communicate massive amounts of information through their fingers by pressing the keys of computers. A modest document of ten pages will have some 40000 characters, while a book such as this has about half a million key presses. There are millions of books and documents which do not exist as computer files, and even those that are on computer are not always accessible. This creates a massive commercial demand for optical character recognition (OCR) and devices which can read characters.

In the last ten years OCR and scanner technologies have advanced to the position that clean black-on-white documents can be read by computer with an error rate of a few percent. This is acceptable for some purposes, but not for others.

172	169	173	168	172	171	175	172	172	169	172	166	166	166	166	167	168	169	166	169	175	171	166	169	168	168	166	171	165	172	171
170	176	173	162	166	170	176	170	168	165	173	172	172	164	170	177	174	171	172	170	166	171	173	174	170	177	169	172	170	172	169
177	169	166	167	176	165	170	167	168	169	171	166	170	171	168	169	173	171	170	171	165	168	167	168	165	173	176	180	169	173	175
166	169	170	161	168	171	171	168	169	163	171	168	166	163	168	169	170	170	165	163	165	166	166	170	166	171	166	169	166	174	171
172	167	170	166	170	168	170	165	170	168	168	165	168	163	165	167	168	169	172	170	164	168	168	169	164	176	175	172	166	168	165
170	169	169	167	168	172	177	166	166	171	177	169	166	165	168	169	178	173	166	163	164	166	163	169	170	171	167	172	166	172	173
171	170	166	166	172	168	171	170	176	171	172	170	174	170	171	174	172	168	166	170	170	169	162	163	168	174	166	168	164	170	168
169	164	163	166	169	167	175	166	165	168	172	167	170	166	164	165	168	171	170	166	164	167	162	164	165	172	165	168	166	170	169
176	171	172	166	172	170	174	169	166	166	172	170	172	168	161	154	160	171	172	173	170	168	161	160	163	168	166	170	168	171	176
174	167	167	165	171	170	174	169	172	169	172	176	172	154	146	142	142	156	174	176	169	170	163	166	169	174	161	169	171	174	170
176	166	168	169	172	166	175	168	167	171	183	179	160	133	115	112	114	131	160	173	169	170	167	168	170	174	172	175	171	174	170
174	170	170	165	167	163	170	166	168	168	175	173	165	124	78	71	83	114	144	163	165	161	161	167	170	175	169	171	167	172	170
169	163	167	163	167	164	170	165	164	166	170	167	156	114	59	45	65	100	139	162	159	163	165	165	167	174	169	171	169	171	166
172	168	169	167	173	168	172	166	167	167	175	168	165	134	66	1	69	130	150	158	165	170	167	168	171	176	173	172	167	173	175
170	167	172	166	174	169	175	172	167	163	176	172	174	150	77	0	74	150	165	163	171	174	168	166	167	173	167	169	171	173	174
171	168	169	166	173	171	178	173	166	166	177	174	173	159	88	0	61	151	168	159	164	172	166	168	166	172	167	166	164	170	168
178	172	172	168	167	163	173	170	170	169	178	181	183	162	88	2	53	152	170	162	167	172	167	165	166	173	170	172	167	169	168
169	170	176	171	177	174	175	171	169	168	177	175	183	171	102	2	51	155	176	166	164	172	173	172	167	175	175	173	171	178	174
174	170	172	168	173	168	177	172	165	170	181	175	177	170	95	1	58	154	172	164	166	172	167	166	165	170	172	173	167	172	173
167	164	167	165	174	169	175	172	170	169	178	174	180	168	99	4	33	150	172	159	162	168	169	170	170	175	176	178	169	167	170
176	170	169	168	176	169	176	171	165	169	178	173	177	166	97	2	43	151	176	164	164	173	173	173	174	176	171	170	167	170	168
172	166	174	171	180	175	181	175	168	168	177	177	183	169	106	0	38	152	175	164	168	178	174	173	171	174	171	173	169	175	172
173	168	167	171	177	168	175	171	171	edges detected to sub-pixel accuracy					98	1	39	150	edges detected to sub-pixel accuracy					173	171	178	171	173	171	175	171
173	167	174	168	170	170	176	166	164						111	1	27	147						167	165	175	173	174	171	177	172
171	166	168	168	173	164	173	169	167						105	3	29	145						166	163	171	169	171	169	173	172
175	171	173	172	174	168	176	172	168	164	173	175	177	173	112	2	29	141	170	161	160	168	169	167	168	176	171	173	170	176	174
175	171	173	171	178	169	172	174	170	167	175	176	180	170	120	7	18	141	176	165	164	172	168	168	165	173	171	175	171	178	174
171	166	168	167	172	165	175	173	170	171	177	175	178	173	116	0	19	140	172	159	158	170	168	168	167	173	169	170	167	170	166
169	171	172	163	170	168	171	169	169	167	173	176	179	172	125	2	12	137	176	160	157	168	167	168	165	175	167	171	173	172	167
177	168	172	173	176	170	178	173	163	169	180	176	179	177	128	3	9	136	174	163	164	174	172	173	170	173	170	173	168	172	170
172	169	172	164	169	170	171	169	170	168	176	175	182	175	130	1	8	137	179	165	161	174	169	165	170	174	165	170	169	172	168

Figure 10.16
Sub-pixel edge detection allows the diameter of the pin to be measured as 2.5 pixels = 0.025 inches (micrometer measurement was 0.026 inches). The image was digitized at 100 dots per inch.

Surprisingly, there are many potential applications in which the images are perfectly readable to human eyes, but not clean or clear enough for current OCR technology to handle.

Typical applications of OCR include reading documents for word processing and editing, reading mail addresses for automatic sorting, reading credit card receipts for automatic banking, reading cheques for automated sorting, and so on.

Although humans see text and handwriting very clearly as 'black on white', digital images do not reflect this intuition. Figure 10.17 shows the digitization of a credit card slip, and an enlargement of the handwritten 'e' at the far right. The pixels here are far from being black or white. At the left of the character the pixels have good contrast but at the top the greyscales fade into the background as the pressure of the writer changes. This kind of variability makes the character more difficult to read automatically. As it happens, people tend to be reasonably consistent in the way they apply pressure differently as they write, and work is under way to exploit this for security applications.

Figure 10.18(a) shows the letters 'DESC' from the beginning of the printed word 'DESCRIPTION' below the date boxes. Although they are more regular than the handwritten characters, it can be seen that there is considerable ambiguity between the pixels of the characters and the background.

Figure 10.18(b) illustrates one of the major problems in reading handwriting, namely that people make characters in a stylistic way which may deviate considerably from the 'norm'. The handwritten description is '5 STAR 204573', but the 'T' in 'STAR' looks more like the Greek letter σ than a T. Also, there is ambiguity between this letter and the printed line on the voucher.

Note also that the '8' is written slightly differently in the two versions of '80.75', and that the '5's appear to be made up of two different strokes. Such idiosyncrasies make the general problem of reading handwriting very difficult. These are reasonably good images for automatic reading: other images can be much worse, with bits of characters missing or obliterated by spurious marks.

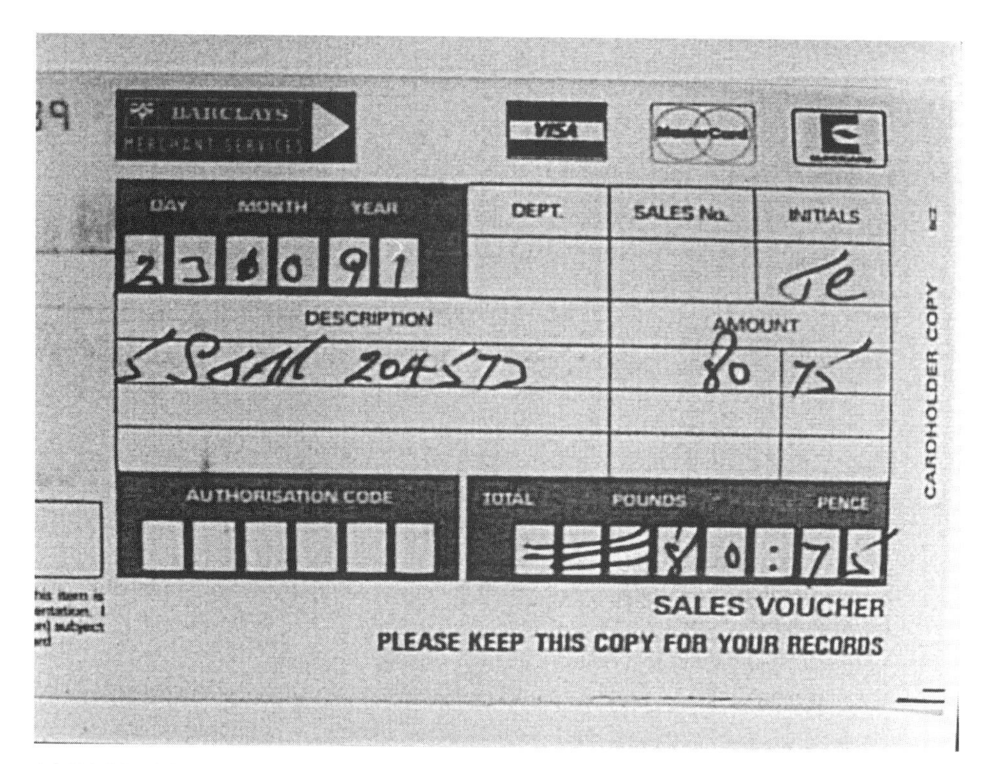
BARCLAYS
MERCHANT SERVICES
VISA
MasterCard
DAY MONTH YEAR
DEPT.
SALES No.
INITIALS
DESCRIPTION
AMOUNT
AUTHORISATION CODE
TOTAL POUNDS PENCE
SALES VOUCHER
PLEASE KEEP THIS COPY FOR YOUR RECORDS
CARDHOLDER COPY

(a) Digitized image of a credit card voucher

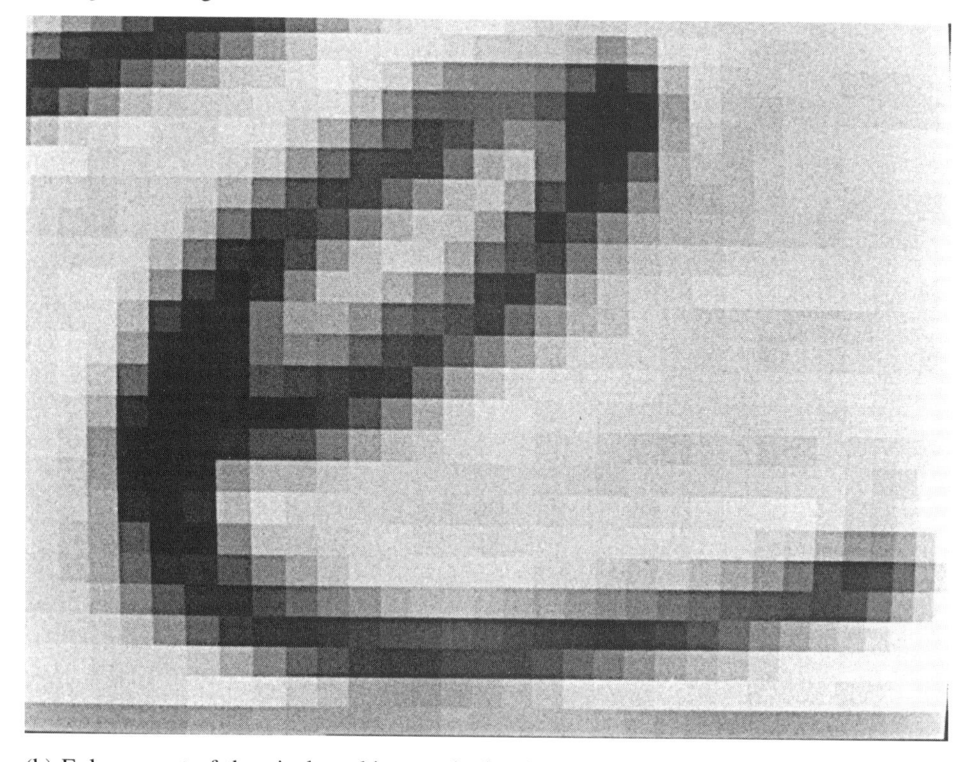

(b) Enlargement of the pixels making up the handwritten 'e' in the 'INITIALS' box

***Figure 10.17* Digitized image of handwriting on a credit card voucher.**

(a) Enlargement of the pixels making up part of the word 'DESCRIPTION' on the credit card voucher of Figure 10.17(a)

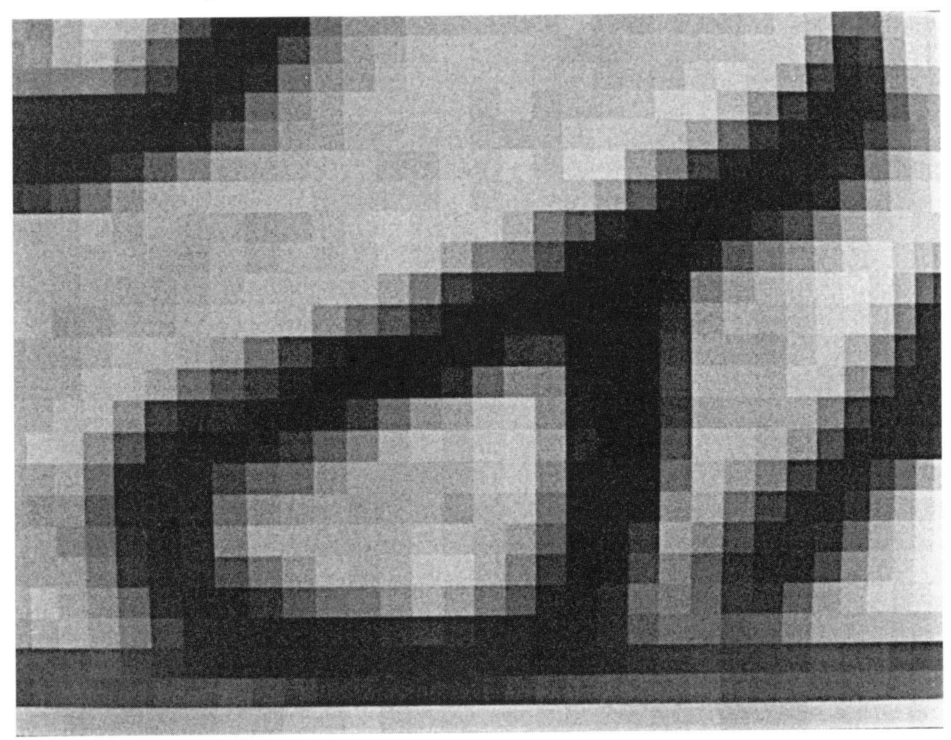

(b) Enlargement of the letter T in the word 'STAR'

***Figure 10.18* Digitized characters from the credit card voucher in Figure 10.17.**

In OCR there are a number of major divisions between the difficulty of the problems. They include:

- reading clear machine-typed black text on white background;
- reading poor machine-typed text on white background;
- reading machine-typed text on a textured or patterned background (e.g. anti-forgery patterns on cheques and bonds);
- reading handwritten text as discretely spaced capitals, small letters, and numbers, in pre-set fields or boxes on pre-printed forms;
- reading carefully handwritten text in which the letters are all clearly written;
- reading carefully written joined-up handwriting;
- reading cursive script (general handwriting).

Of these there is good progress with the first, and some progress with handwritten text on forms. Progress on the others is less good, and the general problems of reading cursive script remain unsolved in 1995. When this problem is solved it will open up many areas for new applications of computer systems, since in many information systems getting information into the system is a major problem. Currently it has to be done by human keyboard operators, and this makes many potential systems uneconomic.

New notepad computers are creating a tremendous drive for recognizing handwritten text. They have the great advantage that the pixels which make up the text are unambiguous, having been defined by the user with the electronic 'pen'. Nevertheless, in 1995, abstracting text information from handwriting remains a very difficult problem, as illustrated by the cartoon strip in Figure 10.19.

Doonesbury BY GARRY TRUDEA

Figure 10.19
Reading handwriting is a difficult problem.

10.5.4 Rejection versus error in pattern recognition

OCR exemplifies an important idea in pattern recognition and computer vision, namely the distinction between *rejection* of the recognition as dubious as opposed to *failure* by the acceptance of a misclassification.

To understand the difference, consider a machine which is sorting cheques in an automatic banking system. Suppose it has to read the amount of money on the cheque. Suppose also that one of the digits is recognized incorrectly as a 7 (seven) instead of a 1 (one). If the amount of money is £1,242,341 it would not be a terrible mistake to recognize it as £1,242,347. However if the amount were recognized as £7,242,341 the machine's failure could be very serious. Usually one has a degree of 'confidence' in a computer's vision recognition, and sometimes it makes sense for the machine to say 'I cannot recognize this with the required confidence and I reject it'. Then the rejected items can be read by humans, who tend to make fewer errors in resolving ambiguous images.

The problem with rejection is that humans usually have to take over, and the higher the rejection rate the more expensive the system becomes to run. However, the rejection rate is inversely related to the failure rate in which the machine makes potentially expensive errors, and so a trade-off has to be made.

10.6 A summary of the basic techniques in computer vision

The previous sections have given a flavour of some of the problems and techniques used in computer vision. After establishing some criteria for success, this section will summarize the basic techniques of machine vision currently available to designers.

10.6.1 Criteria for success in computer vision

Before enumerating the various techniques currently used in computer vision, some criteria will be established for their efficacy. These conditions include:

- acceptably high rates of correct pattern recognition, acceptably low rates of rejection, and acceptably low rates of errors in pattern recognition;
- tolerance to changes in the levels of absolute and ambient illumination;

- invariance to changes in position, size and orientation of the object to the camera or scanner:

    translational invariance,

    scale invariance,

    rotational invariance;

- acceptably high speeds of pattern recognition;
- ease and cost of implementation;
- ease and cost of maintenance;
- acceptable hardware demands;
- acceptable levels of operator skills.

Not all applications will weight these criteria equally. For example, invariance to orientation could be very important in an aircraft detection system. On the other hand, we expressly do not want total orientation invariance in OCR, otherwise we would not be able to discriminate symbols such as + from ×, < from >, d from p, and so on.

The criteria for recognition rates will also vary considerably according to the application. A system which takes an hour to process the information from a medical scanner could be satisfactory, while a system which takes a second to process a sales voucher could be considered to be too slow.

Computer vision techniques vary from 'cheap and cheerful' approaches such as pixel matching to the implementation of very expensive handcrafted methods. Vision systems of any sophistication at all usually involve a lot of highly skilled research and development effort, and their costs reflect this.

Many applications of computer vision require 'real time' processing, which means in practice that results must be delivered within fractions of a second. This can sometimes be achieved by employing powerful but expensive processors, and increasingly it is being achieved by various parallel processing configurations. This includes implementation of 'neural' processors in hardware.

As in other engineering disciplines, computer vision involves selecting the most appropriate approach for the particular application. Sometimes this involves understanding that the present achievements in machine vision are rather limited, and there are many potentially valuable problems which it cannot solve. The machine vision problem is a bottleneck in many applications.

10.6.2 Pixel grid template matching

Pixel grid template matching was discussed in Chapter 4 of Volume 1. It works very well in cases in which the objects always appear in the same place in the image, i.e. they do not change too much in their shape, size, orientation or position. In cases which do not satisfy these criteria, pattern matching gives poor performance and more advanced techniques must be used.

10.6.3 Associative memory

An interesting and successful variant on template-matching approaches is provided by the WISARD system developed in England by Wilke, Stonham and Aleksander at Brunel University in the 1980s. This has been used successfully for banknote recognition and other commercial applications. This vision system works by taking a video image, binarizing it, and using pre-defined random combinations of pixel '1's (white) and '0's (black) to address several banks of memory, where each bank corresponds to a particular class of problem. Essentially this approach works by storing a '1' in each bank of memory at the locations that are addressed by the input images during training. When being used to recognize patterns it uses the new image to address the memory and counts the number of '1's produced at the output of each bank of memory, the sum being the measure of how well the image is recognized.

This is called an *associative memory*, in which the score from each bank is a measure of the association between the current input image and the images used during training. WISARD works in real time due to its special hardware architecture, which allows fast learning and fast response. An amusing and remarkable application of WISARD involves discriminating smiling faces from those that have frowns.

10.6.4 Spectrum histogram and statistical matching

In applications such as remote sensing from satellites, the infra-red spectrum may be divided into many 'bands' with sensors which are especially sensitive to particular parts of the spectrum. For example, the Multispectral Scanning System has four bands while the Thematic Mapper has seven. This means that each pixel in a Thematic Mapper image identifies a point in a seven-dimensional space. A typical application in agricultural planning involves using these data to classify the pixels by crop type, such as 'wheat', 'barley', 'corn', 'sugar beet', 'apples', and so on.

A whole battery of techniques has been developed for these multidimensional data, as illustrated in Chapter 2 on Pattern recognition. Typically they work on spectral histograms and statistical models calibrated from them. The classification paradigm is: 'this pixel of unknown class is close to a pixel of known class, therefore this pixel has that class'.

10.6.5 Binarization of greyscale images and local thresholding

As shown in Section 10.2, one can get somewhere with image recognition through thresholding. However, there is the problem of selecting the threshold. Also, contrast may vary over an image due to different levels of illumination. This could happen, for example, when an image is illuminated from the side.

Although it does not work very well over entire images, the concept of thresholding is not without merit. Clearly, as one goes from a dark object to a light background there will be dark object pixels and light background pixels. The problem is knowing where to set the boundary between them. It seems reasonable that this boundary will be relatively constant locally, even if it should vary over the image. This kind of reasoning has led to a number of techniques for *local thresholding*. The details of any particular technique can be rather involved, but the basic idea is that an appropriate threshold for binarization will depend on the local greyscale statistics. These are computed to provide thresholds which are adapted to local conditions.

10.6.6 Skeletonization

In some applications a further operation of *skeletonization* after thresholding is considered to be useful. For example, in character recognition it is easier to deal with pixel configurations that are one pixel wide rather than those that are perhaps three, four or five pixels wide. Skeletonization algorithms effectively 'eat away' the outside pixels until there is just one pixel left. The results of this for an '8' and an 'S' are illustrated in Figure 10.20. However, as this example shows, skeletonization sometimes creates ambiguity.

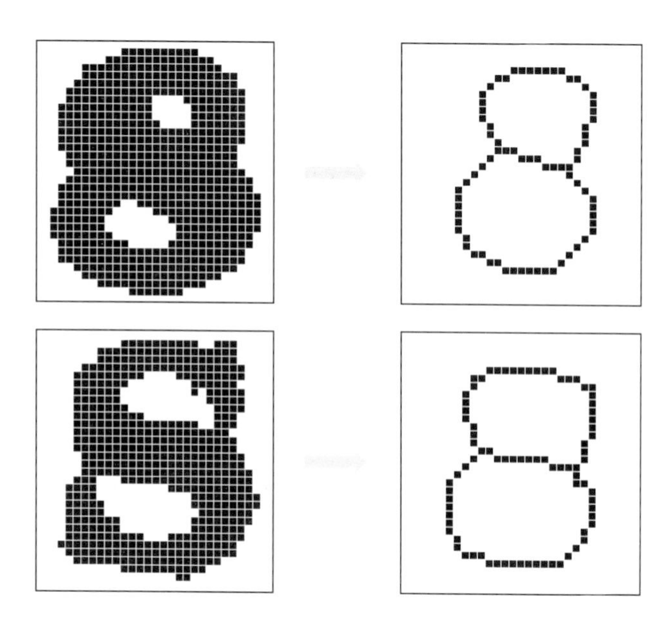

Figure 10.20 ***Skeletonization and its limitations.***

10.6.7 Edge detection

Edge detection is an important technique in computer vision, especially in applications in which a geometric model is used. Section 10.5.2 showed how edge detection can be used in object recognition, and also to make measurements. For example one can use the kind of hierarchical architecture described later in Section 10.7 in which parts of the boundaries of objects are recognized, and assembled by bottom-up processing. Subsequently, some degree of top-down processing might be involved to find missing bits of the object prior to final recognition.

There are many techniques for edge detection in the literature. Most of these are 'filters', i.e. operators which filter out non-edge from edge pixels. As we observed in Volume 1, this leads to edges which are polygons. Also the approach depends on setting parameters, and post-processing operations such as thinning.

10.6.8 Mathematical morphology

A new theory of image processing was suggested by Jean Serra in his book *Image Analysis and Mathematical Morphology* (1982). There he introduced a wide range of operators for image processing. To illustrate one idea, consider a binary image in which the pixels are either black or white. Then let the 'thinning' operator be defined as one which makes every black pixel with a black neighbour into a white pixel. This operator makes objects and lines thinner, as shown in Figures 10.21 (a) and (b). The 'thickening' operator is defined as one which makes black every white pixel with a black neighbour. This operator makes objects and lines thicker. By applying the thinning operator followed by the thickening operator, messy parts of images can be cleaned up as shown in Figure 10.21(c).

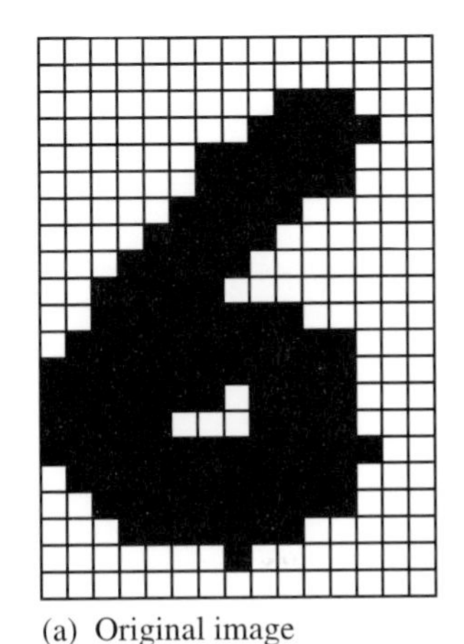

(a) Original image

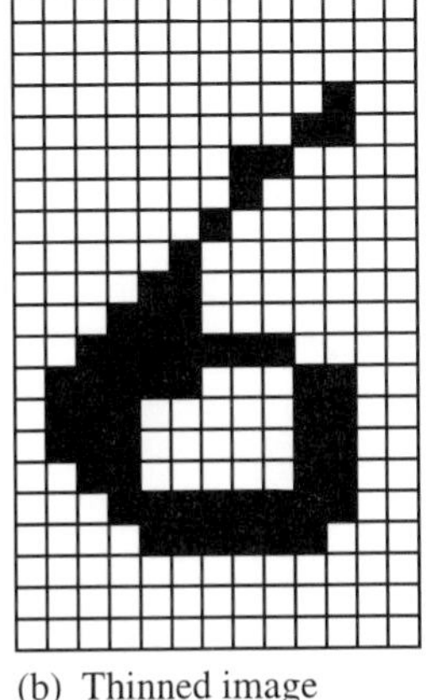

(b) Thinned image

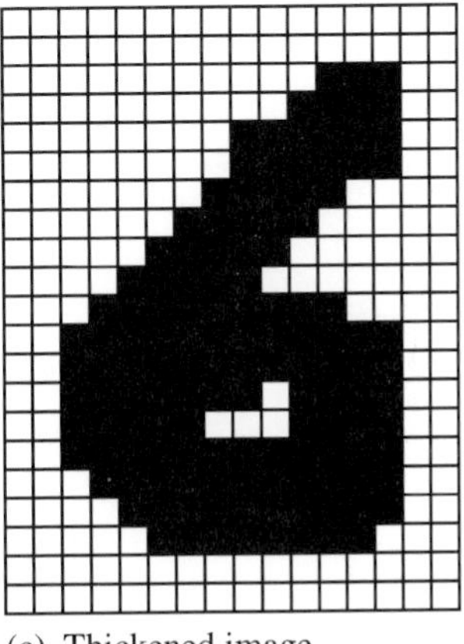

(c) Thickened image

***Figure 10.21* Mathematical morphology: an example of applying 'thinning' and 'thickening' operators to a binary image.**

In some images it is known that the objects of interest are relatively smooth at the given level of resolution. However, digitization noise may cause edges to be 'jaggy'. In such cases the application of the thinning operator followed by the application of the thickening operator may smooth out the jaggy edges and make objects easier to recognize. There are many other operators within mathematical morphology which enhance images according to 'logical' principles and operations.

10.6.9 Neural networks

Neural networks are increasingly being used in computer vision. There are many architectures and variations, but for many of them a lot of work may be required to ensure that appropriate information enters the network in the first place. Thus a lot of expensive work may be necessary to build a *pre-processing* system which, it can be argued, moves the vision problem one step back beyond the network in order to get a solution. This was discussed in Chapter 3 on Search.

Section 10.5.1 showed neural networks applied to a vision problem in which the pre-processing involved collecting data on pixel colour frequencies of insects. In this application the *spatial* element of the problem was ignored. In other applications shape and size may be crucial, and the engineer must ensure that these are encoded in a way which makes them suitable for neural networks to process.

Section 10.7 shows a hierarchical architecture for computer vision. There the higher level objects are assembled from lower level substructures, and it is possible that neural networks could be used for this.

10.6.10 Reasoning in computer vision

Computer vision often occurs when there is a model of the system or its environment. A common problem in computer vision occurs when parts of objects are missing because they are occluded (hidden by other objects) or because the image is poor. Although one cannot create more information than is in the image from the image itself, the information in the image combined with the general model and other expectations may enable strong hypotheses to be made. Usually these will be stated in the form: 'if this is true and that is true, then the image viewed contains a such and such', where the confidence in the hypothesis may be weighted.

In the early days of computer vision, when people were just beginning to learn the power of *If–Then* reasoning through knowledge-based systems and the like, it was felt that higher level reasoning would be very powerful in machine vision. So powerful, in fact, that it would not matter too much if the primitives could not be abstracted with great fidelity. In fact this turned out to be wrong, as illustrated in the field of speech recognition. No amount of reasoning and logic could compensate for rather poorly defined and abstracted phonemes.

10.6.11 Simon's Principle of Robust Primitives

This principle states that, in computer vision, we must seek to abstract primitive objects with great reliability and replicability. These are called ***robust primitives***. Vision techniques which do not begin with robust primitives are likely to fail.

It would be hard to underestimate the practical importance of this principle. The edges mentioned in Section 10.5.2 are robust primitives which have enabled the authors to build very reliable and rugged industrial vision systems which work under a wide range of ambient lighting conditions and operators. The principle of robust primitives is very important in pattern recognition and computer vision.

10.7 A hierarchical architecture for computer vision

10.7.1 Bottom-up processing in computer vision

In the earliest stages of processing image data, computer vision is *bottom-up*. At the bottom of the hierarchy, we have pixels and their greyscale values. These pixels must be combined using various criteria to form classes or structures from which useful information can be abstracted. At this earliest stage in processing it is essential that robust primitives are used. This is illustrated in Figure 10.22(a) in

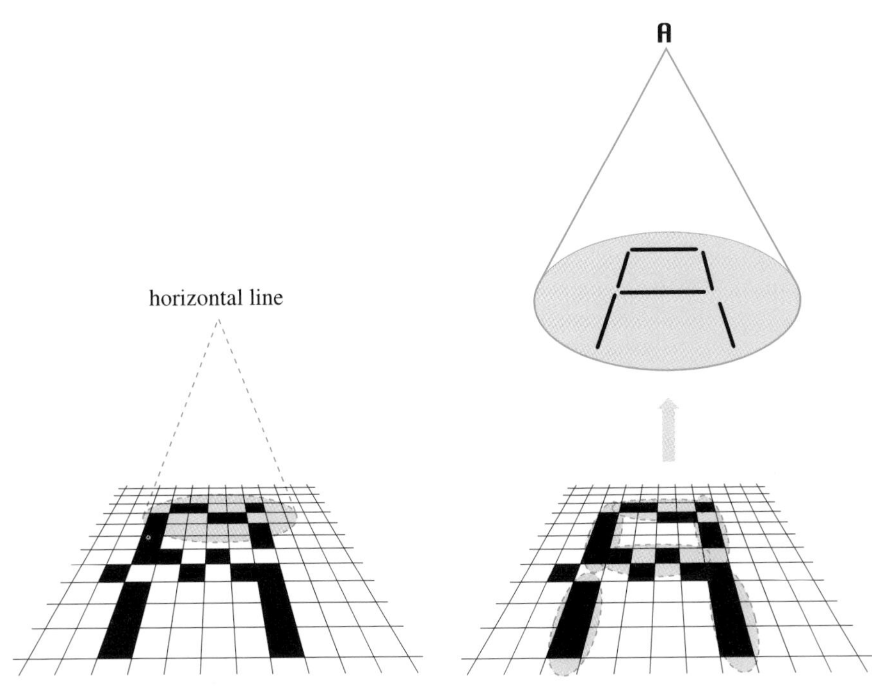

(a) A set of pixels assembled to form a line (b) A set of lines assembled to form a character

***Figure 10.22* Bottom-up aggregation in computer vision.**

which a group of pixels is aggregated to form a line. In this application, detecting the lines is highly replicable and so lines are robust primitives for this application. Figure 10.22(b) shows a set of lines assembled to form a character.

10.7.2 Top-down reasoning in computer vision

In the last section we saw how computer vision begins with a kind of feedforward bottom-up information processing, which builds increasingly more complex objects from simpler objects. Sometimes this will be sufficient by itself for successful pattern recognition. However, it is necessary to use *a priori* knowledge to reason about the objects in the image. This is essentially a *top-down* process in which a given configuration of objects is hypothesized to be an object of interest. If it actually is that object, one can deduce things that ought to be true about the image at lower levels. These hypotheses can then be tested at the lower levels and their outcome can inform the lower level pattern recognition.

10.7.3 Computer vision as an iterative top-down, bottom-up process

It is becoming increasingly accepted that computer vision must involve both of the bottom-up and top-down aspects of information processing discussed in the previous two sections. Thus we come to an architecture for computer vision which combines the two. This means that any vision system must have a *control mechanism* to determine the current modes of the system and transfers between them. In the simplest case this may just be an iterative process in which one goes between bottom-up and top-down until the pattern recognition is made with the required degree of confidence. This general architecture is represented in Figure 10.23.

This architecture has been used in practice in the examples of eye recognition (Volume 1, Chapter 4) and character recognition. In eye recognition, pixel configurations were used which act as extremely robust primitives in this and many other applications. These have been combined to form sub-objects such as the pupils and whites of the eyes, and these in turn have been combined to recognize the eyes in the context of the face (Volume 1, Chapter 4). This higher level recognition enables discrimination of configurations at the lower level which correspond to eyes in the image from other eye-like configurations in swirls of hair and elsewhere.

Apart from the bottom-up aspects of computer vision, there is a *top-down* aspect when there is ambiguity in the recognition. In such a case the system may attempt to resolve the ambiguity by acting in a top-down fashion by seeking specific diagnostic information. For example, the lines which have been abstracted with confidence in Figure 10.23 lead to an ambiguous recognition between an **A** and an **R**. In order to resolve this the machine needs more information about the right side of the character, and so goes down the hierarchy and looks at the pixels 'through a magnifying glass' in order to see more precisely what occurs at the right edge of the character.

In general, computer vision is an iterative process, with information and control moving up and down the hierarchy of representation until recognition is achieved or the attempt is abandoned and the recognition rejected.

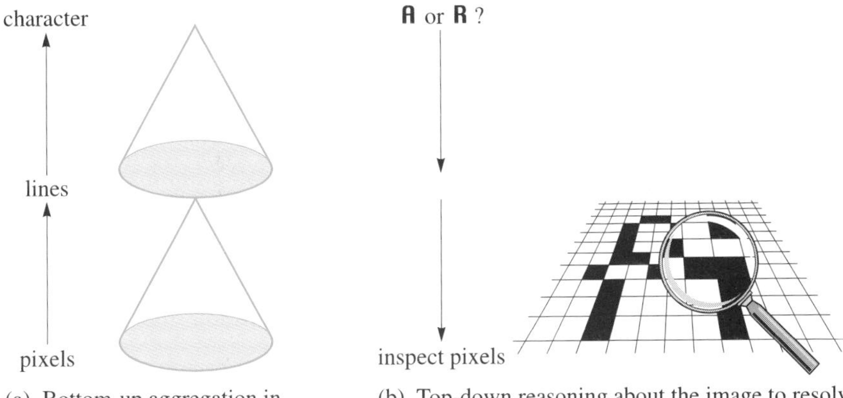

Figure 10.23
The bottom-up, top-down nature of computer vision.

10.8 Conclusion: computer vision in intelligent machines

There is no doubt that computer vision will be one of the most important enabling technologies over the next decade. Current generations of robots have poor visual sensing, and severe limitations in their capabilities as a result. The robot of science fiction which will make your tea and do the housework remains a long way off: current robots find it very difficult to abstract the necessary information from their cameras and other sensors.

Computer vision remains a relatively expensive way of sensing for perception and cognition in intelligent machines. This is true of both the hardware and software, but the situation is beginning to change as the drive for intelligent machines generates new devices and approaches. For example, the University of Edinburgh in Scotland has designed a small and inexpensive camera. No doubt as such devices move to large-scale mass production the price will drop and designers will be able to incorporate many 'electronic eyes' into their systems without making them prohibitively expensive. Similarly, custom-made hardware for processing visual information is emerging and it too can be expected to drop in price as time goes on. Also the price of general-purpose sequential computers continues to drop in price/performance terms.

Software (or more precisely the development of algorithms and procedures) is the other major element in computer vision. This is less likely to decrease in price so quickly. The reason is that there is no general methodology of computer vision to act as a rigorous basis for engineering. Currently applications tend to be one-off solutions to particular problems. Invariably this involves a considerable investment in research and development undertaken by highly skilled engineers, and such computer vision systems tend to be very expensive. The hierarchical architecture described in this chapter is part of an attempt to formulate a general theory of computer vision, and is implicitly or explicitly used by many researchers and practitioners in the field.

It is probably not too optimistic to expect computer vision techniques to improve considerably over the next decade. This will allow us to build machines with capabilities that we can only dream of at present. Areas likely to benefit especially include:

aeronautics and space travel
agriculture through remote sensing
automotive systems
business systems
car and truck design
crime detection and prevention
disaster prediction
domestic consumer goods
industrial and domestic security
industrial inspection
international security
medicine
military systems
paper processing and administration
robotics
scientific research
telecommunications
toys
underwater surveying and mining
weather forecasting

There can be little doubt that human kind has the potential to benefit considerably from new generations of intelligent machines, enabled by computer vision.

Further reading

Levine, D.M. (1985) *Vision in Man and Machines*, McGraw Hill Series in Electrical Engineering, New York. This book gives a comprehensive account of most of the standard techniques used in computer vision.

References

Serra, J.A. (1982) *Image Analysis and Mathematical Morphology*, Academic Press.
Simon, J.C. (1986) *Patterns and Operators: The foundations of data representation*, Tr. J. Howlett, North Oxford Academic, division of Kogan Page.

CHAPTER 11
INTEGRATION

11.1 An introduction to blackboard systems

In this book you have encountered the following concepts and techniques:

- pattern recognition
- search
- neural networks
- scheduling
- reasoning
- rule-based systems
- learning
- intelligent control
- computer vision.

Even if you have mastered all of these components you may have wondered how they can all be brought together in a single 'intelligent' system.

One approach to *integrating* the component parts of intelligent systems involves the concept of *blackboard systems*, and this is the main topic of this chapter. This is not the only approach, but it is simple and sufficiently powerful to enable quite complex machines and systems to be developed.

The central feature of the blackboard system is an area of working memory called the *blackboard*, as shown in Figure 11.1. The knowledge and data stored or *written* 'on the blackboard' is intended to be *public* and accessible to any one of a set of independent *agents*. Any agent can write to the blackboard and read from the blackboard. This public information is not necessarily managed by any of the agents, and may *emerge* from their interaction. This architecture allows each of the agents to do its own business with its only external interface being the blackboard. This greatly simplifies the conceptual nature of each agent, and what one needs to know about its interactions with other agents.

The term *agent* is intended to be very general, but includes ordinary computer programs (both declarative and procedural, and object-oriented modules, which are beyond the scope of this book), rule-based systems and neural networks. To these can be added the sensors which *write* information on the blackboard, and the actuators which *read* information such as control commands and control parameters from the blackboard.

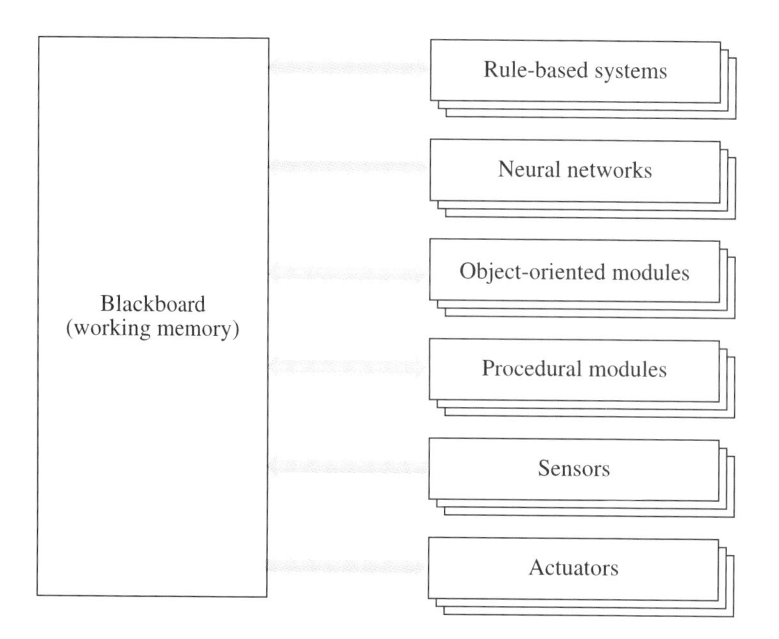

Figure 11.1 The blackboard model (adapted from Hopgood, 1993).

To illustrate the application of the blackboard architecture, consider an autonomous robot which must operate within a changing environment. We will consider how the perception subsystem, the cognition subsystem and the execution subsystem can be implemented within the blackboard architecture. First, however, it is necessary to understand how a blackboard system can itself be implemented as a software environment which enables systems to be developed rapidly without the developer having to start right from the beginning every time.

11.2 The blackboard system as a development environment

In Chapter 7 we explained how rule-based systems are often implemented using a *shell* into which particular facts and rules could easily be introduced using *editors*. Usually the editors check that the format of the facts and rules is syntactically correct and help to overcome errors which would stop the system running. The shell also contains the *inference engine* which operates on the facts database using the rules to produce new facts. This requires that various modes of *conflict resolution* can be set within the shell. In other words, the shell gives the user everything that is needed to create a rule-based system except the facts and rules for a specific domain of application.

Although it is possible to program a blackboard system from first principles for a particular application, this may not be a cost-effective way of proceeding. Often it is better to use a system *development environment* which attempts to provide the 'engine' of the blackboard system and all the necessary editors to input the subsystems or agents. Furthermore, such an environment automatically *integrates* the subsystems by giving them access to the public data on the blackboard, and controlling the way that the agents are distributed over the available hardware and how they are scheduled to run on the hardware. For example, the Open University's SmartLab home experiment laboratory runs on a single personal computer. Since a PC usually has only one processor it means that all the agents have to share this processor as they run. The way this is done is mostly transparent to the user, and this makes system development relatively simple.

Typically, a blackboard system development environment for creating intelligent machines will include:

- a rule-based system shell to create autonomous rule-based systems;
- a neural network editor and controller to create neural subsystems;
- editors, compilers and linkers for creating:

  procedural programs (written in languages such as C and FORTRAN),

  declarative programs (written in languages such as Prolog),

  object-oriented programs (written in languages such as C++ and Smalltalk);
- software interfaces to external sensor hardware;
- software interfaces to external actuator control hardware.

Apart from this functionality, it is important that the environment can be used easily, and this is often achieved using a graphic user interface (GUI). For example, in some systems rules can be defined by manipulating graphic entities such as boxes and arrows, where the boxes represent facts and the arrows represent implication. Also, they sometimes allow the user to zoom in and out to see the system in greater or lesser detail, the former giving insight into the details of subsystems and the latter giving more of an overview.

11.3 Running many rule-based systems in parallel

To understand how many agents can run 'in parallel' on the available hardware, consider a system with many rule-based systems running on a single sequential processor. Since the processor can only execute one instruction at a time, it can only service one rule-based system at a time. This leads to the need to decide how the processor will service each system.

For example, the software managing the execution might allow each rule-based system enough resource to test one rule and move on to the next rule-based system. Alternatively, it might allow each rule-based system to go through its rules once, perform conflict resolution, and fire the selected trigger rule before moving on to the next system. This approach is easy to understand since it means that each rule-based system gets a turn to fire a rule as the control software cycles around.

With a single sequential processor, by definition, nothing happens in parallel. However, by 'slicing' up the processor's time, each rule-based system gets a slice of time and the overall appearance is of many rule-based systems running in parallel. However an important aspect of using a single processor is that only one rule-based system is updating the blackboard at any given time. This can be useful when designing blackboard systems because, although all the agents are independent, it can be useful to know that one system will have updated the facts database if possible before the antecedent facts are examined for a subsequent system.

So, one of the simplest control strategies for multiple rule-based systems is to order them all, and let each rule-based system fire a rule (if it can) when its turn comes round. Since each rule-based system can update the information on the blackboard, the outcome of one rule-based system can affect the conflict set and outcome of the next. Unless the rules are very simple the behaviour of the interacting rule-based system is *emergent*, and cannot necessarily be predicted by the constructor of the blackboard system.

Sometimes the designer of a blackboard system may want to impose a control mechanism on how the rules in the different rule-based systems fire, and how the systems interact. Although the agents of the blackboard system *can* be independent, there is no absolute requirement that they *must* be independent.

As explained in Chapter 7 on Rule-based systems, one sometimes wants a particular rule-based system to give each rule a 'Buggins' Turn' chance of firing. We have found that this can be extended to wanting all the rules in a cycle to fire before control is handed over to the next rule-based system. We have called this blackboard control mechanism 'Buggins' cycle'. For example, the rules of a particular rule-based system may complete a whole task which one wants to be completed before handing on to the next.

It should be clear from this discussion that blackboard systems require control strategies in much the same way that rule-based systems require conflict resolution control strategies. In proprietary software environments the control strategy may be part of the environment, or the system designer may have to select the control mechanism to be applied at any given time.

11.4 Running many agents in parallel

In systems with many processors it is possible for each of the agents to genuinely run in parallel. However, the problem of control remains because a conflict could arise when two completely autonomous agents want to write on the blackboard at the same time. If one agent wants to set a blackboard variable to one thing while another wants to set that variable to something else, there is a conflict which must be resolved. This conflict amounts to deciding which system will be allowed to update the blackboard, and this is similar to deciding which system will get a slice of a single sequential processor (and therefore unrestricted access to the blackboard) at any given time.

11.5 Implementing a perception subsystem

The sensors of a machine form an essential part of its perception subsystem. Usually the sensor data enter the machine through special interfaces and in principle the sensor data can be written on the blackboard to be read by the other agents. By definition the data enter the system at the lowest level in the perception-cognition hierarchy.

Example

For a simple example of how hierarchical pattern recognition might be implemented, consider the problem of reading characters in a grid of pixels. Instead of trying to match the characters directly, it is much more effective to try to recognize their parts using an agent which reads the pixel data from the blackboard and writes the recognized block data onto the blackboard. Then another agent can read the blocks from the blackboard and write the recognized character data on the blackboard.

The first agent might be a neural network operating directly on the black/white binary pixel data, or it might be a rule-based system operating on those data, or it might be a program of some kind. The second agent might also be a neural network, a rule-based system, or a program. In this way a perception subsystem might be implemented through a mixture of information processing techniques.

The important point is that, using the blackboard model, the agents which make up the hierarchical pattern recognition system can be kept reasonably simple and their interface through the blackboard is also simple.

More complex perception subsystems might also have rule-based systems acting in a 'top-down' mode and making deductions about the sensed data on the basis of prior knowledge. These too can be implemented as agents reading and writing data on the blackboard, and again one benefits from the implicit simplicity of the architecture. However, this top-down processing might require some explicit control in the order that the agents operate. For example, a procedural program may be required to execute in its entirety before other agents modify the multiple hierarchical level data on the blackboard and it could force this by setting a flag variable on the blackboard which effectively prevents all the other agents operating until the flag is reset.

Chapter 2 on Pattern recognition and Chapter 10 on Computer vision are concerned with techniques for perception and the interface to the cognition subsystem. The techniques described in those chapters will usually be implemented by a mixture of agents writing data of different hierarchical levels to the blackboard. Typically the cognition subsystem will read the data written at the highest levels by the perception subsystem.

11.6 Implementing a cognition subsystem

As far as we are concerned here, cognition involves the processing of higher level data, usually symbolic or parametric. An intelligent machine operating in real time constantly has to find an answer to the question 'what shall I do next', and this problem is mainly the concern of the cognition subsystem.

For the most general type of intelligent machine the cognition subsystem must fulfil many functions including:

- interfacing to the perception subsystem, including reasoning about the output of the perception subsystem and even controlling it to make it deliver information which is particularly relevant;
- controlling *learning*;
- maintaining a *map* of the relationship between the machine and its environment;
- *searching* for and establishing *goals* and sequences of intermediate goals, and the order in which to attain them;

- *scheduling*: path planning in space and time; also planning sequences of actuator activities including the motion of the whole machine and parts of the machine such as its arms and grippers;
- *pattern recognition* of external states which are particularly desirable or undesirable, possibly using *neural networks*;
- identifying problems in any of the above and finding ways to overcome them.

For mobile machines the cognition subsystem first has to update the 'map' of the machine's physical environment. This may mean recognizing the position of the machine on a given map, or it may require that the machine learns what is in its environment. The map data will probably be implemented as arrays of numbers on the blackboard, or variables which give the x-y positions of objects in the environment as described in Chapter 8 on Learning.

Some machines will have to learn their environment explicitly before they can operate within it. General-purpose machines should learn from how they have solved previous problems and their experiences so that this knowledge can be applied to subsequent situations.

Given that the cognition subsystem has decided where the machine is in relation to its environment, mobile machines must decide the current *goals*, including where they want to be. The criteria for deciding a 'good' position will usually be programmed into the machine as rules in a rule-based system or within a computer program agent. These may involve scheduling criteria such as those discussed in Chapter 5.

Selecting goals and planning how to achieve them is crucial to the *performance* of an intelligent machine. A machine that consistently selects near-optimum goals and achieves them by scheduling near-optimum space–time paths will exhibit good performance in normal usage. A machine which can maintain these characteristics by adapting to unexpected events with new near-optimal goals and schedules will fulfil the concept of an 'intelligent machine' which performs well in the face of uncertainty.

Apart from the more strategic aspects of scheduling, the cognition subsystem must interface to the *execution* subsystem. For example, the sequences of movements of its drive subsystem must be planned, as must sequences of movements of other actuators such as its arms and grippers. One of the main problems with controlling actuators is that complex interactions of forces such as momentum and friction can make it impractical to formulate a precise mathematical model of the dynamics of the system. In such cases intelligent control techniques such as those discussed at the beginning of Chapter 9 are applicable.

11.7 Implementing an execution subsystem

At some stage the cognition subsystem must send control information to the machine's actuators in order to result in execution of its goal-oriented plan. In the simplest case the cognition subsystem can simply write an item of data to the blackboard. For example, the cognition subsystem could change the value of a blackboard variable called 'left_motor' from zero to one. In some blackboard systems it is possible to define special variables such as 'left_motor' which are interfaced to external devices such as switches which supply power. In this case the interface hardware may automatically switch to no power when the variable has value 0, and switch to full power when the variable is 1. Thus the actuators can be activated by writing data on the blackboard. There are many variants of this idea, and the data that are written on the blackboard could be more subtle, including parameters which control the amount of power which is switched and the direction.

In principle it is possible for the cognition subsystem to undertake all the processing necessary to control the actuators of the execution subsystem to give the desired composite action. However, actuators themselves are becoming increasingly intelligent and some of the low-level processing such as deciding the precise sequence of movements of a gripper may be undertaken by the actuator itself. The intelligence in the actuators may be implicit in the mechanical design of its components, or the actuator may even have its own processor(s). Indeed, some grippers are themselves complete mechatronic subsystems, having their own perception–cognition–execution cycle. In this case the cognition subsystem of the master machine has an easier job, since it need only write data about what is wanted onto the blackboard and leave the intelligent actuator to get on with the job.

Whatever the level of intelligence of the actuators, the blackboard architecture gives a conceptually simple way of interfacing the cognition subsystem to the execution subsystem.

11.8 Integration: emergent behaviour and control

The blackboard model is essentially an *integrating* architecture. It is designed to allow components to be implemented as *autonomous agents* which can be designed in relative isolation without the designer having to know the detailed implementation of all the other agents: all the designer needs to know is how the particular agent interacts with the blackboard data.

If a mechatronic system is assembled entirely in terms of autonomous agents there may be *emergent behaviour* which cannot be predicted before the machine runs. The designer may have a model of the whole system which makes the emergent behaviour a logical consequence of the way the agents are implemented with respect to the blackboard.

However, in very complex systems it may be impossible for the designer to predict all the emergent behaviour of a system. For example, road traffic systems are made up of stretches of road supporting many autonomous agents (humans driving vehicles). One way to try to understand the emergent behaviour of this system is to observe it at the side of the road. Recent research at the Los Alamos National Laboratory in the USA has shown that the only way to predict the emergent behaviour of a road system is to *simulate* the interactions of the many thousands of drivers: at present there is no other known way to 'predict' the emergent dynamic behaviour.

In complex machines the simulation may simply involve running the machine and observing its behaviour under specified conditions. This corresponds to the usual testing of systems in all engineering design. However the difference is that one must *expect* emergent behaviour and try to understand it. In this respect the design of intelligent machines may begin as a kind of research exercise which eventually becomes development leading to products as the emergent behaviour is better understood.

The use of autonomous subsystems with a simple blackboard data exchange interface makes it easier to formulate theories as to how the components of the machine are interacting, why the emergent behaviour is as it is, and how the emergent behaviour may be controlled.

One way to control emergent behaviour is to make it impossible. A simple way of doing this is to 'switch off' one or more of the agents which result in behaviour which turns out to be undesirable. But this means that some agents are allowed to control others by switching them on and off, and the controlled agents thereby cease to be autonomous. In fact, it is perfectly possible that the system designer will have all the agents grouped so that the first group executes, enables the second group, and switches itself off; the second group executes, enables the third group, and switches itself off; and so on until the last group executes, enables the first group, and switches itself off. In other words, it is possible that the agents all control each other in some way introduced by the designer. By imposing this kind of control structure on the system the designer may be able to make useful deductions about which data will be on the blackboard when, and thereby make the whole system more predictable.

In summary, the blackboard architecture allows each agent to be totally autonomous, but sometimes the system designer may remove some of that autonomy by making the behaviour of some agents dependent on blackboard data written by other agents. It is the designer's responsibility to ensure that the logic of this is correct, but they are considerably assisted in this by the inherent simplicity of the architecture of the agents being interfaced through the blackboard.

11.9 Blackboard systems and the concepts and techniques of AI

11.9.1 Search

Search is one of the most important theoretical areas of artificial intelligence, especially heuristic search where one has to find good sub-optimal solutions to problems which cannot be solved using optimal methods. Data for the spaces being searched may be encapsulated in particular agents with the results of the search being written on the blackboard, or all the data relating to the search may be publicly available on the blackboard. Search heuristics may be implemented as rule-based systems, programs, and neural network agents.

11.9.2 Pattern recognition

This covers a wide variety of techniques which are used in the design of intelligent machines. Any particular pattern recognition technique might be implemented as a rule-based system, a bespoke computer program, or a neural network agent. In hierarchical pattern recognition, different agents may be implemented to perform the recognition at different levels with each reading its data from the blackboard and writing its results on the blackboard. Pattern recognition implicitly or explicitly involves search.

11.9.3 Neural networks

In Figure 11.1 neural networks were shown as agents which may interact with the blackboard data. In general the network will have to be interfaced to its inputs by a *pre-processor* and its outputs may have to interpreted by a *post-processor*. The pre-processor might be an agent which, for example, reads sensor data from the blackboard, transforms it into a form on which a neural network can operate (e.g. a sequence of numbers) and then writes this back on the blackboard. The network can then read the pre-processed input data from the blackboard and write its outputs on the blackboard. The post-processor may be an agent which reads that output and transforms it to a form which can be used by other agents. The pre-processor and post-processor agents may be implemented as rule-based systems or conventional computer programs.

11.9.4 Scheduling

Scheduling is one of the main tasks of the cognition subsystem. As discussed in Chapter 5, scheduling can be considered to be the ordering of events in space and time. The particular nature of their events and their representation will depend on the particular application. For example, although the scheduling problems are similar, a robot planning its route around a factory will require a different representation to an arm which must position a gripper inside a complex mechanical object. The representation data for a particular scheduling problem may be encapsulated in a single agent, or they may be public data on the blackboard. Scheduling subsystems may be implemented as rule-based systems, programs, and even neural networks (using appropriate pre-processors and post-processors).

11.9.5 Reasoning

Two of the main information processing paradigms discussed in this book are logical deduction through reasoning, and learned classification of data in neural networks. Most human theorizing and technical communication is based on reasoning using explicit vocabularies and rules of inference. The *If–Then* construct is one of the most fundamental ideas in reasoning and it pervades every aspect of designing an intelligent machine. Inevitably designers will have a mental model of the machine and its environment which may be expressed in natural language and mathematical formalism, and they will constantly be musing along the lines that *if* this is so *then* that must follow. Since humans frequently make conceptual and computation errors in reasoning, the designer's conclusions may turn out to be incorrect, as may become apparent when the machine is tested.

Computer programs are usually full of *If–Then* constructs, and they give a very flexible way of encoding the reasoning of the programmer. It is more difficult to make a machine reason for itself, and rule-based systems have been very successful in this respect. Many computer-based systems now exist for the manipulation of knowledge expressed as facts and rules. Some of these systems have the objective of deducing new knowledge from old and ordering this knowledge systematically, while other systems have the objective of the machine *proving* hypothetical results from external sources or which it has hypothesized itself.

The agents which reason in an intelligent machine will be rule-based systems, or other systems implemented as computer programs in languages such as C, C++, LISP and Prolog.

11.9.6 Rule-based systems

The architecture of rule-based systems is one of the triumphs of research into artificial intelligence. Even though this appears to be a standard technology today, research in this area goes back only some thirty or forty years. Rule-based systems are explicitly considered to be agents which can interact with the blackboard data.

11.9.7 Learning

Learning is considered to be one of the most important features in making machines more intelligent and adaptable. Machines do not have anything like the learning capabilities of humans, and this is a very active research area. The agents which learn in a blackboard system will usually be implemented as rule-based systems or programs. The exception is neural networks, which learn from examples during training. The learning process will usually involve the learning agent reading data from the blackboard, processing it, and writing the result back on the blackboard.

As discussed in Chapter 1, the *indexing problem* is one of the most challenging problems in artificial intelligence. It is relatively easy to put huge amounts of information into machines, but it is difficult to synthesize and extract that information in a useful form for any particular task. As described here, the blackboard is a passive information carrier, allowing information to be written to and read from it. It may be that intelligent machines will require pro-active synthesizing agents which transform passive data into useful information, and pro-active indexing agents which post *meta-data* on a *meta-blackboard* enabling other agents to find the information they need when they need it.

11.9.8 Intelligent control

As seen in Chapter 9 on Intelligent control, low-level control such as that for the broom-balancer may be implemented as agents which are rule-based systems, programs and neural networks. In general these agents would read the control data such as the position of the trolley and the angle of the broom from the blackboard. Estimates of the rate of change of the variables might be calculated by an agent which then writes velocity and angular velocity values on the blackboard for other agents to read. The various control strategies will be implemented as agents of appropriate types, and these will write control data to the blackboard. The hardware and software interface of the blackboard system will then switch the motors and apply power according to the parameters stored on the blackboard.

The control of large complex systems is likely to be distributed over many agents in the blackboard system according to the perception, cognition, and execution considerations given above. In particular, some parts of the system may indeed be autonomous agents, and the blackboard architecture allows this since individual agents or groups of agents can be implemented in ways which make them totally independent of all the others, except for their interaction through the blackboard variables.

11.9.9 Computer vision

Computer vision can be considered to be a special case of pattern recognition, and its implementation will be through agents which are rule-based systems, programs and neural networks. For complex scenes computer vision requires a sophisticated representation which usually involves two- or three-dimensional geometry. These data may be encapsulated in specific computer vision agents, or they may be publicly available as blackboard data. Usually the incoming images will be public data which are written on the blackboard by the hardware and software interface. Various perception agents may access these image data, and process them to produce synthesized data to be written on the blackboard to be used by other agents. Computer vision may involve top-down processing, which is usually performed by agents which are rule-based systems or conventional computer programs.

11.10 Conclusion

In this book we have presented some of the most important concepts and techniques of artificial intelligence as they apply to the design of intelligent machines. Each of the concepts needs to be understood in order that the designer has an overview of the technologies available to solve large, complex, and sometimes ill-defined problems. The techniques we have shown can be adapted for particular problems, and again a good understanding of the underlying theory is necessary for this. We have concluded the book by considering how the various concepts and techniques might be implemented, and we have shown that the blackboard system architecture is a simple but powerful way of integrating the various techniques. Other architectures will be appropriate in some situations.

After reading this book you should have a good grasp of the elementary principles of artificial intelligence and be able to implement these ideas in practical systems. Each of the areas that we have covered has its own specialist literature which you should now be able to read and understand in order to extend your knowledge. Our hope is that this book has given you a good foundation from which you can go forward to design and implement your own intelligent machines.

Reference

Hopgood, A.A. (1993) *Knowledge-Based Systems for Engineers and Scientists*, CRC Press: Boca Raton, Florida, USA.

ACKNOWLEDGEMENTS

Grateful acknowledgement is made to the following sources for permission to reproduce material in this text.

Chapter 2

Figure 2.2: Levine, M.M. (1979), *Vision in Man and Machine*, Copyright © 1985 by McGraw-Hill, Inc., Reproduced with the permission of McGraw Hill, Inc.; *Figures 2.3 and 2.4:* Kanizsa, G. (1979), *Organization in Vision*, Copyright © 1979 Praeger, an imprint of Greenwood Publishing Group, Inc., Westport, CT. Reprinted with permission; *Figures 2.25, 2.26, 2.27(a), 2.27(b) and 2.28:* Lillesand, T.M. and Kiefer, R.W. (1979), *Remote Sensing and Image Interpretation*, Copyright © 1979 by John Wiley and Sons Inc., Reprinted by permission of John Wiley and Sons Inc.; *Figure 2.32:* Reprinted from *Artificial Intelligence in Engineering*, **8**, Johnson, J.H., Picton, P.D. and Hallam, N.J., 'Artificial intelligence', pp. 307–313, Copyright 1993, with kind permission from Elsevier Science Ltd, The Boulevard, Langford Lane, Kidlington OX5 1GB, UK; *Figure 2.33:* M.C. Escher's 'Sky and Water I' © 1999 Cordon Art B.V. – Baarn – Holland. All rights reserved.

Chapter 10

Figure 10.8: Scala; *Figure 10.10:* Reproduced with the permission of The Phillips Collection, Washington, D.C.; *Figure 10.11:* Reproduced by permission of Royal Mail; *Figure 10.14:* Heather Angel; *Figure 10.19:* Trudeau, G. (1993) Universal Press Syndicate © 1993 G. B. Trudeau; *Figure 10.20:* Bokser, M. (1992) 'Omnidocument technologies', *Proceedings IEEE*, July 1992, Special Issue on OCR, © 1992 IEEE.

Chapter 11

Figure 11.1: Hopgood, A. A. (1993), Chapter 7 'Systems for interpretation and diagnosis', in *Knowledge-Based Systems for Engineers and Scientists*, © 1993 by CRC Press, Inc. Reprinted by permission of CRC Press, Boca Raton, Florida.

Cover

Computer art courtesy of Dr. Paul Margerison.

INDEX TO VOLUME 2

Page numbers in bold indicate principal references to key terms, which are flagged in the text by the use of bold italic type.